Intelligent Systems, Control and Automation: Science and Engineering

Volume 100

Intelligent Systems, Control and Automation: Science and Engineering book series publishes books on scientific, engineering, and technological developments in this interesting field that borders on so many disciplines and has so many practical applications: human-like biomechanics, industrial robotics, mobile robotics, service and social robotics, humanoid robotics, mechatronics, intelligent control, industrial process control, power systems control, industrial and office automation, unmanned aviation systems, teleoperation systems, energy systems, transportation systems, driverless cars, human-robot interaction, computer and control engineering, but also computational intelligence, neural networks, fuzzy systems, genetic algorithms, neurofuzzy systems and control, nonlinear dynamics and control, and of course adaptive, complex and self-organizing systems. This wide range of topics, approaches, perspectives and applications is reflected in a large readership of researchers and practitioners in various fields, as well as graduate students who want to learn more on a given subject.

The series has received an enthusiastic acceptance by the scientific and engineering community, and is continuously receiving an increasing number of high-quality proposals from both academia and industry. The current Series Editor is Kimon Valavanis, University of Denver, Colorado, USA. He is assisted by an Editorial Advisory Board who help to select the most interesting and cutting edge manuscripts for the series:

Springer and Professor Valavanis welcome book ideas from authors. Potential authors who wish to submit a book proposal should contact Thomas Ditzinger (thomas.ditzinger@springer.com)

Indexed by SCOPUS, zbMATH, SCImago.

More information about this series at https://link.springer.com/bookseries/6259

Paramanand Vivekanand Nandihal ·
Ashish Mohan · Subir Kumar Saha

Dynamics of Rigid-Flexible Robots and Multibody Systems

Springer

Paramanand Vivekanand Nandihal
Department of Robotics and Automation Engineering
Sister Nivedita University
Kolkata, West Bengal, India

Ashish Mohan
Confederation of Indian Industry
New Delhi, Delhi, India

Subir Kumar Saha
Department of Mechanical Engineering
Indian Institute of Technology Delhi
New Delhi, Delhi, India

ISSN 2213-8986 ISSN 2213-8994 (electronic)
Intelligent Systems, Control and Automation: Science and Engineering
ISBN 978-981-16-2797-2 ISBN 978-981-16-2798-9 (eBook)
https://doi.org/10.1007/978-981-16-2798-9

This Springer imprint is published by the registered company Springer Nature Singapore Pte Ltd.
The registered company address is: 152 Beach Road, #21-01/04 Gateway East, Singapore 189721, Singapore

Preface

Writing the preface of work is one of the most revered moments for the authors in their journey of finalization of the manuscript. This is because of two reasons: (1) a preface gives an opportunity to the authors to conversate with the readers and set up the requisite context of the book, and (2) it gives authors an opportunity to share their philosophy and fundamental thought, behind their work. In both ways, it is one of the unique opportunities at hand for the authors.

Context of this book and the subject of its analysis are fairly straightforward and most relevant in light of the arrival of Industry 4.0 and the era of hyper-automation. As the world paces itself into the new age of robotics, there are going to be tremendous aspirations from the controller systems of these robots for higher processing speeds and faster real-time analysis in more complex boundary conditions. This is going to exert unprecedented expectations on the speed, agility, and efficiency of the processor algorithms of these controllers. Indeed, the computational efficiencies and numerical stability of the processing algorithms of the multibody robotic systems are foreseen as the most crucial differentiators determining their real-time control characteristics.

In this book, the dynamic analysis and simulation algorithms for real-time control of rigid–flexible robotic systems are presented. The architectural configuration of robotic systems is taken as serial open-loop as well as closed-loop chains of kinematic linkages. The elasticity components are considered present in the system throughout the analysis. Thus, the book presents comprehensive dynamic analyses of serial open and closed chain robotic systems with rigid and flexible architectural links. The consideration of flexibility/elasticity properties of the links is crucial because of many lightweight applications of the robots in industry these days. The flexibility considerations, however, make the dynamic formulations very complex and intricate.

Dynamic formulation is done based on the unique approach of decoupling the natural orthogonal complement (NOC) of velocity constraints expressions of the system. The NOC is the matrix that relates the angular and linear velocities of the links in the system to the associated joint rates. The NOC is decoupled and expressed as a product of two matrices, of which one is a block triangular, and the other is a block diagonal. Hence, the name "Decoupled NOC" or DeNOC is coined. Unlike the NOC, its modified form, i.e., the DeNOC, allows one to write the expressions of the elements of the associated matrices and vectors in analytical recursive form. These characteristics are not so obvious with the original form of the NOC. The DeNOC matrices lead to recursive, computationally efficient dynamic algorithms for many robotic systems but require the dynamic formulation based on the NE equations. In this book, it is shown that it is also possible to begin with the Euler–Lagrange equations of motion and use the DeNOC matrices to obtain the same independent equations of motion as obtained from the Newton–Euler equations of motion. With the proposed approach, the explicit evaluation of the partial differentiations is avoided. Subsequently, a recursive and efficient simulation algorithm for the rigid link robots is developed, which is also shown to be numerically stable. The main advantage of the approach proposed in this book is that the links are treated in a continuum, and hence, the scheme derived for the rigid robots is extended to the flexible link robots with ease. The DeNOC-based formulation of dynamic equations of the robotic systems is proposed to be a unique contribution of this book. The dynamic formulation can be easily modified to fit in any architectural configuration—serial open or closed chain systems. Formulation of a computationally efficient and numerically stable algorithm for dynamic analysis of serial open and closed chain robotic systems with rigid and flexible links forms the core kernel of the book.

Contents of this book are broadly classified into the following parts:

I. Dynamic modeling of open-loop serial-chain systems with rigid and flexible links.
II. Dynamic modeling of closed-loop systems with rigid and flexible links.
III. Study of computational efficiency and numerical stability of dynamic algorithm.
IV. Experimental study of flexible link robotic systems.

The assumed mode method is used to discretize the link deflection for the flexible links. The DeNOC formulation, as mentioned above, gives rise to the analytical expressions for the matrices and vectors associated with the dynamic equations of motion. Geometric stiffness is introduced in these DeNOC-based formulations to deal with the high-speed, large deformation problem.

The proposed approach actually allows one to solve the closed chain systems using the knowledge of serial-chain robot dynamics. Computationally efficient forward dynamics algorithms for serial-chain robots and closed-loop systems with rigid or flexible or rigid–flexible links are then developed. As a result, we present in the book the following:

- an $O(n)$—n being the total number of links in the serial-chain robot—recursive, numerically stable, and computationally efficient inverse and forward dynamics algorithms for serial-chain rigid link robots and
- a recursive $O(n)+O(m^3/3)$–m being the maximum number of flexible link modes of vibration–numerically stable and computationally efficient forward dynamics algorithm for serial-chain flexible link robots.

Inverse dynamics algorithms are also presented in the book, along with the illustrations of algorithm simulations with different serial and closed-loop systems of both planar and spatial in nature.

Next in series, the numerical stability of the forward dynamics algorithms is investigated using different methods. A new criterion of testing the numerical stability based on the zero eigenvalue problems is proposed and shown to be versatile and precise. The proposed algorithms are compared with the others available in the literature and are shown to be computationally efficient for rigid as well as flexible body serial robots and closed-loop multibody systems.

Finally, a series of experiments are carried out on a single- and two-link flexible link arm to validate the simulated results obtained from the proposed algorithms. Inclusion of the damping in the proposed algorithm is also illustrated through these experiments.

The book is deemed to be useful in introducing some of the novel and unique concepts on the DeNOC-based formulation of rigid–flexible robotic systems. It should be handy and useful for teaching graduate-level (Master's and Ph.D.) courses specialized in robot and machine dynamics. The work presented in this book will be useful for researchers working in the area of flexible robotic and multibody systems. It is envisaged that the book will also be highly useful for researchers working in robotic applications in defense, space, atomic, and ocean application sectors.

Bringing up work with a comprehensive and detailed analysis of robotic systems for wide applications like those mentioned above was also part of the fundamental philosophy of the authors behind this work. Another fundamental thought of the authors was to illustrate the application of proposed simulation algorithms on a number of popular robots in industries. Also, the wish list included elucidating the theoretical concepts through the demonstration on experimental setups and showcases the objective gains achieved in real-time control of the robotic systems through the proposed algorithms. Fortunately, and with the support of its patrons, authors have been successful in including these concepts, experiments, and comparative studies in the manuscript. The inclusion of theoretical, simulation, and experimental work in a single manuscript makes it a very comprehensive work and extremely fulfilling for the aspirations of the authors of the book.

We wish our readers a happy and insightful journey into the vast world of dynamics of robotic systems. We would look forward to any feedback and suggestions that our readers and reviewers would like to give us for further enhancing and enriching the contents of the book.

Kolkata, India — Paramanand Vivekanand Nandihal
New Delhi, India — Ashish Mohan
New Delhi, India — Prof. Subir Kumar Saha

Acknowledgements

A book may be the work of its authors, but it certainly encompasses the efforts of many more people who are associated with the authors. These people are the professional colleagues, institutional patrons, family members, and friends of the authors. These people have made huge contributions to the making of the manuscript—through their expert advice, comments, and feedbacks (by the professional colleagues) and patiently providing time to the authors for completing the manuscript (by the family members and friends). A book can never be complete without the due acknowledgments to the efforts of all these people. In this sense, the authors consider this page and section of acknowledgments as the most important section of the book.

On top of this list is the patron institution of the three authors—the Indian Institute of Technology Delhi. The authors are truly blessed to have been associated with this great institute in various capacities and at various points of time in their professional careers. Thus, while the first two authors have completed their doctorates from the institution, the third author has held various faculty positions and also the position of Associate Dean of Students and Head of the Department of Mechanical Engineering of the institution. First two authors have indeed completed their doctorate under the guidance of the third author. The institute has thus extended enormous support to all the three authors in terms of its research infrastructure, laboratories, and experimental facilities. The authors owe wide-ranging acknowledgments to IIT Delhi and would also be indebted to it for being their institutional patron.

The book presents some experimental work, for which the hardware setup was partially funded by the Department of Science and Technology, Govt. of India. The authors express their thankfulness to the DST for this institutional support.

Next, the authors are indebted to a large number of research colleagues and professionals at IIT Delhi and at JCB India and Confederation of Indian Industries, CII (where the second author has been employed during the course of writing of this book). The insightful and detailed discussions with these researchers on the subject of robotics have been very instrumental in formulating the framework of the book and setting up the relevance of its subject in the current times of Industry 4.0. Our deepest regards and gratitude to these people for providing us detailed insights on the subject and making the overall thought process more comprehensive.

From the professional networks, the authors would like to express their thankfulness and gratitude to the members of Mechatronics and Program for Autonomous Robots Laboratories (for the first author), and Mr. D. Jaitly who helped in conducting the experiments, and Prof. S. P. Singh who was the Coprincipal Investigator of the DST-funded project (for the third author). These people have given immense support to the authors in various capacities, which has helped the author broaden the scope of this book.

In particular, both the first two authors, Paramanand Vivekanand Nandihal and Ashish Mohan, are deeply thankful to the third author Prof. Subir Kumar Saha, for having guided them throughout the work and giving rightful directions both in research content and in the subject perspective.

Similarly, there are numerous people who have given their professional reviews and comments over time which has now culminated in the form of this book. We express our thanks to all of them.

Now, the most important acknowledgments—to the families. All three authors are most deeply grateful for the superb support given to them by their respective families during writing of this book. No words can duly represent the appreciation which the authors have for their respective families for giving the authors space and time out of their personal times in order to complete this work. We would rather like to thank them mentioning their names here. Thus, we would like to thank Smt. Mahadevi Nandihal (mother) of Paramanand Vivekanand Nandihal; Smt. Kiran Srivastava (mother), Ms. Purnima Srivastava (wife), and Aditi and Atharv (children) of Ashish Mohan; and Smt. Bulu Saha (wife) and Ms. Esha Saha (daughter) of Subir Kumar Saha for their patience and understanding while this book was under preparation. Thank you, families. We owe it to you all!

Finally, we would like to thank the reviewers and Springer (publisher) for their support in organizing the work and bringing the manuscript into its final form of this book.

Paramanand Vivekanand Nandihal
Ashish Mohan
Subir Kumar Saha

Units and Notation

The International System of Units (SI) is used throughout this book. The boldface Latin/Greek letters in lower and upper cases denote vectors and matrices, respectively, whereas light face Latin/Greek letters in lower case with italic font denote scalars. The upper-case letters *R* and *F* with italic font represent the rigid and flexible links, whereas the non-italic letters R and P represent revolute and prismatic joints. In the case of any deviation in the above definitions, an entity is defined as soon as it appears in the text.

Contents

About the Authors

Dr. Paramanand Vivekanand Nandihal completed his Ph.D. from the department of Mechanical Engineering, Indian Institute of Technology (IIT) Delhi in 2018. His expertise and core-research domain subjects are Robotics, multibody dynamics, kinematics, and Finite element analysis (FEA) with key focus on harmonics impact on structural dynamics. He has worked on key projects on advanced engineering research and analytics related to rigid-flexible robotics with the Department of Science and Technology, Government of India. He is currently working as an assistant professor in the department of Robotics and Automation Engineering at Sister Nivedita University, Kolkata, India. He has a M.Tech. in Machine Design (2009), and Bachelor of Engineering in Mechanical Engineering (2007) from BEC Bagalkot affiliated to Visveswaraiah Technological University (VTU), India. Dr. Paramanand has published several papers in journals and conferences of national and international repute.

Dr. Ashish Mohan completed his doctorate in robotics and mechatronics from the Indian Institute of Technology (IIT) Delhi in 2006. He has a total industry experience of around 24 years which includes working with JCB India, Tata Motors, Hi-tech Robotic Systems Ltd., and a brief research project at IIT Delhi for ISRO and Department of Science and Technology, Government of India. Dr. Mohan is currently working at the Confederation of Indian Industry and heads its portfolios of Technology, Advance Design, Engineering Research, Innovation, Science, IP Creation and Entrepreneurship. His core areas of expertise include new technology development and advanced engineering domains like robotics, multibody dynamics, kinematics & mechatronics, and, establishment of industry best innovation ecosystems, innovation incubation proto-labs, and IP delivery processes. In last 5 years, Dr. Mohan was part of the leadership team at JCB India which led it to win six Innovation leadership awards. He has over 15 publications in various international journals and conferences. A mechanical engineer by graduation from National Institute of Technology Bhopal (1997), he also holds a Masters in Technology Management from IIT Delhi (2012). Dr. Mohan has also been visiting IIT Delhi, IIT Guwahati, and IIT Jodhpur as guest industry speaker, visiting/faculty and external industry expert.

Prof. Subir Kumar Saha received his mechanical engineering degree from RE College (now NIT), Durgapur, India, in 1983, prior to completing his M.Tech. at the Indian Institute of Technology (IIT) Kharagpur, India, and his Ph.D. at McGill University, Canada. After his studies, he joined Toshiba Corporation's R&D Center in Japan. After four years of work experience in Japan, he has been with IIT Delhi since 1996, and is currently a Professor in the Department of Mechanical Engineering. Prof. Saha is additionally the Project Director of the newly formed Rs. 170 crores (USD 20.0 millions) non-profit company of IIT Delhi I-Hub Foundation for Cobotics (IHFC) funded by Department of Science and Technology, Government of India. For his research contributions, he has been awarded the 2020 Distinguished Alumnus Award by NIT Durgapur in January 2021. He also held the Naren Gupta Chair Professorship at IIT Delhi for 10 years since 2010. Prof. Saha awarded with the Humboldt Fellowship by the AvH Foundation, Germany, in 1999 to conduct research at the University of Stuttgart. Later, he has been visiting Canada, Australia, and Italy for short-term research assignments. Prof. Saha has authored more than 200 research publications in reputed journals/conference proceedings, and delivered equal number of invited/keynote lectures globally.

Chapter 1
Introduction

In this chapter, several robotic and multibody systems, which are used in practical applications, are introduced. The need for dynamic modeling and the consideration of flexibility in a link are explained. Several modeling techniques and other issues related to dynamic modeling are also highlighted. Lastly, the chapter is concluded along with the contributions and the outline of this book.

1.1 Background

Over the last two decades, applications of multibody dynamics have spread over the fields of robotics, automobile, aerospace, biomechanics, and many more. With continuous development in the fields mentioned above, many complex multibody systems have evolved, and dynamics play a pivotal role in the analysis of those systems. Computer-aided dynamic analysis of multibody systems has been a prime motive of engineers since the evolution of high-speed computing facilities using computers. In order to perform computer-aided dynamic analysis, the actual system is represented by its dynamic model, which has information in terms of link parameters, joint variables, and constraints. A dynamic model of a multibody system is represented by a set of equations governed by physical laws of motions. For a system with few links, it is easier to obtain explicit expressions for the equations of motion. However, finding equations of motion for complex systems with many links is not an easy task. Sometimes even with 4 or 5 links, say, a five-bar mechanism, it is difficult to find an explicit expression for the system inertia in terms of the link length, mass, and joint angle. Hence, the development of the equations of motion is an essential step for dynamic analysis.

In serial multibody systems, the links are concatenated such that each link is coupled to a predecessor and a successor link, except for the first and the last one which are coupled, respectively, to a successor and a predecessor only. For dynamic

P. V. Nandihal et al., *Dynamics of Rigid-Flexible Robots and Multibody Systems*,
Intelligent Systems, Control and Automation: Science and Engineering 100,
https://doi.org/10.1007/978-981-16-2798-9_1

(a) Simulation of the flexible helicopter fan blade [122] (b) Canadarm 2 (source: https://www.nasa.gov)

Fig. 1.1 Flexible multibody systems

analysis, these links have generally been treated as rigid bodies. However, due to the recent requirements of higher productivity and applications in the areas of space, ocean, and nuclear installations, there is a great emphasis on the high speed and lightweight of the manipulators. If various links of a manipulator are considered rigid, they must be structurally stiff, and this leads to heavy and massive designs. If speed is not to be sacrificed, the powerful and heavy actuators with high energy consumptions are required to move these arms. Hence, the requirements of lightweight and high precision for mechanical systems such as space robots and high-speed machinery create the need for accurate simulation of flexible multibody systems, creating the need for efficient dynamic algorithms. The flexibility of the links must be included in the dynamic model so that their vibrations are predicated and avoided using a suitable control scheme, particularly in position control. The time-optimal simulation of these flexible robots has been an area of extensive research for more than a decade and is still an active area for research. Figure 1.1 shows the examples of flexible multibody systems.

1.2 Dynamics of Multibody Systems

Multibody systems can be classified based on their architecture as open-loop and closed-loop systems. The open-loop systems basically have serial and tree-type architectures, as shown in Fig. 1.2a–b, respectively. In serial systems, the links are concatenated such that each link is coupled to a predecessor and a successor link, except for the first and the last one, which are coupled respectively to a successor and a predecessor only, see Fig. 1.2a. On the other hand, a tree-type system has the links that are connected in the form of branches, as shown in Fig. 1.2b. Each link in a tree-type system is connected to a predecessor link called parent link and one or more successor links known as child links except for the first and the last one. The closed-loop systems have closed architectures with single or multiloops. Generally, the first link of a closed-loop system is coupled to the last link, as depicted in

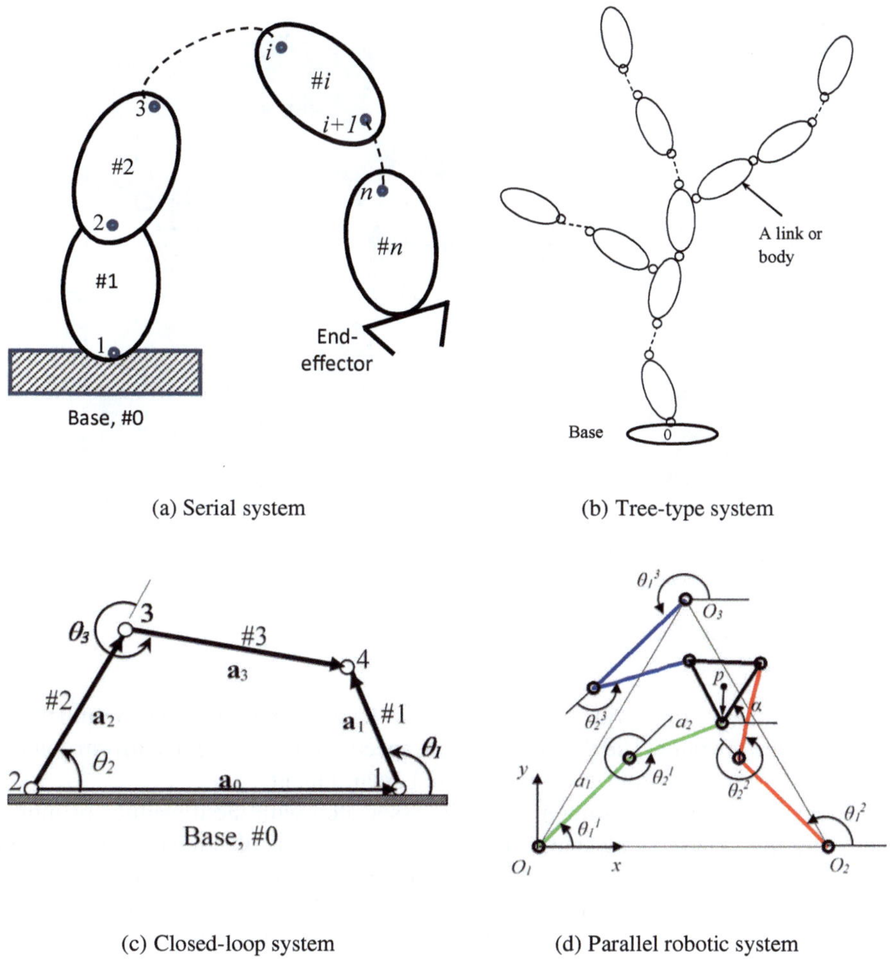

Fig. 1.2 Different architecture of multibody systems

Fig. 1.2c. A closed-loop system having parallel architectures, as shown in Fig. 1.2d, is called parallel system.

In a serial system, each link is connected by an active joint, i.e., a joint which is actuated by an actuator, whereas in closed-loop systems, links are connected by both active as well as passive joints. Passive joints are those which have no actuator connected to them. Figure 1.3 shows two robotic devices. Figure 1.3a shows a robotic gripper (robotic hand), which is an example of a tree-type system, whereas Fig. 1.3b shows the Stewart platform, an example of a parallel system. The accurate dynamic modeling of a robotic gripper is required to calculate the force/torque requirement on each finger to hold the objects ranging from softball to a thin glass tube. On the other hand, the Stewart platform can be used as a roller-coaster simulator, flight, car,

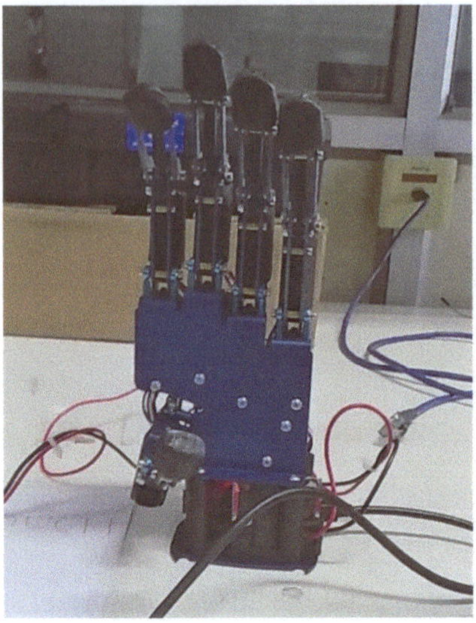

(a) AR10 Robotic Hand
(source: Mechatronics Lab, IIT Delhi)

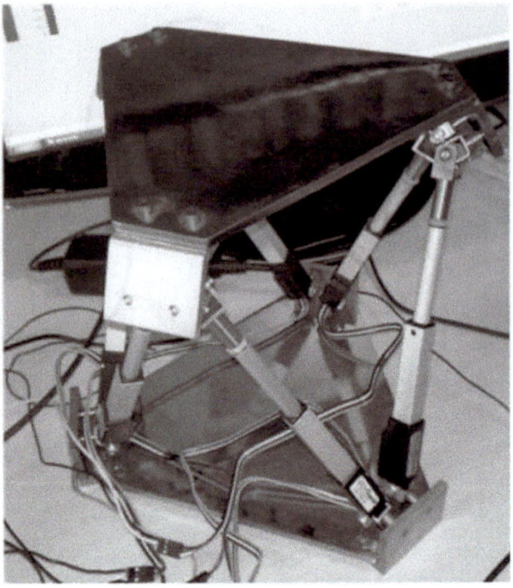

(b) Stewart Platform
(source: PAR lab, IIT Delhi)

Fig. 1.3 Two robotic devices

or ship simulator along with a virtual reality (VR) headset to provide an immersive feeling to the operator. The dynamic motion, speed calculation, and platform control are necessary to synchronize with the virtual environment.

There are two most common problems associated with the dynamics of multibody systems. They are inverse and forward dynamics, as illustrated in Fig. 1.4. In inverse dynamics, the torque or force requirements at different actuated joints of a given robot are determined for a set of known time histories of the joint parameters. The forward dynamics, on the other hand, is an initial value problem in which the

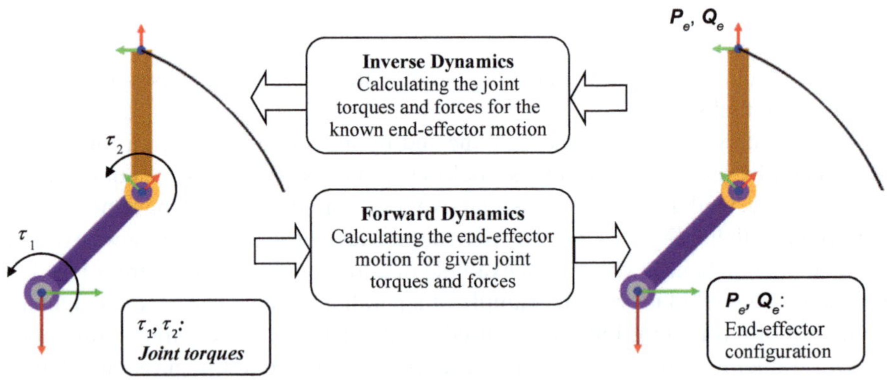

Fig. 1.4 Inverse and forward dynamics

joint accelerations, velocities, and positions are determined for a given set of joint torques and forces, and initial configuration of the robot. The dynamic analysis of multibody systems, i.e., inverse or forward, requires the determination of the equations of motion, i.e., dynamic model of the system at hand. The equations of motion can take the form of a set of pure differential equations or a set of mixed differential and algebraic equations.

These equations must be solved numerically, although it might be possible to obtain closed-form solutions for some simple systems. Note that no numerical method is capable of finding exact solutions. Hence, a numerical error defined as the difference between the actual and numerical solutions is encountered. Such errors consist of two components, namely truncation errors and round-off errors. A truncation error depends on the nature of the numerical algorithm used in computing the state vector. In order to minimize the truncation error, it is desired to have an algorithm that determines the state vector in a minimum number of computations. A round-off error, on the other hand, is caused due to the limitation on the memory allocation of the computing system and, in combination with the truncation error, it affects the numerical stability of the behavior of the physical system under study. Thus, the study of computational efficiency and numerical stability of various algorithms for the dynamic analysis of multibody systems is crucial.

Research work in the past on various issues relating to the dynamics of multibody systems is reviewed. In order to know the various aspects of dynamic modeling of rigid–flexible or flexible multibody systems, it is required to review the dynamic formulations of systems with both rigid as well as flexible links. Hence, the literature survey on the dynamic formulations of both rigid and flexible systems are discussed.

The aim of this book is broadly classified into three parts: (1) dynamic modeling of serial and closed-loop systems with only rigid and rigid–flexible links; (2) analysis of computational efficiency and the numerical stability of the algorithms proposed for the dynamic analysis; and (3) experimental verification of few theoretical results.

1.3 Dynamic Modeling

Literature in the category of dynamic modeling is subclassified into dynamics of rigid systems and flexible systems.

1.3.1 Dynamics of Rigid Systems

Recent trends in dynamic modeling of rigid systems and a pervasive review of various modeling techniques can be found in various literature [98, 110, 113, 46]. A review of some recently published formulations for increasing efficiency of the dynamic simulation of a rigid system is given in Eberhard and Schiehlen [43] and Saha et al. [108]. Most of the literature reports the selection of dynamic modeling techniques, mainly, for simulation purposes, based on computational efficiency and numerical stability of the algorithm. However, there are several fundamental methods [52] to formulate the equations of motion. For example, Newton–Euler (NE) formulation, Euler–Lagrange principle, Gibbs–Appel approach, Kane's method, and D'Alembert's principle. All the approaches mentioned above, when applied to multibody systems, have their own advantages and disadvantages. Newton–Euler (NE) approach is one of the classical methods for dynamic formulation. This approach is based on the concept of free body diagrams. If it is constrained, associated forces of constraints are included in the free bodies with those which are externally applied. Mathematically, the NE equations of motion lead to two vector equations, which in the scalar form are equivalent to three translational equations of motion of the center of mass (COM) and three equations determining the rotational motion of the rigid body. The NE equations are related simply by the constraint forces, and hence, for open-loop multibody systems, they can easily be solved recursively. For closed-loop systems, however, some of the NE equations need to be solved simultaneously to obtain the constraint and driving forces. Hence, the use of the NE equations of motion for closed-loop systems is not as efficient as those for open-loop systems. Euler–Lagrange (EL) formulation is another classical approach that is widely used for dynamic modeling. The EL formulation uses the concept of generalized coordinates instead of Cartesian coordinates. It is based on the minimization of a functional called 'Lagrangian', which is nothing but the difference between the kinetic energy and potential energy of the system at hand. For open-loop multibody systems, where typically the number of coordinates equals the degrees of freedom of the system, the constraint forces do not appear in the equations of motion. For the closed-loop multibody systems, however, forces of constraints appear as Lagrange's multipliers. Kane's formulation [70], which is the same as Lagrange's form of D'Alembert's principle, has also been used by many researchers to develop the equations of motion. Kane's formulation uses the concept of the partial velocity vector to formulate the final equations of motion of a system at hand.

Several other methods of dynamic formulations were also proposed in the literature. For example, Khatib [72] presented the operational space formulation. Angeles and Lee [5] presented the natural orthogonal complement (NOC) to obtain a set of constrained equations of motion from those which are unconstrained. Blajer and Schiehlen [17] also derived another orthogonal complement to derive the constrained equations of motion for a multibody system. Park et al. [96] presented dynamics using a Lie group formulation, while Stokes and Brockett [127] derived the equations of motion of a kinematic chain using concepts associated with a special Euclidean group. McPhee [84] showed the use of linear graph theory in multibody system dynamics. Cameron and Book [22] described a technique based on Boltzmann–Hamel equations to derive the dynamic equations of motion. The comprehensive discussion on dynamic formulations can be found in several textbooks, for example, by Wittenburg [136], Roberson and Schwertassek [100], Schiehlen [109], and Shabana [111].

1.3.1.1 The Natural Orthogonal Complement (NOC)

It is pointed out here that the NE equations of motion are still found popular in the literature of dynamic modeling and analysis. However, it requires the solution of the constraint forces and moments which do not play any role in the motion of a system. Hence, in motion studies, extra calculations are required. To avoid such extra calculations, a series of different formulations proposed in the literature where the equations of motion in the EL form are obtained from the NE equations. Huston and Passerello [62] were the first to introduce a computer-oriented method to reduce the dimension of the unconstrained NE equations by eliminating the constraint forces. Later, Kim and Vanderploeg [74] derived the equations of motion in terms of relative joint coordinates from Cartesian coordinates through the use of the velocity transformation matrix (VTM). A VTM relates the translational and angular velocities of links with joint velocities. It is worth noting that the vector of constraint forces and moments is orthogonal to the columns of the VTM. More precisely, the null space of the transpose of a VTM is an orthogonal complement to the column space of the VTM. Hence, the VTM is also referred to as the orthogonal complement matrix (OCM). The pre-multiplication of the transpose of an OCM with the unconstrained dynamic equations of motion vanishes the constrained moments and forces. As a result, a set of independent dynamic equations of motion, which are ordinary differential equations (ODE), is obtained. The ODEs are known to provide numerically stable algorithms compared to the differential-algebraic equations (DAE), representing the same system dynamics [95]. The phrase 'orthogonal complement' was first coined by Hemami and Weimer [58] for the modeling of non-holonomic systems. Note that an OCM is not unique. In some approaches, it was obtained numerically using singular value decomposition or as an eigenvalue problem [69, 81, 135] which are computationally inefficient.

Alternatively, Angeles and Lee [5] presented a methodology where they derived an OCM naturally from the velocity constraints. Hence, the word 'natural' was added to the orthogonal complement matrix. The natural orthogonal complement (NOC) matrix, when combined with the unconstrained NE equations of motion, leads to the minimal order constrained dynamic equations of motion by eliminating the constraint forces, as explained in the previous paragraph. Hence, this facilitates the representation of the equations of motion in Kane's form that is suitable for recursive computation in inverse dynamics or in the EL form that is suitable for forward dynamics and integration. Later, Angeles and Ma [6], Cyril [36], Angeles et al. [7] and Saha and Angeles [106] showed the effectiveness of the use of the NOC matrix while applied to systems with holonomic and non-holonomic constraints.

1.3.1.2 The Decoupled Natural Orthogonal Complement (DeNOC)

The decoupled form of the NOC for serial multibody systems was introduced by Saha [102, 104]. The decoupled NOC (DeNOC) matrices have two block matrices, namely a lower block triangular matrix and a block diagonal matrix. In contrast to the NOC, the DeNOC matrices allowed one to recursively obtain the analytical expressions of the vectors and matrices appearing in the equations of motion [103]. This, in turn, helped one to analytically decompose the generalized inertia matrix (GIM) of the system at hand, allowing one to obtain a recursive algorithm for forward dynamics [105] and the analytical inverse of the GIM [103]. Inspired by the concept of the DeNOC matrices, Dimitrov [41] used a method for dynamic analysis, trajectory planning, and control of space robots. Mohan and Saha [89, 90] showed that the DeNOC-based modeling for rigid–flexible multibody systems provides not only efficient but also numerically stable algorithms. More recently, Shah et al. [116, 117] presented the concept of 'kinematic modules' to obtain module-level DeNOC matrices leading to intra-modular and inter-modular recursive algorithms for analysis and control of legged robots. The concept also allowed macroscopic purview of the multibody systems. Gupta et al. [53] recently extended this modeling method to systems where each link and body is represented with the help of point masses. Then, they optimized the required torques and forces at the joints of a serial robot moving along a specified trajectory.

1.3.2 Dynamics of Flexible Systems

Research on the dynamics of flexible multibody systems is started in the international arena in the early 1970s. Book [18], Baruh and Tadikonda [12], Huston [61], Shabana [113] , Bauchau [13, 14], Wasfy and Noor [134], and Dwivedy and Eberhard [42] give comprehensive reviews of various techniques available for the modeling of link flexibility. Generally, the dynamic modeling of flexible multibody systems is done by superposition of the large rigid body motion and the small elastic deformation. The Ritz method approximates the flexible deformation as a product of known shape functions and the unknown time-dependent flexible coordinates. The structural flexibility in multibody systems is mainly due to both link and joint deflections. However, depending on a system under study, contributions of these deflections vary considerably. The serial robots like space shuttle arm, because of its long slender links, display dominant link compliance [2] compared to industrial robots whose link deflections are generally small because of short link lengths [49]. The dynamic model of a flexible system is a set of highly coupled and nonlinear equations which involve inertial, Coriolis, centrifugal, and elastic forces.

Different beam theories, namely Euler–Bernoulli beam theory, Timoshenko beam theory, and Euler–Bernoulli beam model with rotary inertia have been used by different researchers for dynamic analysis of a flexible beam. In Euler–Bernoulli beam theory [85], the assumption is made that the beam's cross-section always remains normal to its neutral axis (before and after deformation). In this theory, the transverse shear deformation is neglected. Midha et al. [86], Jain and Rodriguez [66], Boyer et al. [19] and Chandrashaker and Ghosal [25] have obtained the dynamic model of a serial flexible system considering each link as Euler–Bernoulli beam. The

Euler–Bernoulli model provides results with good accuracy for beams with cross-sectional dimensions less than one-tenth of the beam length. The flexible multibody formulation using Timoshenko beam theory is provided by Bakr and Shabana [10], Iura and Atluri [65], Wang et al. [133], and Shabana [114]. This theory takes into account both the transverse shear deformation and the rotary inertia. The model gives results with good accuracy for thick beams with lengths more than three times the cross-sectional dimensions.

In the flexible modeling, the kinematic description of the finite discretization of each link is very crucial. The most common approaches are lumped parameters [49, 137], finite element method (FEM) [3, 51, 55, 56, 92, 93, 114, 130], and the assumed mode method (AMM) [85, 89, 114, 129].

In the lumped-parameter approach, deflection of the flexible link is described by an appropriate number of fictitious/pseudo-joints. Each non-actuated pseudo-joint is accompanied by a linear spring to restrict the joint motion and model the flexibility. Additional fictitious joints may be added to represent the effect of rotational vibrations. Fresonke et al. [49] developed an end-effector deflection prediction scheme in terms of geometrically dependent influence coefficients that determine the stiffness of the linear springs with the pseudo-joints. Yoshikawa and Hosoda [137] have described the flexible deflection using the concept of virtual links and passive joints and modeled each link individually in a method which closely resembles the elastodynamics modeling technique [121]. However, the lumped-parameter method is generally not preferred by the researchers because of difficulty in determining the spring constants for the pseudo-joints. Moreover, only very small elastic deflections can be modeled accurately using this technique.

Hence, the other two methods, namely finite element method (FEM) and assumed mode method (AMM), are more commonly used by the researchers to model flexible deflection of the links. In FEM, each flexible link is modeled by a finite number of elements. The shape, number, and deformation properties of these elements depend not only on the construction of the link but also on the dynamic requirements imposed on the control system [121]. A beam-like finite element provides the mathematical description of the displacement, force, and moment acting at the ends, whereas a mass matrix added to each element accounts for the inertia effect. Mathematical formulations of the properties of a number of different element types are available and can be used for a computer-aided formulation [121]. In the AMM formulation, the link flexibility is usually represented by the truncated finite modal series in terms of spatial mode eigenfunctions and time-varying model amplitude. Shabana [114] has shown that the FEM and AMM are analogous and comparable to each other. Usoro et al. [131], Naganathan and Soni [91], Chang and Shabana [27], Chedmail et al. [30], Ambrosio [3], Shyu and Gill [121], Geradin and Cardona [50], Lim and Taylor [78], and Shabana [114] have used the FEM approach to model flexible deflection of the links. Forward dynamics algorithm for flexible manipulators using FEM-based kinematic discretization of link deflection are given by Ryu et al. [101], Surdilović and Vukobratović [128], and Znamenáček and Valášek [138]. On the other Book [18], Cetinkunt and Book [23] Cyril [36], Asada et al. [8], De Luca and Siciliano [38], Cetinkunt and Ittoop [24], Li and Sankar [77], D'Eleuterio and Barfooy [38], Martins et al. [82] and Hwang [63] have used AMM approach to formulate link deflection and obtain the dynamic model of flexible link robots. Efficient simulation algorithms for flexible manipulators using AMM-based kinematic discretization of link deflection is given by Cyril [36], De Luca and Siciliano [39], Jain and Rodriguez [66], Theodore and Ghosal [129], and others. Comprehensive comparisons of the computational efficiency and stability of the algorithms using FEM and AMM formulation are made by Sharf and Damaren [119] and Theodore and Ghosal [129]. The finite element methodology (FEM) has advantages of its generality and ease of implementation. In particular, the flexible links with non-uniform mass and stiffness distributions can be readily treated, and the scheme of FEM allows one to discretize bodies of complex shapes also. However, in the context of manipulator dynamics, the generality of the FEM scheme is usually a redundant feature, as the flexible links of manipulators, which are long and slender, are well approximated by the Euler–Bernoulli beam [85]. Moreover, in FEM, more than one element are required in order to obtain an acceptable approximation of a particular frequency. This leads to the introduction of extra nodal coordinates, along with high and spurious frequencies, which have several detrimental effects on the dynamic simulation, such as degradation of computational and integration efficiency, the increased dimensionality of the state vector, unstable closed-loop response and the numerical noise [129]. In order to overcome these limitations and reduce the overall flexible degree of freedom of the system at hand, the FEM is used by the researchers along with different modal condensation methods, namely static condensation [54], component mode synthesis (CMS) method [60], and a variant of the CMS method, Craig–Bampton method [33]. The above condensation methods are based on the partitioning of generalized coordinates into two sets of desired and superfluous coordinates, use of Ritz vectors for the transformation of the system at hand into a group of subsystems of reduced DOF, followed by the solution of the models of these subsystems. The three methods differ in the transformation matrix used to reduce the system mass and stiffness matrices. Shabana and Wehage [115], Agrawal and Shabana [1], Ambrosio and Nikravesh [4], and Cuadrado et al. [35] have presented the dynamic analysis of flexible links using component mode synthesis methods. In the assumed mode method (AMM), the link deflections are discretized using the Rayleigh–Ritz method into appropriately chosen shape functions and time-dependent amplitudes of vibrations [36, 39, 66, 85, 118, 129]. Thus, by using a specific number of eigenfunctions, one is assured of reproducing the corresponding natural frequency exactly. Moreover, in manipulator dynamics, higher modes can be dispensed with as the consideration of more than first two modes do not have significant effect on the results [36, 39]. Hence, although efficient dynamic models based on FEM approach of kinematic discretization are present in the literature [47], AMM is preferred by researchers to develop recursive and efficient algorithms for robots with flexible links. However, proper selection of the link boundary conditions and the mode eigenvalues is vital. Methods to select the most appropriate mode shapes are given in Sharf [118], Craig and Kurdila [34] and Mohan [88].

Numerous efficient dynamic algorithms for flexible multibody systems based on different approaches are available in the literature. Most of the algorithms found the difficulty of expressing the dynamic equations for the flexible links in NE form. On the other hand, the equations can be readily expressed in EL form. Although recursive algorithm based on the NE form of dynamic equations for flexible multibody systems has been proposed by Shabana [112], the researchers tend to prefer readily available EL form of the dynamic model [14, 73]. One of the approaches proposed in the literature to exploit the advantages of both the NE and EL formulations is the DeNOC-based approach. In that first, the dynamic model is expressed in readily available EL form and then, using the equivalence of NE and EL formulations, the dynamics model is written in the form as one writes the NE equations. Mohan and Saha [90] followed this approach to obtain the recursive algorithms for rigid–flexible multibody systems using the DeNOC matrices and the AMM discretization.

1.3.3 Dynamics of Closed-Loop Systems

The closed-loop systems have recently inspired many researchers for their superior features in many industrial and commercial applications, such as high structural stiffness, higher load-carrying capacity, and good precision. A large number of formulations are also available for closed-loop systems. With the advent of high-speed computing technology, computer-aided analysis of multibody systems emerged as an essential tool for the simulation of such systems. Mainly, two types of formulations are used for the closed-loop systems. One is to write the equations of motion of the system in terms of active and passive coordinates. Then convert all equations in the form of passive joint variables to those in terms of active joint variables only. For that, one needs to solve the kinematic equations to obtain the passive variables in terms of the active joint variables. Hence, the equations of motion get very complicated. Using such an approach, Saha and Schiehlen [107] obtained recursive dynamic algorithms for closed-loop systems using the DeNOC matrices. Later, Khan et al. [71] illustrated the effectiveness of the DeNOC-based method in modeling a parallel manipulator. The second method which is called as augmented formulation, where the system is converted into an equivalent open-loop subsystems by placing cuts at appropriate joints. The cut joint is replaced by suitable constraint forces to represent the actual presence of those joints. Such constraint forces are known in the literature as Lagrange multipliers. Such an approach, along with the concept of the DeNOC matrices was proposed by Chaudhary and Saha [29], where they introduced the philosophy of 'determinate' and 'indeterminate' subsystems. This approach leads to achieving the subsystem-level recursions for the inverse dynamics of closed-loop systems. Later, Koul et al. [76] provided the reduced-order formulations for the closed-loop systems, which avoided the inversion of the large generalized inertia matrix by calculating the constraint forces, i.e., the Lagrange multipliers, separately.

1.4 Numerical Stability

The phenomenon of numerical stability is due to the magnification of the errors caused by a coupled effect of the truncation and round-off errors. In context of the forward dynamics and simulation, the numerical stability of the algorithm becomes crucial and needs to be investigated appropriately. Since a numerical integration algorithm yields only an approximate solution to the exact response, the computed values of rate of displacement and displacement vectors contain some numerical error which compounds with every step and reduces numerical stability of the system at hand. An initial approach, to control the accumulation of errors and overcome the instability inherent with the numerical integration, is the Baumgarte stabilization [16], in which an artificial feedback, namely position and velocity terms, is added in the second derivative of the constraint equation of the system. Nikravesh [95] studied the numerical integration of the equations of motion with dependent coordinates without stabilization and with Baumgarte stabilization comparatively. The disadvantage of this method is that there is no reliable method for selecting the coefficients of the position and velocity terms. Improper selection of these coefficients can lead to erroneous results. Chang and Nikravesh [26] proposed different methods for selecting these coefficients for the artificial feedback to the constraint differential equations. The correct choice of coefficients for the integrations based on the Runga–Kutta method has been proposed by Lin and Huang [79]. It should be noted, however, that Baumgarte stabilization does not solve all possible instabilities, such as those arising due to near kinematic singular configurations [57]. This aspect works in favor of another stabilization method, namely augmented Lagrangian formulation. A brief overview of various stabilization methods is given in Neto and Ambrôsio [94] and Nikravesh [95].

Different researchers have devised various tests to investigate the numerical stability of their algorithms; however, there appears to be no standard investigation technique to test the numerical stability of an algorithm. The phenomenon of error magnification is available in Stewart [125]. Ider [64] has used the Jacobian matrix and investigated its rank for the conditions close to the singularity. Sharf and Damaren [119] have taken the example of a Canadarm robot with flexible links and compared four different models to investigate their simulation characteristics. They studied the drift in the energy of the system to test the numerical stability of the algorithms. Ellis et al. [44] have studied the numerical stability of the forward dynamic algorithms using methods based on energy conservation. Cloutier et al. [32] and Ascher et al. [9] have investigated the numerical stability aspects using the formulation stiffness phenomenon. They have used the condition number of the inertia matrix as an indication of the stability characteristics of the algorithms and compared the composite body method and the articulated body inertia methods. It is shown that a variant of the articulated body inertia method is better suited to deal with certain types of ill-conditioning than the composite rigid body method. Jain and Rodriguez [67, 68] have focused on the computation of the sensitivity of the mass matrix and developed an analytical expression for the same using spatial operator algebra. Stability aspects of a mobile robot based on the posture velocity error dynamics are proposed by Shim and Sung [120]. The linearized stability analysis of dynamical systems modeled using the finite element-based multibody formulations is done by Bauchau and Wang [15]. The emphasis is, however, more on the geometric stability of the systems and numerical stability aspects of dynamic algorithms is not covered. Moreover, most of the study on the stability of numerical algorithms, etc., is limited to the rigid multibody systems. The effect of the flexibility of links on the numerical aspects of forward dynamics algorithms still remains a relatively less explored area. Ching and Wang [31] studied the stability of a single flexible link in collision, which is, however, limited to structural stability of the link only and not extended to the numerical algorithm for analysis of flexible link. Pascal [97], on the other hand, has studied the dynamics and stability of a two degree of freedom oscillator under impact conditions. In this research, the numerical stability aspects of the proposed forward dynamics algorithm are studied, for both rigid and flexible link robotic systems. Two schemes: one based on power drift of the system, and other using the time duration for which simulation results match the desired results, are adopted to test the stability of the proposed algorithm. It is also shown that the consideration of flexibility of links has a significant effect on the stability of the algorithms, and it becomes an important criterion for the selection of any particular algorithm.

1.5 Computational Efficiency

In this section, computational efficiencies of the DeNOC-based algorithms are investigated and compared with those available in the literature. Figure 1.5 compares computational complexities required by several inverse dynamics algorithms when a system has 1-DOF (Fig. 1.5a), 2-DOF (Fig. 1.5b), 3-DOF (Fig. 1.5c), and equal numbers of 1-, 2-, and 3-DOF joints (Fig. 1.5d). It may be seen that the recursive inverse dynamics algorithm given in Shah et al. [116] outperforms all except the one by Balafoutis and Patel [11] when the system has only 1-DOF joints, as evident from Fig. 1.5a. This is clear from Fig. 1.5b–d. From Fig. 1.6, it is also clear that the forward dynamics algorithm explained in Shah et al. [116] performs better than any other algorithm available in the literature. This is mainly due to the implicit inversion of the GIM based on the reverse Gaussian elimination [102] of the GIM and simplification of the expressions associated with the multiple DOF joints [113]. In the tree-type robotic systems, such as biped and quadruped, where the DOF of the system is more than 30, and the system consists of many multiple DOF joints, the DeNOC-based algorithms significantly improve the computational efficiency.

It has been observed in Sect. 1.3 that the researchers have always strived for more and more computationally efficient dynamics algorithms. Generally, the floating-point operations in terms of multiplications and additions are counted in order to obtain the complexity of an algorithm. For example, Tables 1.1 and 1.2 report the

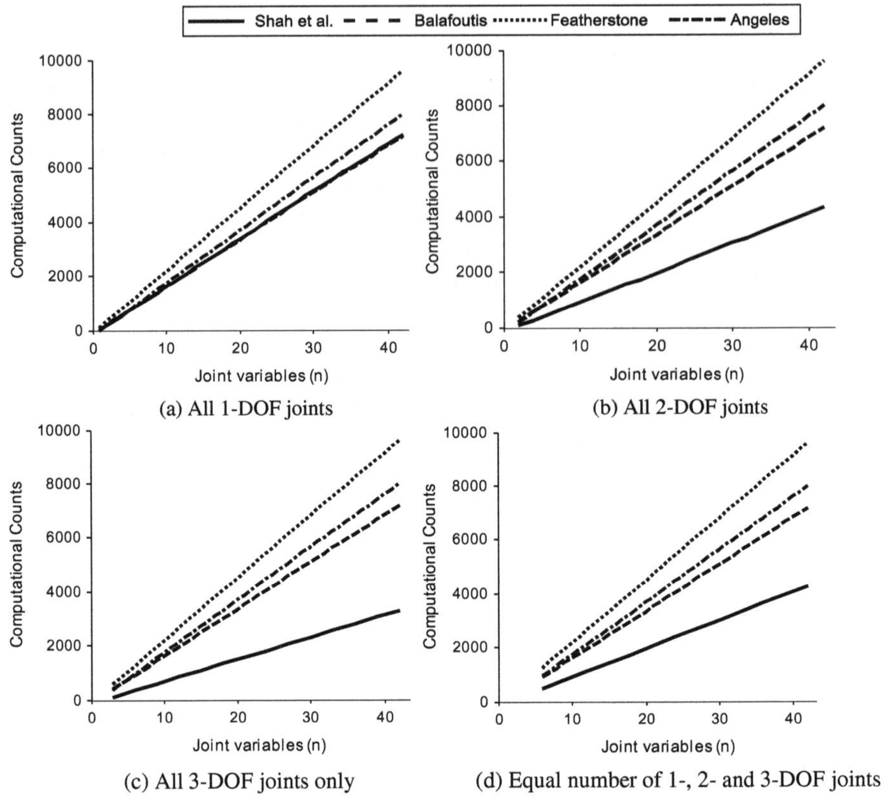

Fig. 1.5 Performance of several inverse dynamics algorithms [108, 116]

computational complexity of the dynamics algorithms for rigid and flexible systems, respectively.

In Table 1.2, numerical computations required for determination of the GIM **I**, and the terms associated with the centrifugal and Coriolis accelerations, i.e., $\boldsymbol{\varphi}$, for a six-link manipulator with each link vibrating in two modes in bending and one mode in torsion are given. In the simulation, the computational counts only do not provide the complete picture, as it comprises of two steps, forward dynamics and numerical integration, and the researchers report only the complexity of forward dynamics algorithm. Even though the complexity of numerical integration is known, the total number of iterations required to obtain the simulation results is not known in advance. Hence, the best forward dynamics and integration algorithms may perform worst if they require many iterations before the integrated values converge to a solution, thereby taking more CPU time of a computer. Such a phenomenon is associated with the numerical stability of an algorithm. What is emphasized here is that, in addition to the computational counts, one must study the CPU times required by a simulation algorithm for a particular application.

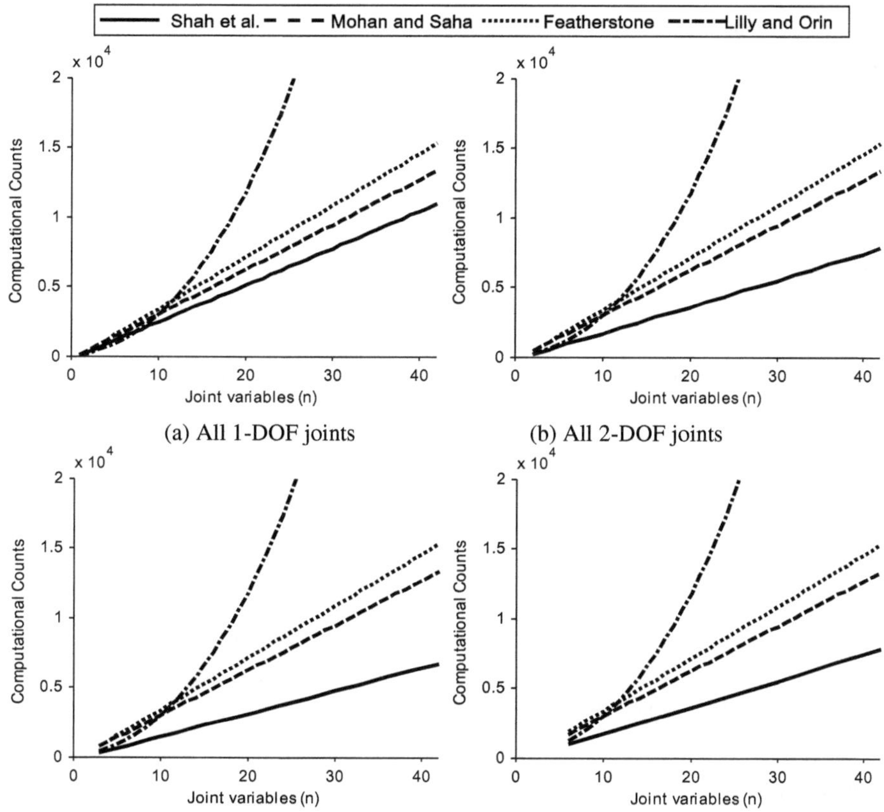

(a) All 1-DOF joints (b) All 2-DOF joints

(c) All 3-DOF joints only (d) Equal number of 1-, 2- and 3-DOF joints

Fig. 1.6 Performance of forward dynamics algorithms [108, 116]

1.6 Experimental Work

Experiments have generally been conducted by the researchers to quantify the physical parameters which cannot be easily modeled in the dynamic equations, namely damping, friction, etc. For example, Chapnik et al. [28] have experimentally investigated the projectile impact dynamics of an unactuated flexible beam. Mingli et al. [87] have conducted experiments on a two flexible link arm to identify the friction coming on the robot. Feliu et al. [48] have conducted experiments on a three degree of freedom robot to control its position using sensors and feedback loops. Mavroidis et al. [83] and Stieber et al. [126] have conducted experiments on the multilink robot whose end-effector is supported on a long and flexible link. Similarly, De Luca and Siciliano [40] have used the experimental results to study the regulation of the flexible arms under gravity. Another objective of the researchers to conduct experiments is to validate their theoretical models. Thus, Stanway et al. [123] have validated the dynamic simulation of the flexible link manipulators presented in Damaren and

Table 1.1 Computational complexity for rigid serial robots

Algorithm	Multiplication	Additions
(a) Inverse Dynamics		
Shah et al. [116] (DeNOC based)	$97n_r + 58n_u + 46\,ns$-81	$82n_r + 48n_u + 36\frac{2}{3}n_s - 75$
Saha [104] (DeNOC based)	$120n$-44	$97n$-55
Walker and Orin [132]	$137n$-22	$101n$-11
Luh et al. [80]	$150n$-48	$131n$-48
Hollerbach [59]	$412n$-277	$320n$-201
(b) Forward Dynamics		
Shah et al. [116] (DeNOC based)	$135n_r + 99n_u + 87\,ns$-116	$131n_r + 92n_u + 79\frac{2}{3}n_s - 123$
Mohan and Saha [89] (DeNOC based)	$173n$-128	$150n$-133
Featherstone [45]	$199n$-198	$174n$-173
Walker and Orin [132] (as implemented by Featherstone)	$\frac{1}{6}n^3 + \frac{23}{2}n^2 + \frac{115}{3}n - 47$	$\frac{1}{6}n^3 + 7n^2 + \frac{233}{6}n - 46$
Stejskal and Valášek [124]	$226n$-343	$206n$-345
Brandl et al. [20]	$250n$-222	$220n$-198

$n = n_r + n_u + n_s$, where n_r, n_u, and n_s, are the total DOF associated with all revolute, universal, and spherical joints, respectively

Table 1.2 Computations required for determination of GIM and $\boldsymbol{\phi}$ of a six-flexible link robot (each link vibrating in two modes in bending and one mode in torsion)

Algorithm	Multiplication	Additions
Book [18]	34,593	33,653
King et al. [75]	9,799	8,064
Cyril [36]	7,612	6,526
Mohan and Saha [89]	3,635	2,893

Sharf [37]. Both the time and frequency domain analyses were presented. Similarly, Brogliato et al. [21], Queiroz et al. [99] and Feliu et al. [48] have validated the dynamic models proposed by them using experimental verifications and extended the results for feedback control. Various experimental methodologies vary in their approaches, e.g., the type of sensors to measure vibrations, the architecture of the robot, initial conditions, etc., and make trade-offs among the contradicting requirements. For example, while the frequency of rotational joint oscillations is moderate and the frequency of vibrations of the flexible link is very high, it is difficult to measure both using a single sensor. Consequently, various researchers have adopted a different combination of sensors, data acquisition systems, and actuators to study the tip performance characteristics. Moreover, a study conducted under open-loop control provides a better scenario for simulation validation since the use of feedback tends to mask some of the flexibility effects present in the system [123]. In the present book, experiments have been conducted, (i) to estimate the damping to be incorporated in the model, and (ii) to validate the simulation results.

1.7 Important Features of the Book

Computational efficiency is one of the important criteria for selecting a dynamic modeling technique. For flexible systems, the computational efficiency and simulation time gained prime importance due to the need for real-time controls. There are several numerical issues present associated with the computer simulation of flexible multibody systems. These issues are related to the use of detailed finite element models that contain high frequencies as well as the need to efficiently solve differential and algebraic equations of the flexible multibody systems, etc. Hence for flexible systems, still there is a need for an efficient simulation technique for real-time applications.

The contents of this book are broadly classified into the following categories:

1. Dynamic modeling of open-loop serial-chain systems with rigid and flexible links.
2. Dynamic modeling of closed-loop systems with rigid and flexible links.
3. Study of computational efficiency and numerical stability of dynamic algorithm based on DeNOC-based methodology.
4. Experimental study of flexible system.

The important contributions in the book are outlined below:

1. Dynamic modeling of serial-chain robots with only rigid and rigid–flexible links robots using hybrid formulation and the DeNOC matrices.
2. Physical interpretations of many matrices and vectors associated with dynamic models of serial flexible robots.
3. An O(n)–n being the total number of links in the serial-chain robot–recursive, numerically stable, and computationally most efficient inverse and forward dynamics algorithms for serial-chain rigid link robots.
4. Recursive O(n) + O(m^3/3)–m being the maximum number of flexible link modes of vibration–numerically stable, and computationally efficient forward dynamics algorithm for serial-chain flexible link robots.
5. The effectiveness of the above formulations in items 1 and 3 is shown by analyzing two popular industrial robots, namely, spatial serial-chain PUMA robot and Stanford arm, whereas the formulations in items 1 and 4 are applied to (a) single-link, two-link, and three-link robot arms with one and all links flexible; (b) two-link robot arms with its inner link rigid and the outer one flexible (RF system), and vice-versa, i.e., the inner link is flexible and the outer one is rigid (FR system); and (c) Space Shuttle Remote Manipulator System (SSRMS).

6. A dynamic formulation suitable for the use of existing dynamics algorithms for serial-chain systems by the closed-loop system is developed.
7. The addition of geometric stiffness to the existing formulation of rigid–flexible systems by extending its capability to solve for the high speed and large deflection problems.
8. The algorithms for closed-loop rigid–flexible systems for improved efficiency.
9. Computationally efficient forward dynamics algorithm for closed-loop systems with rigid or flexible or rigid–flexible links based on the reverse Gaussian elimination (RGE) of the generalized inertia matrix (GIM). This step computes the generalized accelerations from the equations of motion.
10. An alternative approach for inverse dynamics algorithm for closed-loop systems with rigid or flexible or rigid–flexible links which simplifies the closed-loop formulations. It uses the information of passive joints to obtain the constraint forces and the active joints for joint torques/forces.
11. Dynamic analyses and simulations of several open-loop, closed-loop systems with single and multiloop systems, namely rigid, rigid–flexible, and full flexible (i) four-bar mechanisms, (ii) five-bar mechanisms, (iii) three-DOF parallel manipulators, and (iv) a spatial 4-bar mechanism which would help researchers and practicing engineers to use these results for design, simulation and control of those systems.

1.8 Organization of the Book

The book has seven chapters and an appendix. They are organized as follows:

Chapter 1: Introduction
The significance of the study on flexible system dynamics and its computational complexity is explained. Research work in the past on various issues relating to the dynamics of multibody systems is reviewed. The literature survey is divided into the following parts: dynamic modeling of rigid open-loop serial-chain systems, flexible systems, closed-loop systems, and computational efficiency. Modeling techniques used by other researchers to achieve high computational efficiency are also reported. Various discretization techniques used to kinematically model the deflections of flexible links are compared. Criteria used by various researchers to investigate the computational efficiency of simulation algorithms are reviewed as well. Finally, based on the advantages of the DeNOC-based methodology over others are described.

Chapter 2: Decoupled Natural Orthogonal Complement (DeNOC)-Based Dynamic Formulation
In this chapter, the dynamic modeling of serial rigid link robots is presented. The dynamic modeling based on the hybrid formulation and the decoupled natural orthogonal complement (DeNOC) matrices is presented. First, some definitions are given, and DeNOC matrices for the rigid robots are introduced. The recursive relations for the elements of the matrices and vectors associated with the equations

of motion are presented. Also, a recursive forward dynamic algorithm for serial rigid robots and its computational complexity are presented.

Chapter 3: Dynamics of Serial Rigid–Flexible Robots

The dynamic modeling of serial-chain rigid–flexible systems comprising of both rigid and flexible links is presented in this chapter. The dynamic equations of motion for serial-chain rigid–flexible systems are obtained using the decoupled natural orthogonal complement (DeNOC) matrices. The analytical expressions for the matrices and the vectors associated with the dynamic equations of motion of the system are also obtained. The effect of geometric stiffness is incorporated in the existing DeNOC-based formulation to deal with the high speed and large displacement problems. The recursive forward dynamic algorithm for serial-chain rigid–flexible robots and its computational complexity are presented. Furthermore, The algorithm was illustrated with the examples of different rigid–flexible systems.

Chapter 4: Dynamics of Six-Link Spatial Robot Arm

Simulation results of the six-link spatial industrial and space robots, namely, PUMA robot, Space Shuttle Remote Manipulator System (SSRMS), and Stanford arm, with only rigid and rigid–flexible links, are presented in this chapter.

Chapter 5: Dynamics of Closed-loop Systems

This chapter presents the formulation of dynamic equations of motion of closed-loop rigid–flexible multibody systems using the DeNOC matrices. The methodology allows one to exploit the existing dynamic algorithms for the open-loop serial-chain systems to solve for the closed-loop systems. To illustrate the concept and methodology, several planar closed-loop systems, with single and multi-closed-loops, namely four-bar, five-bar mechanisms, and the three-DOF parallel manipulator, were undertaken. The results were validated with the results from a commercial software RecurDyn. An alternative method for inverse dynamics is also presented in this chapter. Further, the forced simulation is carried out to study the effect of flexibility in a system. It is illustrated with the three-DOF parallel manipulator.

Chapter 6: Dynamics of Spatial Four-bar Mechanism

The formulation of dynamic equations of motion of the spatial closed-loop rigid–flexible multibody system using DeNOC matrices is presented in this chapter. It is based on the same cut joint methodology adopted for the closed-loop systems of Chapter 4 to form several open-loop subsystems and subsequently use existing dynamics algorithms available for the open-loop systems.

Chapter 7: Numerical Stability and Efficiency

In this chapter, the numerical stability aspects and the computational efficiencies of the proposed forward dynamics algorithms for the rigid and rigid–flexible serial robotic systems are analyzed. First, the importance of the study of numerical stability characteristics is underlined, and various criteria for testing it are listed. Then, using a series of numerical experiments, their behaviors are compared. Examples of the spatial six-link rigid PUMA robot, the planar three-link rigid–flexible Canadarm, and the spatial six-link rigid–flexible Space Shuttle Remote

Manipulator System are considered to demonstrate the stability and efficiency of the proposed algorithms.

Chapter 8: Experimental Results

A series of experiments were conducted on a single flexible link compound pendulum and a robot arm with two flexible links. The experiments have been conducted (i) to validate the theoretical model and (ii) to incorporate the damping into the model. The modifications in the dynamic model due to the incorporation of the joint and structural damping factors are presented. The experimental results are then compared with the simulation results to verify the proposed modeling.

1.9 Summary

This chapter briefly introduces the dynamics of multibody systems, contribution and contents are highlighted. Research work in the past on various issues relating to the dynamics of multibody systems is reviewed. The literature survey is carried out in the following topics: dynamic modeling of rigid open-loop serial-chain systems, flexible systems, closed-loop systems, and computational efficiency. Criteria used by various researchers to investigate the computational efficiency of simulation algorithms are reviewed as well. Finally, based on the advantages of the DeNOC-based methodology over others are described.

Bibliography

1. Agrawal OP, Shabana AA (1985) Dynamic analysis of multibody systems using component modes. Comput Struct 21(6):1303–1312. https://doi.org/10.1016/0045-7949(85)90184-1
2. Alberts TE, Xia H, Chen Y (1992) Dynamic analysis to evaluate viscoelastic passive damping augmentation for the space shuttle remote manipulator system. J Dyn Syst Meas Contr 114(3):468–475. https://doi.org/10.1115/1.2897370
3. Ambrosio J (1996) Dynamics of structures undergoing gross motion and nonlinear deformations: a multibody approach. Comput Struct 59(6):1001–1012
4. Ambrosio JAC, Nikravesh PE (1992) Elasto-plastic deformations in multibody dynamics. Nonlinear Dyn 3(2):85–104
5. Angeles J, Lee SK (1988) Formulation of dynamical equations of holonomic mechanical systems using a natural orthogonal complement. J Appl Mech, Trans ASME 55(1):243–244
6. Angeles J, Ma O (1988) Dynamic simulation of n-axis serial robotic manipulators using a natural orthogonal complement. Int J Robot Res 7(5):32–47
7. Angeles J, Ma O, Rojas A (1989) An algorithm for the inverse dynamics of n-axis general manipulators using Kane's equations. Comput Math Appl 17(12):1545–1561
8. Asada H, Ma Z-D, Tokumaru H (1990) Inverse dynamics of flexible robot arms: modeling and computation for trajectory control. J Dyn Syst Meas Contr 112(2):177–185. https://doi.org/10.1115/1.2896124
9. Ascher UM, Pai DK, Cloutier BP (1997) Forward dynamics, elimination methods, and formulation stiffness in robot simulation. Int J Robot Res 16(6):749–758
10. Bakr EM, Shabana AA (1987) Timoshenko beams and flexible multibody system dynamics. J Sound Vib 116(1):89–107

11. Balafoutis CA, Patel RV (1991) Dynamic analysis of robot manipulators: a Cartesian tensor approach, vol 131. Springer Science & Business Media
12. Baruh H, Tadikonda SSK (1989) Issues in the dynamics and control of flexible robot manipulators. J Guid Control Dyn 12(5):659–671. https://doi.org/10.2514/3.20460
13. Bauchau OA (1998) Computational schemes for flexible, nonlinear multi-body systems. Multibody Syst Dyn 2(2):169–225
14. Bauchau OA (2000) On the modeling of prismatic joints in flexible multi-body systems. Comput Meth Appl Mech Eng 181(1–3):87–105
15. Bauchau OA, Wang J (2006) Stability analysis of complex multibody systems. J Comput Nonlinear Dyn 1(1):71–80
16. Baumgarte J (1972) Stabilization of constraints and integrals of motion in dynamical systems. Comput Methods Appl Mech Eng 1(1):1–16
17. Blajer W, Schiehlen W (1992) Walking without impacts as a motion/force control problem. J Dyn Syst, Meas Control, Trans ASME 114(4):660–665
18. Book WJ (1984) Recursive lagrangian dynamics of flexible manipulator arms. Int J Robot Res 3(3):87–101
19. Boyer F, Glandais N, Khalil W (2002) Flexible multibody dynamics based on a non-linear Euler-Bernoulli kinematics. Int J Numer Meth Eng 54(1):27–59
20. Brandl H, Johanni R, Otter M (1986) A very efficient algorithm for the simulation of robots and similar multibody systems without inversion of the mass matrix. IFAC Proc Volumes 19(14):95–100
21. Brogliato B, Rey D, Pastore A, Barnier J (1998) Experimental comparison of nonlinear controllers for flexible joint manipulators. Int J Robot Res 17(3):260–281. https://doi.org/10.1177/027836499801700304
22. Cameron JM, Book WJ (1997) Modeling mechanisms with nonholonomic joints using the Boltzmann-Hamel equations. Int J Robot Res 16(1):47–59
23. Cetinkunt S, Book WJ (1987) Symbolic modeling of flexible manipulators, pp 2074–2080
24. Cetinkunt S, Ittoop B (1992) Computer-automated symbolic modeling of dynamics of robotic manipulators with flexible links. IEEE Trans Robot Autom 8(2):94–105. https://doi.org/10.1109/70.127243
25. Chandrashaker M, Ghosal A (2006) Nonlinear modeling of flexible manipulators using nondimensional variables. J Comput Nonlinear Dyn 1(2):123–134
26. Chang B, Nikravesh P (1985) An adaptive constraint violation stablisation method for dynamic analysis of mechanical systems. Trans ASME Appl Mech 104:488–492
27. Chang B, Shabana AA (1990) Nonlinear finite element formulation for the large displacement analysis of plates. J Appl Mech 57(3):707–718. https://doi.org/10.1115/1.2897081
28. Chapnik BV, Heppler GR, Aplevich JD (1991) Modeling impact on a one-link flexible robotic arm. IEEE Trans Robot Autom 7(4):479–488. https://doi.org/10.1109/70.86078
29. Chaudhary H, Saha SK (2007) Constraint force formulation for closed-loop multibody systems. Trans ASME J Mech Des 129:1234–1242
30. Chedmail P, Aoustin Y, Chevallereau C (1991) Modelling and control of flexible robots. Int J Numer Meth Eng 32(8):1595–1619
31. Ching FMC, Wang D (2003) Exact solution and infinite-dimensional stability analysis of a single flexible link in collision. IEEE Trans Robot Autom 19(6):1015–1020. https://doi.org/10.1109/TRA.2003.819716
32. Cloutier BP, Pai DK, Ascher UM (1995) Formulation stiffness of forward dynamics algorithms and implications for robot simulation, pp 2816–2822
33. Craig RR, Bampton MCC (1968) Coupling of substructures for dynamic analyses. AIAA J 6(7):1313–1319. https://doi.org/10.2514/3.4741
34. Craig RR, Kurdila AJ (2006) Fundamentals of structural dynamics. John Wiley & Sons
35. Cuadrado J, Cardenal J, García de Jalón J (1996) Flexible mechanisms through natural coordinates and component synthesis: an approach fully compatible with the rigid case. Int J Numer Meth Eng 39(20):3535–3551
36. Cyril X (1988) Dynamics of flexible-link manipulators. McGill University, Canada

37. Damaren C, Sharf I (1995) Simulation of flexible-link manipulators with inertial and geometric nonlinearities. J Dyn Syst Meas Contr 117(1):74–87. https://doi.org/10.1115/1.2798525
38. D'Eleuterio GMT, Barfooy TD (1999) Just a second, we'd like to go first: a first-order discreized formulation for structural dynamics. In: Proceedings of the fourth international conference on dynamics and controls, London, pp 1–24
39. De Luca A, Siciliano B (1991) Closed-form dynamic model of planar multilink lightweight robots. IEEE Trans Syst Man Cybern 21(4):826–839
40. De Luca A, Siciliano B (1993) Regulation of flexible arms under gravity. IEEE Trans Robot Autom 9(4):463–467. https://doi.org/10.1109/70.246057
41. Dimitrov D (2005) Dynamics and control of space manipulators during a satellite capturing operation. Tohoku University, Graduate School of Engineering, Japan
42. Dwivedy SK, Eberhard P (2006) Dynamic analysis of flexible manipulators, a literature review. Mech Mach Theory 41(7):749–777
43. Eberhard P, Schiehlen W (2006) Computational dynamics of multibody systems: history, formalisms, and applications. J Comput Nonlinear Dyn 1(1):3–12
44. Ellis RE, Ismaeil OM, Carmichael IH (1992) Numerical stability of foward-dynamics algorithms, pp 305–311
45. Featherstone R (1983) Calculation of robot dynamics using articulated-body inertias. Int J Robot Res 2(1):13–30
46. Featherstone R, Orin D (2000) Robot dynamics: equations and algorithms. In: IEEE international conference on robotics and automation 2000, vol 821. pp 826–834. https://doi.org/10.1109/robot.2000.844153
47. Feliu JJ, Feliu V, Cerrada C (1999) Load adaptive control of single-link flexible arms based on a new modeling technique. IEEE Trans Robot Autom 15(5):793–804. https://doi.org/10.1109/70.795785
48. Feliu V, García A, Somolinos JA (2001) Gauge-based tip position control of a new three-degree-of-freedom flexible robot. Int J Robot Res 20(8):660–675. https://doi.org/10.1177/02783640122067598
49. Fresonke D, Hernandez E, Tesar D (1988) Deflection prediction for serial manipulators. In: IEEE international conference on robotics and automation, pp 482–487
50. Geradin M, Cardona A (2001) Flexible multibody dynamics: a finite element approach. John Wiley and Sons, New York
51. Geradin M, Cardona A, Doan DB, Duysens J (1994) Finite element modeling concepts in multibody dynamics. In: Computer-aided analysis of rigid and flexible mechanical systems, pp 233–284
52. Greenwood DT (1988) Principles of dynamics. Prentice-Hall of India, New Delhi
53. Gupta V, Chaudhary H, Saha SK (2015) Dynamics and actuating torque optimization of planar robots. J Mech Sci Technol 29(7):2699–2704. https://doi.org/10.1007/s12206-015-0517-z
54. Guyan RJ (1965) Reduction of stiffness and mass matrices. AIAA J 3(2):380–380. https://doi.org/10.2514/3.2874
55. Hamper MB, Recuero AM, Escalona JL, Shabana AA (2011) Modeling rail flexibility using finite element and finite segment methods
56. Hamper MB, Recuero AM, Escalona JL, Shabana AA (2012) Use of finite element and finite segment methods in modeling rail flexibility: a comparative study. J Comput Nonlinear Dyn 7(4)
57. Haug EJ (1989) Computer aided kinematics and dynamics of mechanical systems. Basic methods, vol 1. Allyn & Bacon, Inc.
58. Hemami H, Weimer F (1981) Modeling of nonholonomic dynamic systems with applications. In: American Society of Mechanical Engineers and American Society of Civil Engineers, Joint Applied Mechanics, Fluids Engineering, and Bioengineering Conference, University of Colorado, Boulder, CO
59. Hollerbach JM (1980) A recursive lagrangian formulation of maniputator dynamics and a comparative study of dynamics formulation complexity. IEEE Trans Syst Man Cybern 10(11):730–736

60. Hurty WC (1965) Dynamic analysis of structural systems using component modes. AIAA J 3(4):678–685. https://doi.org/10.2514/3.2947
61. Huston RL (1991) Multibody dynamics-modeling and analysis methods. Appl Mech Rev 44(3):109–117
62. Huston RL, Passerello CE (1974) On constraint equations-a new approach. ASME J Appl Mech 41:1130–1131
63. Hwang YL (2005) A new approach for dynamic analysis of flexible manipulator systems. Int J Non-Linear Mech 40(6):925–938. https://doi.org/10.1016/j.ijnonlinmec.2004.12.001
64. Ider SK (1990) Stability analysis of constraints in flexible multibody systems dynamics. Int J Eng Sci 28(12):1277–1290
65. Iura M, Atluri S (1995) Dynamic analysis of planar flexible beams with finite rotations by using inertial and rotating frames. Comput Struct 55(3):453–462
66. Jain A, Rodriguez G (1992) Recursive flexible multibody system dynamics using spatial operators. J Guid Control Dyn 15(6):1453–1466
67. Jain A, Rodriguez G (2000) Sensitivity analysis for multibody systems using spatial operators. In: International conference (VI) on methods and models in automation and robotics, pp 30–31
68. Jain A, Rodriguez G (2003) Multibody mass matrix sensitivity analysis using spatial operators. Int J Multiscale Comput Eng 1(2–3)
69. Kamman JW, Huston RL (1984) Dynamics of constrained multibody systems. J Appl Mech, Trans ASME 51(4):899–903
70. Kane TR, Levinson DA (1983) The use of Kane's dynamical equations in robotics. Int J Robot Res 2(3):3–21
71. Khan WA, Krovi VN, Saha SK, Angeles J (2005) Modular and recursive kinematics and dynamics for parallel manipulators. Multibody Sys Dyn 14(3–4):419–455
72. Khatib O (1987) A unified approach for motion and force control of robot manipulators: the operational space formulation. IEEE J Robot Autom 3(1):43–53. https://doi.org/10.1109/jra.1987.1087068
73. Kim SS, Haug EJ (1988) A recursive formulation for flexible multibody dynamics, part I: open-loop systems. Comput Meth Appl Mech Eng 71(3):293–314
74. Kim SS, Vanderploeg MJ (1986) A General and efficient method for dynamic analysis of mechanical systems using velocity transformations. J Mech Trans Autom Des 108(2):176–182
75. King JO, Gourishankar VG, Rink RE (1987) Lagrangian dynamics of flexible manipulators using angular velocities instead of transformation matrices. IEEE Trans Syst Man Cybern 17(6):1059–1068. https://doi.org/10.1109/TSMC.1987.6499316
76. Koul M, Shah SV, Saha SK, Manivannan M (2014) Reduced-order forward dynamics of multiclosed-loop systems. Multibody Syst Dyn 31(4):451–476. https://doi.org/10.1007/s11044-013-9379-2
77. Li C-J, Sankar TS (1993) Systematic methods for efficient modeling and dynamics computation of flexible robot manipulators. IEEE Trans Syst Man Cybern 23(1):77–94
78. Lim H, Taylor RL (2001) An explicit–implicit method for flexible–rigid multibody systems. Finite Elem Anal Des 37(11):881–900
79. Lin S-T, Huang J-N (2002) Stabilization of Baumgarte's method using the Runge-Kutta approach. J Mech Des 124(4):633–641
80. Luh JYS, Walker MW, Paul RPC (1980) On-line computational scheme for mechanical manipulators. J Dyn Syst, Meas Control, Trans ASME 102(2):69–76
81. Mani N, Haug E, Atkinson K (1984) Application of singular value decomposition for analysis of mechanical system dynamics. J Mech Trans Autom Des 107(1):82–87
82. Martins JM, Miguel AB, da Costa J (2003) On the formulation of flexible multibody systems with constant mass matrix. In: Proceedings of the 9th ECCOMAS thematic conference on multibody dynamics, Lisbon, July 1–4
83. Mavroidis C, Rowe P, Dubowsky S (1995) Inferred end-point control of long reach manipulators. In: Proceedings 1995 IEEE/RSJ international conference on intelligent robots and systems. Human robot interaction and cooperative robots, 5–9, vol 72, pp 71–76. https://doi.org/10.1109/IROS.1995.526141

84. McPhee JJ (1996) On the use of linear graph theory in multibody system dynamics. Nonlinear Dyn 9(1):73–90
85. Meirovitch L (1967) Analytical methods in vibration. The Mcmillan Company, New York, NY
86. Midha A, Erdman AG, Frohrib DA (1978) Finite element approach to mathematical modeling of high-speed elastic linkages. Mech Mach Theory 13(6):603–618. https://doi.org/10.1016/0094-114X(78)90028-9
87. Mingli B, Dong Hua Z, Schwarz H (1999) Identification of generalized friction for an experimental planar two-link flexible manipulator using strong tracking filter. IEEE Trans Robot Autom 15(2):362–369. https://doi.org/10.1109/70.760359
88. Mohan A (2007) Dynamic analysis of flexible multibody robotic systems. IIT, Delhi
89. Mohan A, Saha SK (2007) A recursive, numerically stable, and efficient simulation algorithm for serial robots. Multibody Sys Dyn 17(4):291–319
90. Mohan A, Saha SK (2009) A recursive, numerically stable, and efficient simulation algorithm for serial robots with flexible links. Multibody Sys Dyn 21(1):1–35
91. Naganathan G, Soni AH (1987) Coupling effects of kinematics and flexibility in manipulators. Int J Robot Res 6(1):75–84. https://doi.org/10.1177/027836498700600106
92. Nagarajan S, Turcic DA (1990) Lagrangian formulation of the equations of motion for elastic mechanisms with mutual dependence between rigid body and elastic motions: part I—element level equations. J Dyn Syst Meas Contr 112(2):203–214. https://doi.org/10.1115/1.2896127
93. Nagarajan S, Turcic DA (1990) Lagrangian formulation of the equations of motion for elastic mechanisms with mutual dependence between rigid body and elastic motions: part II—system equations. J Dyn Syst Meas Contr 112(2):215–224
94. Neto MA, Ambrôsio J (2003) Stabilization methods for the integration of DAE in the presence of redundant constraints. Multibody Sys Dyn 10(1):81–105
95. Nikravesh PE (1988) Computer-aided analysis of mechanical systems. Prentice-Hall Englewood Cliffs, New Jersey
96. Park FC, Bobrow JE, Ploen S (1995) A Lie group formulation of robot dynamics. Int J Robot Res 14(6):609–618
97. Pascal M (2005) Dynamics and stability of a two degree of freedom oscillator with an elastic stop. J Comput Nonlinear Dyn 1(1):94–102. https://doi.org/10.1115/1.1961873
98. Paul B (1975) Analytical dynamics of mechanisms—a computer oriented overview. Mech Mach Theory 10(6):481–507. https://doi.org/10.1016/0094-114X(75)90005-1
99. Queiroz MSd, Dawson DM, Agarwal M, Zhang F (1999) Adaptive nonlinear boundary control of a flexible link robot arm. IEEE Trans Robot Autom 15(4):779–787. https://doi.org/10.1109/70.782034
100. Roberson RE, Schwertassek R (1988) Dynamics of multibody systems, vol 18. Springer-Verlag, Berlin
101. Ryu J, Sung-Soo K, Sang-Sup K (1992) An efficient computational method for dynamic stress analysis of flexible multibody systems. Comput Struct 42(6):969–977
102. Saha SK (1997) A decomposition of the manipulator inertia matrix. IEEE Trans Robot Autom 13(2):301–304
103. Saha SK (1999) Analytical expression for the inverted inertia matrix of serial robots. Int J Robot Res 18(1):116–124
104. Saha SK (1999) Dynamics of serial multibody systems using the decoupled natural orthogonal complement matrices. J Appl Mech, Trans ASME 66(4):986–995
105. Saha SK (2003) Simulation of industrial manipulators based on the UDUT decomposition of inertia matrix. Multibody Sys Dyn 9(1):63–85
106. Saha SK, Angeles J (1991) Dynamics of nonholonomic mechanical systems using a natural orthogonal complement. J Appl Mech, Trans ASME 58(1):238–243
107. Saha SK, Schiehlen WO (2001) Recursive kinematics and dynamics for parallel structured closed-loop multibody systems. Mech Struct Mach 29(2):143–175
108. Saha SK, Shah SV, Nandihal PV (2013) Evolution of the DeNOC-based dynamic modeling for multibody systems. Mech Sci 4(1):1–20. https://doi.org/10.5194/ms-4-1-2013

109. Schiehlen W (1990) Multibody systems handbook. Springer-Verlag, New York Inc., New York, NY (USA)
110. Schiehlen W (1997) Multibody system dynamics: roots and perspectives. Multibody Sys Dyn 1(2):149–188
111. Shabana A (2001) Computational dynamics. Wiley-Interscience
112. Shabana AA (1990) Dynamics of flexible bodies using generalized Newton-Euler equations. J Dyn Syst, Meas Control, Trans ASME 112(3):496–503
113. Shabana AA (1997) Flexible multibody dynamics: review of past and recent developments. Multibody Sys Dyn 1(2):189–222
114. Shabana AA (2005) Dynamics of multibody systems. Cambridge University Press
115. Shabana AA, Wehage RA (1983) Coordinate reduction technique for dynamic analysis of spatial substructures with large angular rotations. J Struct Mech 11(3):401–431
116. Shah S, Saha S, Dutt J (2011) Modular framework for dynamic modeling and analyses of legged robots. Mech Mach Theory
117. Shah SV, Nandihal PV, Saha SK (2012) Recursive dynamics simulator (ReDySim): a multibody dynamics solver. Theor Appl Mech Lett 2(6):14–063011. https://doi.org/10.1063/2.1206311
118. Sharf I (1999) Nonlinear strain measures, shape functions and beam elements for dynamics of flexible beams. Multibody Sys Dyn 3(2):189–205
119. Sharf I, Damaren C (1992) Simulation of flexible-link manipulators: basis functions and nonlinear terms in the motion equations, pp 1956–1962
120. Shim HS, Sung YG (2004) Stability and four-posture control for nonholonomic mobile robots. IEEE Trans Robot Autom 20(1):148–154
121. Shyu YJ, Gill KF (1997) Dynamic modelling of planar flexible manipulators: computational and algorithmic efficiency. Proc Inst Mech Eng C J Mech Eng Sci 211(2):119–133. https://doi.org/10.1243/0954406971521700
122. Sousa PJ, Barros F, Tavares PJ, Moreira PMGP (2019) Displacement analysis of rotating RC helicopter blade using coupled CFD-FEA simulation and digital image correlation. Procedia Struct Integrity 17:812–821. https://doi.org/10.1016/j.prostr.2019.08.108
123. Stanway J, Sharf I, Damaren C (1996) Validation of a dynamics simulation for a structurally flexible manipulator. In: Proceedings of IEEE international conference on robotics and automation, 22–28 April 1996, vol 1953, pp 1959–1965. https://doi.org/10.1109/ROBOT.1996.506159
124. Stejskal V, Valášek M (1996) Kinematics and dynamics of machinery. M. Dekker
125. Stewart GW (1973) Introduction to matrix computations. Academic Press
126. Stieber ME, McKay M, Vukovich G, Petriu E (1999) Vision-based sensing and control for space robotics applications. IEEE Trans Instrum Meas 48(4):807–812. https://doi.org/10.1109/19.779178
127. Stokes A, Brockett R (1996) Dynamics of kinematic chains. Int J Robot Res 15(4):393–405
128. Surdilović D, Vukobratović M (1996) One method for efficient dynamic modeling of flexible manipulators. Mech Mach Theory 31(3):297–315
129. Theodore RJ, Ghosal A (1995) Comparison of the assumed modes and finite element models for flexible multilink manipulators. Int J Robot Res 14(2):91–111
130. Thompson B, Sung C (1986) A survey of finite element techniques for mechanism design. Mech Mach Theory 21(4):351–359
131. Usoro PB, Nadira R, Mahil SS (1986) Finite element/Lagrange approach to modeling lightweight flexible manipulators. J Dyn Syst, Meas Control, Trans ASME 108(3):198–205
132. Walker MW, Orin DE (1982) Efficient dynamic computer simulation of robotic mechanisms. J dyn syst meas control trans asme 104(3):205–211
133. Wang D, Meng M, Liu Y (1999) Influence of shear, rotary inertia on the dynamic characteristics of flexible manipulators. In: IEEE pacific rim conference on communications, computers and signal processing, pp 615–618
134. Wasfy TM, Noor AK (2003) Computational strategies for flexible multibody systems. Appl Mech Rev 56(6):553–613

135. Wehage RA, Haug EJ (1982) Generalized coordinate partitioning for dimension reduction in analysis of constrained dynamic systems. J Mech Des 104(1):247–255
136. Wittenburg J (1977) Dynamics of systems of rigid bodies. Teubner Stuttgart
137. Yoshikawa T, Hosoda K (1996) Modeling of flexible manipulators using virtual rigid links and passive joints. Int J Robot Res 15(3):290–299
138. Znamenáček J, Valášek M (1998) An efficient implementation of the recursive approach to flexible multibody dynamics. Multibody Sys Dyn 2(3):227–251

Part I
Open-Loop Serial-Chain Systems

Chapter 2
Dynamic Formulation Using the Decoupled Natural Orthogonal Complement (DeNOC)

In this chapter, dynamic formulation, i.e., the derivation of the dynamic equations of motion, is presented using the decoupled natural orthogonal complement (DeNOC). First, some definitions of the DeNOC matrices for rigid robots are given. The dynamic modeling presented here is based on the hybrid formulation. The use of the DeNOC matrices presented here shows the recursive relations for the elements of the matrices and vectors associated with the equations of motion. The associated physical interpretations for the matrices and vectors are also presented. The recursive relations are then used along with the reverse Gaussian elimination (Strang 1980) scheme to factorize the generalized inertia matrix into $\mathbf{UDU}^T$ form–$\mathbf{U}$ and $\mathbf{D}$ being the upper triangular and diagonal matrices. A recursive, $O(n)$ forward dynamics algorithm is then presented for the serial rigid link robots. The computational complexity of the algorithm is compared with that of the algorithms available in the literature, and it is shown that the proposed algorithm is computationally more efficient.

2.1 Kinematics

In this section, some definitions are introduced, and the DeNOC matrices for rigid robots are derived. A serial robotic system having a fixed base and n-moving rigid links is shown in Fig. 2.1. The links are numbered #1, ..., #n, which are coupled by n-one-degree-of-freedom kinematic pairs, revolute or prismatic, denoted by 1, ..., n. The shape of the ith link is shown in Fig. 2.2. Its geometry is defined as per the definitions of the Denavit and Hartenberg (DH) parameters, as presented in Appendix A. Note here that, any other link shape can have an equivalent shape shown in Fig. 2.2, which can be easily determined using the equivalence of the mass, center of mass, and moment of inertia of the link. Moreover, the effect of motor/actuator at the joint is modeled by considering hub inertia at the joint, as indicated in the figure.

P. V. Nandihal et al., *Dynamics of Rigid-Flexible Robots and Multibody Systems*, Intelligent Systems, Control and Automation: Science and Engineering 100,
https://doi.org/10.1007/978-981-16-2798-9_2

Fig. 2.1 A typical n-link serial robot

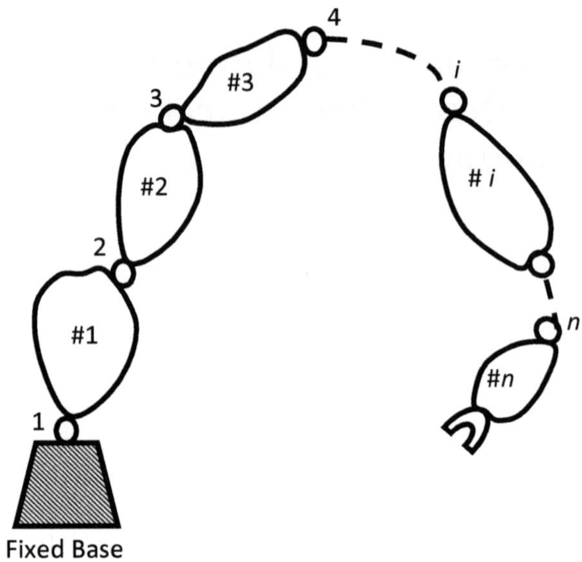

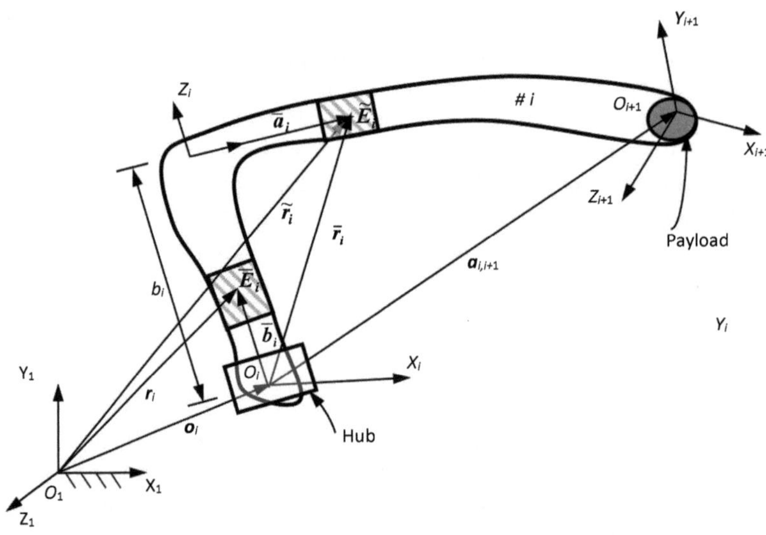

Fig. 2.2 ith rigid link

Furthermore, a point mass is assumed at the tip of the ith link, i.e., at O_{i+1}, to include the effect of any payload on it.

2.1.1 Some Definitions

Referring to the ith link, the following terms are now defined that will be used to derive the kinematic equations of the robotic system at hand:

- $\mathbf{t}_i$ and $\mathbf{w}_i$: The 6-dimensional twist and wrench of the ith link, respectively, i.e.,

$$\mathbf{t}_i \equiv \begin{bmatrix} \mathbf{v}_i^T & \boldsymbol{\omega}_i^T \end{bmatrix}^T; \mathbf{w}_i \equiv \begin{bmatrix} \mathbf{f}_i^T & \mathbf{n}_i^T \end{bmatrix}^T \tag{2.1}$$

where $\mathbf{v}_i$ and $\boldsymbol{\omega}_i$ are the 3-dimensional vectors of velocity of the point on the ith link where it is in contact with the (i-1)st link, i.e., O_i of Fig. 2.2, and its angular velocity, respectively. The vectors, $\mathbf{f}_i$ and $\mathbf{n}_i$, are accordingly the 3-dimensional vectors of force at O_i and the moment about O_i of the ith link, respectively. In this book, unlike Saha [4, 5] and others, the twist is defined with respect to point O_i instead of the mass center of the ith body, and putting the linear velocity first followed by the angular velocity, as in Eq. (2.1). Such a choice of the twist definition provides a simple structure for the various matrices and vectors associated with the dynamic equations of motion and is easily extendable to the flexible link robot dynamics, as shown in Chap. 3.

- $\mathbf{t}$ and $\mathbf{w}$: The $6n$-dimensional vectors of generalized twist and wrench, respectively, which are defined as

$$\mathbf{t} \equiv \begin{bmatrix} \mathbf{t}_1^T & \mathbf{t}_2^T & \cdots & \mathbf{t}_n^T \end{bmatrix}^T; \mathbf{w} \equiv \begin{bmatrix} \mathbf{w}_1^T & \mathbf{w}_2^T & \cdots & \mathbf{w}_n^T \end{bmatrix}^T \tag{2.2}$$

  where $\mathbf{t}_i$ and $\mathbf{w}_i$, for $i = 1, 2, \ldots, n$, are given in Eq. (2.1).

- $\dot{\boldsymbol{\theta}}$: The n-dimensional vector of joint rates, i.e.,

$$\dot{\boldsymbol{\theta}} \equiv \begin{bmatrix} \dot{\theta}_1 & \dot{\theta}_2 & \cdots & \dot{\theta}_n \end{bmatrix}^T \tag{2.3}$$

  where $\dot{\theta}_i$ is the time-rate change of the rotational or translational displacement of the ith joint, θ_i, depending on its type, i.e., revolute or prismatic, respectively.

2.1.2 The DeNOC Matrices

Referring to Fig. 2.2, the steps to obtain the decoupled natural orthogonal complement (DeNOC) matrices for serial rigid robots [4, 5] are outlined below:

(1) The twist of the ith link is expressed in terms of the (i-1)st link as

$$\mathbf{t}_i = \mathbf{R}_{i,i-1}\mathbf{t}_{i-1} + \mathbf{p}_i\dot{\theta}_i \tag{2.4}$$

where $\mathbf{t}_i$ is the 6-dimensional twist vector of the ith body defined in Eq. (2.1), and $\dot{\theta}_i$ is the ith joint rate. Moreover, the 6×6 matrix, $\mathbf{R}_{i,i\text{-}1}$, and the 6-dimensional vector, $\mathbf{p}_i$, are defined as follows:

$$\mathbf{R}_{i,i-1} \equiv \begin{bmatrix} \mathbf{1} & \mathbf{a}_{i,i-1} \times \mathbf{1} \\ \mathrm{O} & \mathbf{1} \end{bmatrix}; \mathbf{p}_i \equiv \begin{bmatrix} \mathbf{0} \\ \mathbf{z}_i \end{bmatrix} \text{ for revolute; } \mathbf{p}_i \equiv \begin{bmatrix} \mathbf{z}_i \\ \mathbf{0} \end{bmatrix} \text{ for prismatic} \quad (2.5)$$

in which, $\mathbf{a}_{i,i-1} \equiv -\mathbf{a}_{i-1,i}$, is the position vector of the point, O_i of ith link, from O_{i-1} of the (i-1)st link. A similarly defined vector, $\mathbf{a}_{i,i+1}$, is shown in Fig. 2.2. Moreover, the 3×3 cross product tensor, $\mathbf{a}_{i,i-1} \times \mathbf{1}$, associated with the vector, $\mathbf{a}_{i,i-1}$ [5], is defined such that $(\mathbf{a}_{i,i-1} \times \mathbf{1})\mathbf{x} = \mathbf{a}_{i,i-1} \times \mathbf{x}$, for any 3-dimensional Cartesian vector $\mathbf{x}$. Furthermore, $\mathbf{z}_i$ is the 3-dimensional unit vector along the axis of rotation of a revolute joint or along the direction of a prismatic joint, whereas $\mathbf{O}$ and $\mathbf{0}$ are, respectively, the 3×3 zero matrix, and the 3-dimensional vector of zeros, and $\mathbf{1}$ is the 3×3 identity matrix. Matrix $\mathbf{R}_{i,i\text{-}1}$ and the vector, $\mathbf{p}_i$, have the following physical interpretations [4–6]:

- If $\dot{\theta}_i = 0$, i.e., when the (i-1)st and ith links are rigidly connected, $\mathbf{R}_{i,i-1}\mathbf{t}_{i-1}$ propagates the twist, $\mathbf{t}_{i\text{-}1}$, to $\mathbf{t}_i$. This implies that the angular velocity of the ith body remains same as that of (i-1)st one, namely $\boldsymbol{\omega}_i = \boldsymbol{\omega}_{i-1}$, while the linear velocity of the ith link is, $\mathbf{v}_i = \mathbf{v}_{i-1} + \mathbf{a}_{i,i-1} \times \boldsymbol{\omega}_{i-1}$, which is the formula for rigid body velocity transfer. Hence, the 6×6 matrix, $\mathbf{R}_{i,i\text{-}1}$, is termed as the twist propagation matrix, which has the following properties:

$$\mathbf{R}_{i,j} = \mathbf{R}_{j,i}; \ \mathbf{R}_{i,i} = \mathbf{1}; \ \text{and } \mathbf{R}_{i,j}^{-1} = \mathbf{R}_{j,i} \quad (2.6)$$

- If $\dot{\theta}_i \neq 0$, i.e., when ith joint motion is allowed, $\mathbf{p}_i\dot{\theta}_i$ adds to the components of $\mathbf{R}_{i,i-1}\mathbf{t}_{i-1}$, thus, giving rise to the actual twist of the ith link, $\mathbf{t}_i$. Hence, the 6-dimensional vector, $\mathbf{p}_i$, is termed as the joint-rate propagation vector of the ith joint.

(2) Now, for the n-link serial robot, Fig. 2.1, the generalized twist, $\mathbf{t}$ of Eq. (2.2), can be expressed using Eq. (2.4), for $i = 1, \ldots, n$, as

$$\mathbf{t} = \mathbf{A}\mathbf{t} + \mathbf{N}_d\dot{\boldsymbol{\theta}} \quad (2.7)$$

where the $6n$-dimensional vector, $\mathbf{t}$, and the n-dimensional vector, $\dot{\boldsymbol{\theta}}$, are defined in Eqs. (2.2) and (2.3), respectively. The $6n \times 6n$ matrix, $\mathbf{A}$, and the $6n \times n$ matrix, $\mathbf{N}_d$, are given as

$$\mathbf{A} \equiv \begin{bmatrix} \mathrm{O} & \cdots & \cdots & \mathrm{O} \\ \mathbf{R}_{21} & \mathrm{O} & \cdots & \mathrm{O} \\ \vdots & \vdots & \ddots & \vdots \\ \mathrm{O} & \cdots & \mathbf{R}_{n,n-1} & \mathrm{O} \end{bmatrix}; \mathbf{N}_d \equiv \begin{bmatrix} \mathbf{p}_1 & \mathbf{0} & \cdots & \mathbf{0} \\ \mathbf{0} & \mathbf{p}_2 & \cdots & \mathbf{0} \\ \vdots & \vdots & \ddots & \vdots \\ \mathbf{0} & \mathbf{0} & \cdots & \mathbf{p}_n \end{bmatrix} \tag{2.8}$$

where **O** and **0** are the 6 × 6 matrix, and the 6-dimensional vector of zeros, respectively. Henceforth, ***O*** and **0** should be understood as of compatible sized based on where they appear.

(3) Equation (2.7) is re-arranged as

$$\mathbf{t} = \mathbf{N}_l \mathbf{N}_d \dot{\boldsymbol{\theta}} \tag{2.9}$$

where $\mathbf{N}_l \equiv (\mathbf{1} - \mathbf{A})^{-1}$, and can be shown to have the following expression [9]:

$$\mathbf{N}_l \equiv \begin{bmatrix} \mathbf{1} & \mathrm{O} & \cdots & \mathrm{O} \\ \mathbf{R}_{21} & \mathbf{1} & \cdots & \mathrm{O} \\ \vdots & \vdots & \ddots & \\ \mathbf{R}_{n1} & \mathbf{R}_{n2} & \cdots & \mathbf{1} \end{bmatrix} \tag{2.10}$$

In Eq. (2.10), **1** denotes the 6 × 6 identity matrix. Henceforth, **1** should be understood as of compatible size based on where it appears. The product of matrices, $\mathbf{N}_l$ and $\mathbf{N}_d$, namely$\mathbf{N} \equiv \mathbf{N}_l\mathbf{N}_d$, is nothing but the natural orthogonal complement (NOC) matrix for the serial robot, as introduced in Angeles and Lee [1]. The matrices $\mathbf{N}_l$ and $\mathbf{N}_d$ are the desired decoupled natural orthogonal complement (DeNOC) matrices for the rigid robot under study, which allow one to write the matrix and vector elements associated with the dynamic equations of motion in analytical form leading to recursive forward dynamics algorithm.

2.2 Dynamic Modeling of Rigid Robots

The dynamic modeling of the rigid robot shown in Fig. 2.1 is now derived using the equivalence of the EL with the NE equations of motion and the DeNOC matrices. The steps are outlined below:

(1) Referring to Fig. 2.2, the position vectors of the elements, $\overline{E}_i$, $\widetilde{E}_i$ and payload of mass m_{pi} on the ith link, namely $\mathbf{r}_i$, $\tilde{\mathbf{r}}_i$, and $\mathbf{r}_{pi}$, are, respectively, given by,

$$\mathbf{r}_i = \mathbf{o}_i + \overline{\mathbf{b}}_i, \text{where } \overline{\mathbf{b}}_i = \overline{b}_i \mathbf{z}_i$$
$$\tilde{\mathbf{r}}_i = \mathbf{o}_i + \overline{\mathbf{r}}_i, \text{ where } \overline{\mathbf{r}}_i = b_i \mathbf{z}_i + \overline{a}_i \mathbf{x}_{i+1}, \text{ and}$$

$$\mathbf{r}_{pi} = \mathbf{o}_i + \overline{\mathbf{r}}_{pi}, \text{ where } \overline{\mathbf{r}}_{pi} = b_i\mathbf{z}_i + a_i\mathbf{x}_{i+1} \tag{2.11}$$

where the non-negative distance, a_i, and the offset of the axis X_{i+1} from the origin of the ith frame, b_i, are the DH parameters of the link, as defined in Appendix A, $\overline{b}_i$ is the axial distance of element $\overline{E}_i$ along Z_i from O_i, and $\overline{a}_i$ is the axial distance of element $\tilde{E}_i$ along X_{i+1} from $O_i^{'}$, as shown in Fig. 2.2. Note that $\overline{\boldsymbol{b}}_i$ is the position vector of element $\overline{E}_i$ along Z_i from O_i, whose magnitude is $\overline{b}_i$. The term, $\overline{b}_i$, varies from 0 to b_i, and $\overline{a}_i$ varies from 0 to a_i. Moreover, the unit vectors along Z_i and X_{i+1}-axes are denoted with $\mathbf{z}_i$ and $\mathbf{x}_{i+1}$, respectively,

(2) The kinetic energy, T_i, for the ith link is then given by

$$T_i = \frac{1}{2}\int_0^{b_i} \rho_i \dot{\mathbf{r}}_i^T \dot{\mathbf{r}}_i d\overline{b}_i + \frac{1}{2}\int_0^{a_i} \rho_i \dot{\tilde{\mathbf{r}}}_i^T \dot{\tilde{\mathbf{r}}}_i d\overline{a}_i + \frac{1}{2} m_{pi}\dot{\mathbf{r}}_{pi}^T \dot{\mathbf{r}}_{pi} + T_{hi} \tag{2.12}$$

where ρ_i is the mass per unit length of the ith link, and the vectors, $\dot{\mathbf{r}}_i$, $\dot{\tilde{\mathbf{r}}}_i$, and $\dot{\mathbf{r}}_{pi}$, are the velocities of the elements, $\overline{E}_i$, $\tilde{E}_i$ and payload, respectively, which can be written from Eq. (2.11) as

$$\begin{aligned} \dot{\mathbf{r}}_i &= \mathbf{v}_i + \boldsymbol{\omega}_i \times \overline{\mathbf{b}}_i; \\ \dot{\tilde{\mathbf{r}}}_i &= \mathbf{v}_i + \boldsymbol{\omega}_i \times \overline{\mathbf{r}}_i + \dot{b}_i\mathbf{z}_i; \text{ and} \\ \dot{\mathbf{r}}_{pi} &= \mathbf{v}_i + \boldsymbol{\omega}_i \times \overline{\mathbf{r}}_{pi} + \dot{b}_i\mathbf{z}_i \end{aligned} \tag{2.13a}$$

where $\mathbf{v}_i$ is substituted for $\dot{\mathbf{o}}_i$, i.e., $\mathbf{v}_i \equiv \dot{\mathbf{o}}_i$. Moreover, $\dot{\overline{a}}_i = \dot{a}_i = 0$ is used in Eq. (2.13a) as the link is rigid. Moreover, $\dot{b}_i$ represents the linear joint rate in the presence of prismatic joint. If the ith joint is revolute then $\dot{b}_i = 0$. For that, Eq. (2.13a) is modified as

$$\begin{aligned} \dot{\mathbf{r}}_i &= \mathbf{v}_i + \boldsymbol{\omega}_i \times \overline{\mathbf{b}}_i; \\ \dot{\tilde{\mathbf{r}}}_i &= \mathbf{v}_i + \boldsymbol{\omega}_i \times \overline{\mathbf{r}}_i; \text{ and} \\ \dot{\mathbf{r}}_{pi} &= \mathbf{v}_i + \boldsymbol{\omega}_i \times \overline{\mathbf{r}}_{pi} \end{aligned} \tag{2.13b}$$

in which $\boldsymbol{\omega}_i$ is the angular velocity of the ith link. Finally, the term, T_{hi} is the kinetic energy due to the hub at the joint, which includes the effect of motors/actuators at the joint, and is given by

$$T_{hi} = \frac{1}{2}\boldsymbol{\omega}_i^T \mathbf{I}_{hi}\boldsymbol{\omega}_i \tag{2.14}$$

(3) The Euler–Lagrange (EL) equations of motion for the whole system are then given by Meirovitch [3]:

$$\frac{d}{dt}\left(\frac{\partial T}{\partial \dot{\theta}_j}\right) - \frac{\partial T}{\partial \theta_j} = \tau_j, \text{ for } j = 1, \ldots, n \tag{2.15a}$$

where n is the total number of bodies in the system, which is same as the total number of joints in the serial robotic system, Fig. 2.1. Moreover,θ_j is the generalized coordinate, and $\tau_j = \tau_j^E + \tau_j^G$, in which τ_j^E is the generalized force, corresponding to the jth generalized coordinate, due to external forces and moments on the whole system and τ_j^G is the generalized forces due to gravity. The generalized forces due to gravity, τ_j^G, have the following form:

$$\tau_j^G = \frac{\partial V}{\partial \theta_j} \tag{2.15b}$$

where V is the potential energy of the system at hand due to gravity. However, in the proposed algorithm, τ_j^G need not be calculated using the partial differential of Eq. (2.15b). The effect of the τ_j^G is included in the algorithm efficiently using the methodology proposed by Walker and Orin [11]. Thus, the terms corresponding to the negative of the acceleration due to gravity are added to the linear velocity component of the twist vector for the first link, i.e., $\mathbf{t}_1$, Eq. (2.1).

Note that in Eq. (2.15a) for a revolute joint, the generalized coordinate, θ_j, represents joint rotation, and for a prismatic joint, it should be understood as joint displacement.

Now, for the total kinetic energy of the system, $T = \sum_{i=1}^{n} T_i$—i being the subscript for the bodies—the dynamic equations of motion, as derived in Appendix B, are given as

$$\left[\left(\frac{\partial \mathbf{t}_1}{\partial \dot{\theta}_j}\right)^T \cdots \left(\frac{\partial \mathbf{t}_n}{\partial \dot{\theta}_j}\right)^T\right]\begin{bmatrix}\mathbf{w}_1^* \\ \vdots \\ \mathbf{w}_n^*\end{bmatrix} = \tau_j, \; j = 1, \ldots, n \tag{2.16}$$

where the 6-dimensional vectors, $\partial \mathbf{t}_i / \partial \dot{\theta}_j$ and $\mathbf{w}_i^*$, for $i, j = 1,.., n$, are defined in eq. (B.11), and reproduced here as

$$\frac{\partial \mathbf{t}_i}{\partial \dot{\theta}_j} \equiv \begin{bmatrix}\frac{\partial \mathbf{v}_i}{\partial \dot{\theta}_j} \\ \frac{\partial \boldsymbol{\omega}_i}{\partial \dot{\theta}_j}\end{bmatrix};$$

$$\mathbf{w}_i^* \equiv \begin{bmatrix} \int_0^{b_i} \rho_i \ddot{\mathbf{r}}_i \mathrm{d}\overline{b}_i + \int_0^{a_i} \rho_i \ddot{\overline{\mathbf{r}}}_i \mathrm{d}\overline{a}_i + m_{pi}\ddot{\mathbf{r}}_{pi} \\ \int_0^{b_i} \rho_i \overline{b}(\mathbf{z}_i \times \ddot{\mathbf{r}}_i)\mathrm{d}\overline{b}_i + \int_0^{a_i} \rho_i \left(\overline{\mathbf{r}}_i \times \ddot{\overline{\mathbf{r}}}_i\right)\mathrm{d}\overline{a}_i + m_{pi}\left(\overline{\mathbf{r}}_{pi} \times \ddot{\mathbf{r}}_{pi}\right) + (\mathbf{I}_{hi}\dot{\boldsymbol{\omega}}_i + \\ \boldsymbol{\omega}_i \times \mathbf{I}_{hi}\boldsymbol{\omega}_i) \end{bmatrix} \tag{2.17}$$

(4) In Eq. (2.17), vector $\mathbf{w}_i^*$ can be physically interpreted as the inertia wrench in line with the definition of the wrench defined in Eq. (2.1). It is interesting to note here that the expression of $\mathbf{w}_i^*$, Eq. (2.17), has the following representation:

$$\mathbf{w}_i^* = \mathbf{M}_i\dot{\mathbf{t}}_i + \boldsymbol{\gamma}_i \tag{2.18}$$

where the 6-dimensional vector, $\dot{\mathbf{t}}_i \equiv \left[\dot{\mathbf{v}}_i^T \ \dot{\mathbf{w}}_i^T\right]^T$, and the 6 × 6 mass matrix, $\mathbf{M}_i$, and the 6-dimensional vector, $\boldsymbol{\gamma}_i$, are defined by

$$\mathbf{M}_i \equiv \int_0^{b_i} \rho_i \begin{bmatrix} \mathbf{1} & -\overline{\mathbf{b}}_i \times \mathbf{1} \\ \text{sym} & -\overline{\mathbf{b}}_i \times (\overline{\mathbf{b}}_i \times \mathbf{1}) \end{bmatrix} \mathrm{d}\overline{b}_i + \int_0^{a_i} \rho_i \begin{bmatrix} \mathbf{1} & -\overline{\mathbf{r}}_i \times \mathbf{1} \\ \text{sym} & -\overline{\mathbf{r}}_i \times (\overline{\mathbf{r}}_i \times \mathbf{1}) \end{bmatrix} \mathrm{d}\overline{a}_i$$
$$+ m_{pi} \begin{bmatrix} \mathbf{1} & -\overline{\mathbf{r}}_{pi} \times \mathbf{1} \\ \text{sym} & -\overline{\mathbf{r}}_{pi} \times (\overline{\mathbf{r}}_{pi} \times \mathbf{1}) \end{bmatrix} + \begin{bmatrix} \mathbf{O} & \mathbf{O} \\ \mathbf{O} & \mathbf{I}_{hi} \end{bmatrix} \tag{2.19a}$$

$$\gamma_i \equiv \int_0^{b_i} \rho_i \begin{bmatrix} \overline{\boldsymbol{\omega}}_i \\ \overline{\mathbf{b}}_i \times \overline{\boldsymbol{\omega}}_i \end{bmatrix} \mathrm{d}\overline{b}_i + \int_0^{a_i} \rho_i \begin{bmatrix} \overline{\overline{\boldsymbol{\omega}}}_i \\ \overline{\mathbf{r}}_i \times \overline{\overline{\boldsymbol{\omega}}}_i \end{bmatrix} \mathrm{d}\overline{a}_i + m_{pi} \begin{bmatrix} \overline{\overline{\boldsymbol{\omega}}}_{pi} \\ \overline{\mathbf{r}}_{pi} \times \overline{\overline{\boldsymbol{\omega}}}_{pi} \end{bmatrix} + \begin{bmatrix} \mathbf{0} \\ \boldsymbol{\omega}_i \times \mathbf{I}_{hi}\boldsymbol{\omega}_i \end{bmatrix} \tag{2.19b}$$

in which $\overline{\boldsymbol{\omega}}_i \equiv \boldsymbol{\omega}_i \times (\boldsymbol{\omega}_i \times \overline{\mathbf{b}}_i)$, $\overline{\overline{\boldsymbol{\omega}}}_i \equiv \boldsymbol{\omega}_i \times (\boldsymbol{\omega}_i \times \overline{\mathbf{r}}_i)$ and $\overline{\overline{\boldsymbol{\omega}}}_{pi} \equiv \boldsymbol{\omega}_i \times (\boldsymbol{\omega}_i \times \overline{\mathbf{r}}_{pi})$, and 'sym' denotes the symmetric elements of the matrix, $\mathbf{M}_i$. Furthermore, Eq. (2.18) has the same structure as one would derive from the Newton–Euler (NE) equations of motion of the ith rigid link [4–6, 8]. Hence, the word hybrid is used in this book, which implies that the Euler–Lagrange (EL) equations of motion can be written in a form, namely Eq. (2.16), whose $6n$-dimensional generalized inertia wrench is the 6-dimensional inertia wrenches, $\mathbf{w}_i^*$, for the n-bodies, that has the same form as can be obtained from the uncoupled NE equations of motion.

(5) Writing Eq. (2.16) for all n-links, i.e., $j = 1,\ldots, n$, one obtains

$$\begin{bmatrix} \left(\frac{\partial \mathbf{t}_1}{\partial \dot{\theta}_1}\right)^T & \cdots & \left(\frac{\partial \mathbf{t}_n}{\partial \dot{\theta}_1}\right)^T \\ \vdots & & \\ \left(\frac{\partial \mathbf{t}_1}{\partial \dot{\theta}_n}\right)^T & \cdots & \left(\frac{\partial \mathbf{t}_n}{\partial \dot{\theta}_n}\right)^T \end{bmatrix} \begin{bmatrix} \mathbf{w}_1^* \\ \vdots \\ \mathbf{w}_n^* \end{bmatrix} = \begin{bmatrix} \tau_1 \\ \vdots \\ \tau_n \end{bmatrix} \tag{2.20}$$

which is expressed as

$$\left(\frac{\partial \mathbf{t}}{\partial \dot{\boldsymbol{\theta}}}\right)^T (\mathbf{M}\dot{\mathbf{t}} + \boldsymbol{\gamma}) = \boldsymbol{\tau} \tag{2.21}$$

where the $6n \times 6n$ matrix, $\mathbf{M}$, the $6n$-dimensional vector, $\boldsymbol{\gamma}$, and the n-dimensional vector, $\boldsymbol{\tau}$, have the following definitions:

$$\mathbf{M} \equiv \text{diag}\left[\, \mathbf{M}_1 \ \cdots \ \mathbf{M}_n \,\right];\ \boldsymbol{\gamma} \equiv \left[\, \boldsymbol{\gamma}_1^T \ \cdots \ \boldsymbol{\gamma}_n^T \,\right]^T;\ \text{and}\ \boldsymbol{\tau} \equiv \left[\, \tau_1 \ \cdots \ \tau_n \,\right]^T \tag{2.22}$$

and the vectors, $\mathbf{t}$ and $\mathbf{w}$, are defined in Eq. (2.2). Moreover, the $6n \times n$ matrix, $\frac{\partial \mathbf{t}}{\partial \dot{\boldsymbol{\theta}}}$ of Eq. (2.21), is nothing but

$$\frac{\partial \mathbf{t}}{\partial \dot{\boldsymbol{\theta}}} \equiv \begin{bmatrix} \frac{\partial \mathbf{t}_1}{\partial \dot{\theta}_1} & \cdots & \frac{\partial \mathbf{t}_1}{\partial \dot{\theta}_n} \\ \vdots & & \vdots \\ \frac{\partial \mathbf{t}_n}{\partial \dot{\theta}_1} & \cdots & \frac{\partial \mathbf{t}_n}{\partial \dot{\theta}_n} \end{bmatrix} \tag{2.23}$$

(6) From Eq. (3.9), it is clear that

$$\frac{\partial \mathbf{t}}{\partial \dot{\boldsymbol{\theta}}} = \mathbf{N}_l \mathbf{N}_d \tag{2.24}$$

Hence, Eq. (2.21) is written as:

$$\mathbf{N}_d^T \mathbf{N}_l^T (\mathbf{M}\dot{\mathbf{t}} + \boldsymbol{\gamma}) = \boldsymbol{\tau} \tag{2.25}$$

(7) Differentiating Eq. (2.9), with respect to time, i.e.,

$$\dot{\mathbf{t}} = \mathbf{N}_l \mathbf{N}_d \ddot{\boldsymbol{\theta}} + \mathbf{N}_l \dot{\mathbf{N}}_d \dot{\boldsymbol{\theta}} + \dot{\mathbf{N}}_l \mathbf{N}_d \dot{\boldsymbol{\theta}} \tag{2.26}$$

and substituting the resulting expression, Eq. (2.26), into Eq. (2.25), the independent set of constrained dynamic equations of motion is finally obtained as

$$\mathbf{I}\ddot{\boldsymbol{\theta}} = \varphi \tag{2.27}$$

where the $n \times n$ generalized inertia matrix (GIM), $\mathbf{I}$, is given by,

$$\mathbf{I} \equiv \mathbf{N}_d^T \widetilde{\mathbf{M}} \mathbf{N}_d, \text{ where } \widetilde{\mathbf{M}} \equiv \mathbf{N}_l^T \mathbf{M} \mathbf{N}_l \tag{2.28}$$

and the n-dimensional vector, φ, is as follows:

$$\varphi = \boldsymbol{\tau} - \mathbf{N}_d^T \mathbf{N}_l^T \left[\mathbf{M}(\mathbf{N}_l \dot{\mathbf{N}}_d + \dot{\mathbf{N}}_l \mathbf{N}_d)\dot{\boldsymbol{\theta}} + \boldsymbol{\gamma} \right] \tag{2.29}$$

As shown in Saha [4–6], the 6×6 block matrix, $\widetilde{\mathbf{M}}_i$, Eq. (2.28), can be evaluated recursively as

$$\widetilde{\mathbf{M}}_i \equiv \mathbf{M}_i + \mathbf{R}_{i+1,i}^T \widetilde{\mathbf{M}}_{i+1} \mathbf{R}_{i,i+1} \tag{2.30}$$

in which $\widetilde{\mathbf{M}}_{n+1} \equiv \boldsymbol{O}$, because there is no $(n+1)$st link in the chain. Hence, $\widetilde{\mathbf{M}}_n \equiv \mathbf{M}_n$. Accordingly, the elements of the GIM, $\mathbf{I}$ of Eq. (3.27), are expressed as

$$i_{ij} = i_{ji} \equiv \mathbf{p}_i^T \widetilde{\mathbf{M}}_i \mathbf{R}_{ij} \mathbf{p}_j; \text{ for } i = 1, \ldots, n; \; j = 1, \ldots, i. \tag{2.31}$$

Equations (2.30)–(2.31) represent the analytical and recursive expressions for the elements of the GIM. The $6n \times 6n$ matrix, $\widetilde{\mathbf{M}}_i$, is physically interpreted as the mass matrix of the ith composite body, which is also reported in Saha [5, 6], where it was derived starting from the NE equations of motion. Here, it is shown how the same can be derived starting from the EL equations of motion, Eq. (2.15a). The advantage of the present formulation is that one can easily accommodate the flexible links in the system without requiring any modification in the dynamic formulation as the rigid bodies are treated here as continuous systems, unlike in Saha [4–6] where the rigid bodies were treated as discrete. Moreover, a computationally more efficient forward dynamics algorithm for the rigid robots is also obtained, as presented next.

2.3 Forward Dynamics Algorithm for Rigid Robots

This section presents a recursive, computationally efficient, and stable algorithm for the forward dynamics of serial rigid robotic systems. First, the recursive algorithm is presented in Sect. 2.3.1. The computational efficiency of the algorithm is then demonstrated in Sect. 2.3.2 by comparing it with those available in the literature. Numerical stability aspects of the algorithm are analyzed separately in Chap. 7.

2.3.1 Recursive Algorithm

In order to obtain an algorithm to calculate the joint accelerations, $\ddot{\boldsymbol{\theta}}$, from Eq. (3.27), the GIM, **I** of Eq. (3.27), is factorized using the reverse Gaussian elimination Strang [7] as

$$\mathbf{I} \equiv \mathbf{U}\mathbf{D}\mathbf{U}^T \tag{2.32}$$

where **U** and **D** are, respectively, the upper and diagonal matrices [7]. Equation (2.32) is then substituted in Eq. (2.27), yielding the following:

$$\mathbf{U}\widehat{\varphi} = \varphi, \text{ where } \mathbf{D}\widetilde{\varphi} = \widehat{\varphi} \text{ and } \mathbf{U}^T\ddot{\boldsymbol{\theta}} = \widetilde{\varphi} \tag{2.33}$$

where φ is the n-dimensional vector of the centrifugal and Coriolis terms given by Eq. (2.29). For the forward dynamics algorithm, vector φ is assumed to be known from any recursive, efficient algorithm. Since the $\mathbf{I} \equiv \mathbf{U}\mathbf{D}\mathbf{U}^T$ factorization is done analytically, expression of each element of the upper triangular matrix **U** is available. The solution of the joint accelerations, $\ddot{\boldsymbol{\theta}}$ of Eq. (2.27), is now obtained in the following steps:

Step A. Solution for $\widehat{\varphi}$: The solution, $\widehat{\varphi} = \mathbf{U}^{-1}\varphi$, is evaluated for $i = n$-1, ..., 1 as

$$\widehat{\varphi}_i = \varphi_i - \mathbf{p}_i^T \boldsymbol{\eta}_{i,i+1} \tag{2.34}$$

where $\widehat{\varphi}_n = \varphi_n$, $\mathbf{p}_i$ is the joint-rate propagation vector defined in Eq. (2.5) and the 6-dimensional vector, $\boldsymbol{\eta}_{n,n+1}$, is obtained recursively as

$$\boldsymbol{\eta}_{i,i+1} = \mathbf{R}_{i+1,i}^T \boldsymbol{\eta}_{i+1}, \text{ and } \boldsymbol{\eta}_{i+1} = \boldsymbol{\psi}_{i+1}\widehat{\boldsymbol{\tau}}_{i+1} + \boldsymbol{\eta}_{i+1,i+2} \tag{2.35}$$

in which $\boldsymbol{\eta}_{n,n+1} = \mathbf{0}$. Moreover, the 6 × 6 twist propagation matrix, $\mathbf{R}_{i+1,i}$, is defined similar to $\mathbf{R}_{i,i-1}$, Eq. (2.5). Furthermore, the 6-dimensional vector, $\boldsymbol{\psi}_i$, is obtained from

$$\boldsymbol{\psi}_i = \frac{\mathbf{R}_{i+1,i}\widehat{\boldsymbol{\psi}}_i}{\hat{i}_i}, \text{ where } \widehat{\boldsymbol{\psi}}_i \equiv \widehat{\mathbf{M}}_i\mathbf{p}_i \text{ and } \hat{i}_i \equiv \mathbf{p}_i^T\widehat{\boldsymbol{\psi}}_i \tag{2.36}$$

Note that the 6 × 6 matrix, $\widehat{\mathbf{M}}_i$, can also be obtained recursively as

$$\widehat{\mathbf{M}}_i = \mathbf{M}_i + \mathbf{R}_{i+1,i}^T\overline{\mathbf{M}}_{i+1}\mathbf{R}_{i+1,i}, \text{ where } \overline{\mathbf{M}}_i \equiv \widehat{\mathbf{M}}_i - \widehat{\boldsymbol{\psi}}_i\boldsymbol{\psi}_i^T \tag{2.37}$$

for $i = n$-1, ..., 1, and $\widehat{\mathbf{M}}_n \equiv \mathbf{M}_n$.

Step B. The expression for each element of $\widetilde{\varphi}$ is now found from, $\widetilde{\varphi} = \mathbf{D}^{-1}\widehat{\varphi}$, i.e.,

$$\widetilde{\varphi}_i = \frac{\widehat{\varphi}_i}{\hat{i}_i} \tag{2.38}$$

for $i = 1, \ldots, n$, where $\widehat{\varphi}_i$ is available from Step A.

Step C. Finally, the expression for each element of the joint acceleration vector, $\ddot{\boldsymbol{\theta}} = \mathbf{U}^{-1}\widetilde{\varphi}$, is given by

$$\ddot{\theta}_i = \widetilde{\varphi}_i - \boldsymbol{\psi}_i^T \boldsymbol{\mu}_{i,i-1} \tag{2.39}$$

for $i = 2, \ldots, n$. Note that, $\ddot{\theta}_1 \equiv \widetilde{\varphi}_1$, and the 6-dimensional vector, $\boldsymbol{\mu}_{i,i-1}$, is obtained similar to $\boldsymbol{\eta}_{i,i+1}$ as

$$\boldsymbol{\mu}_{i,i-1} \equiv \mathbf{R}_{i,i-1}\boldsymbol{\mu}_{i-1}, \text{ where } \boldsymbol{\mu}_{i-1} \equiv \mathbf{p}_{i-1}\ddot{\theta}_{i-1} + \boldsymbol{\mu}_{i-1,i-2} \tag{2.40}$$

and $\boldsymbol{\mu}_{10} \equiv \mathbf{0}$.

2.3.2 Computational Complexity

The total number of computational counts required to determine the joint acceleration vector, $\ddot{\theta}$, using the algorithm presented in Sect. 2.3.1, for a serial n-link rigid robot are performed. The counts are performed in terms of the number of divisions/multiplications (M), subtractions/additions (A), and trigonometric operations (T). The total computational count of the proposed algorithm for the rigid robots, not including the effects of payload and hub inertia, is (173n-133)M, (150n-133)A, and 2nT. The details of the implementation, computational complexity, and the counts are given in Appendix C. Note that the effect of the payload and hub are not considered while computing the computational complexity, as it would enable to compare the final complexity with those where these effects were not reported in the literature. The comparison of the computational complexity for the proposed algorithm with the best non-recursive algorithm and some of the best recursive algorithms is given in Table 1.1. It is clear from Table 1.1 that the DeNOC based forward dynamics algorithm presented in Shah et al. [10] and one presented in this chapter are the best compared to other recursive algorithms. Even compared with the non-recursive algorithms, which had been known to perform better for smaller number of links, the present recursive algorithm has reduced that gap. The proposed algorithm starts benefiting for $n = 12$ and above if the multiplications and additions are both considered. This is a significant achievement compared to those that appeared during the early stages of development of the recursive algorithms, e.g., Featherstone [2]. The algorithm by Featherstone [2] benefits when $n = 14$ and above if both multiplications and additions are considered. If only multiplications are considered, the proposed algorithm starts benefiting for $n = 10$ and above, while that of Featherstone [2] starts benefiting for $n = 12$ and above. Even compared to Saha [5, 6], which also starts benefiting for $n = 10$ and above, if only multiplications are considered, the proposed algorithm is better. The efficiency is mainly achieved due to the consideration of robot links as slender rods rather than of a very general shape, whose lengths are equivalent to the DH parameters, as shown in Fig. 2.2. The assumption avoids matrix multiplications associated with the general 3×3 inertia tensors of the links, thereby reducing the overall number of computations required. Note that the assumption of slender rods is quite practical as it closely resembles real robot links used in the industries.

2.4 Summary

In this chapter, an alternative dynamic modeling approach for serial-chain robots with rigid links, namely based on the Euler–Lagrange equations of motion and the decoupled natural orthogonal complement (DeNOC) matrices, is proposed. First, the Euler–Lagrange (EL) equations of motion of the system are written. Then, using the equivalence of EL and Newton–Euler (NE) equations, and the DeNOC matrices associated with the velocity constraints of the connecting rigid links, the analytical and recursive expressions for the matrices and vectors appearing in the independent dynamic equations of motion are obtained. The analytical expressions allow one to obtain a recursive forward dynamics algorithm. The algorithm is shown to be computationally more efficient than other algorithms in the literature. The computational efficiency is mainly achieved due to some realistic and simplifying assumptions, which reduce the dynamic computations. The algorithm is also numerically more stable than the conventional algorithms, as shown in Chap. 7. One of the advantages of the modeling technique is that the links are treated as continuous systems, and hence, the modeling technique can be easily extended to accommodate flexibility characteristics of the links, which is taken up in Chap. 3. Moreover, physical interpretations of many matrices and vectors associated with the dynamic model of the system are also presented.

Bibliography

1. Angeles J, Lee SK (1988) Formulation of dynamical equations of holonomic mechanical systems using a natural orthogonal complement. J Appl Mech Trans ASME 55(1):243–244
2. Featherstone R (1983) Calculation of robot dynamics using articulated-body inertias. Int J Robot Res 2(1):13–30
3. Meirovitch L (1967) Analytical methods in vibration. The Mcmillan Company, New York, NY

4. Saha SK (1997) A decomposition of the manipulator inertia matrix. IEEE Trans Robot Autom 13(2):301–304
5. Saha SK (1999) Analytical expression for the inverted inertia matrix of serial robots. Int J Robot Res 18(1):116–124
6. Saha SK (1999) Dynamics of serial multibody systems using the decoupled natural orthogonal complement matrices. J Appl Mech Trans ASME 66(4):986–995
7. Saha SK (2003) Simulation of industrial manipulators based on the $\mathbf{UDU}^T$ decomposition of inertia matrix. Multibody Syst Dyn 9(1):63–85
8. Saha SK, Angeles J (1991) Dynamics of nonholonomic mechanical systems using a natural orthogonal complement. J Appl Mech Trans ASME 58(1):238–243
9. Saha SK, Schiehlen WO (2001) Recursive kinematics and dynamics for parallel structured closed-loop multibody systems. Mech Struct Mach 29(2):143–175
10. Shah S, Saha S, Dutt J (2011) Modular framework for dynamic modeling and analyses of legged robots. Mech Mach Theo 49:234–255
11. Walker MW, Orin DE (1982) Efficient dynamic computer simulation of robotic mechanisms. J Dyn Syst Meas Control Trans ASME 104(3):205–211

Chapter 3
Dynamics of Serial Rigid–Flexible Robots

In this chapter, dynamic modeling of serial robots comprising both rigid and flexible links is presented. Such robots are referred here as rigid–flexible robots. First, their kinematics is described, followed by the scheme used to discretize the deflection of each flexible link. Then, some definitions, similar to those defined for rigid robots in Chap. 2, are also introduced for the rigid–flexible robots. The decoupled natural orthogonal complement (DeNOC) matrices for systems with mixed rigid and flexible links are defined next, followed by the derivation of the dynamic equations of motion based on the hybrid formulation presented in Chap. 2. The analytical expressions for the matrices and vectors associated with the dynamic equations of motion of the system are also obtained, and their physical interpretations are presented. It is seen that the recursive relations for the flexible and rigid link robots are exactly similar in structure except that the dimensions of the associated matrices and vectors are different. The similarity in the analytical expressions is exploited to factorize the generalized inertia matrix into $\mathbf{UDU}^T$ form–$\mathbf{U}$ and $\mathbf{D}$ being the upper block triangular and block-diagonal matrices–on the similar lines as done for rigid bodies using the reverse Gaussian elimination scheme. A recursive, forward dynamics algorithm is then presented for the serial rigid–flexible robots. Further, computational complexity for only flexible link robots is compared with some of the algorithms available in the literature. It is observed that the proposed algorithm is computationally more efficient. The numerical stability of the algorithm is also investigated in Chap. 7. The effect of geometric stiffness is also incorporated in the DeNOC-based formulation to deal with the high-speed, large displacement problem which is illustrated with the spinning cantilever example. Also, in this chapter, a series of simulation results are presented to validate the dynamic modeling technique and the forward dynamics algorithms proposed in this book, namely in Chaps. 2 and 3, using the examples of simple single-link, two-link arm, and three-link planar robotic arms.

P. V. Nandihal et al., *Dynamics of Rigid-Flexible Robots and Multibody Systems*, Intelligent Systems, Control and Automation: Science and Engineering 100, https://doi.org/10.1007/978-981-16-2798-9_3

3.1 Kinematics

Consider a serial robotic system having a fixed base and n-moving links, and the links can be either rigid or flexible. The links are coupled by n-one-degree-of-freedom kinematic pairs, say, revolute or prismatic. Thus, the serial robot under study is shown in Fig. 2.1, whose links are either rigid or flexible. Figure 3.1 shows a typical flexible link, say, the i-th one. For the i-th flexible link, it is assumed that the portion, $O_i'\widehat{O}_{i+1}$, is much longer than the portion, O_iO_i'. Hence, the former portion is structurally flexible, which can deflect, whereas the latter one cannot. Typically, the flexible portion of the links has a smaller cross-sectional dimension compared to its length. As a result, the Euler–Bernoulli beam theory [18] can be applied for the vibration analysis of the flexible part of the link. The flexible part is assumed to vibrate in space due to bending and torsion. For the simplification of representation, it is assumed that the flexible part of the links vibrates in bending, in Y- and Z-directions in the equal number of modes, m_i. However, the methodology and formulation of the dynamic model proposed below are well applicable to a more general case of vibrations where the beam vibrates in different numbers of modes in its Y- and Z-directions, as explained later in Sect. 3.1.2, after Eqs. (3.3c) and (3.4a). Moreover, the flexible part of the links is assumed to be vibrating in $\overline{m}_i$ modes in torsional vibrations.

The degree of freedom (DOF) of the rigid–flexible serial robot is then calculated as, $\overline{n} \equiv n + \sum_{i=1}^{n_f} (3m_i + \overline{m}_i)$, where $n \equiv n_r + n_f$ is the total number of links or joints in which n_r is the number of rigid links or the number of joints preceding to them, and n_f is the number of flexible links or the number of joints preceding them.

Moreover, the number 3 in the formula for $\overline{n}$ stands for X-, Y-, and Z-components of the spatial 3-dimensional deflection of each flexible link, namely, vector $\mathbf{u}_i$ of Fig. 3.1.

3.1.1 Rotation Matrix

Considering the i-th flexible link, as shown in Fig. 3.1, the coordinate system (X_{i+1}, Y_{i+1}, Z_{i+1}) is defined at the distal end (tip of the link) of the i-th link, where it is attached. The orientation of the frame and the DH notations are defined similarly to the rigid link systems, as explained in Appendix A. The axis of the i-th joint at O_i is thus coincident with the axis of Z_i. In addition to the rigid link coordinate frames, as defined in Fig. 2.2, an additional coordinate frame ($\widehat{X}_{i+1}$, $\widehat{Y}_{i+1}$, $\widehat{Z}_{i+1}$) is defined here for the flexible link i at $\widehat{O}_{i+1}$, as shown in Fig. 3.1. The point, $\widehat{O}_{i+1}$, is the distal end of the i-th link in its undeformed state. The origin of the $(i + 1)^{st}$ joint, O_{i+1}, thus coincides with $\widehat{O}_{i+1}$ when the link is rigid or when it is undeformed. However, for the flexible links, the orientation matrix representing the transformation of the coordinate system from frame ($\widehat{X}_{i+1}$, $\widehat{Y}_{i+1}$, $\widehat{Z}_{i+1}$) to frame ($\widehat{X}_i$, $\widehat{Y}_i$, $\widehat{Z}_i$) denoted as $\widetilde{\mathbf{Q}}_i$, is given by,

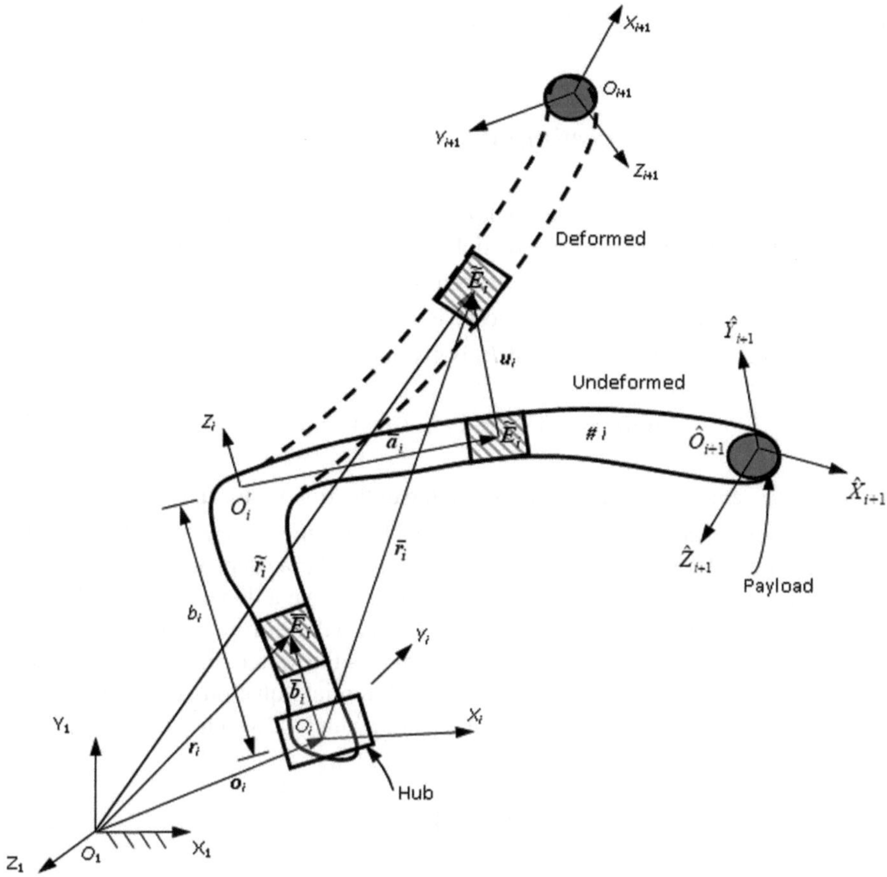

Fig. 3.1 *i*-th flexible link

$$\widetilde{\mathbf{Q}}_i \equiv \mathbf{Q}_i \widehat{\mathbf{Q}}_{i-1} \tag{3.1}$$

where $\mathbf{Q_i}$ is the 3×3 rigid rotation matrix that defines the orientation of frame $(\widehat{X}_{i+1}, \widehat{Y}_{i+1}, \widehat{Z}_{i+1})$ with respect to frame (X_i, Y_i, Z_i) and is given after C.2 of Appendix C. The orientation of frame (X_i, Y_i, Z_i) with respect to frame $(\widehat{X}_i, \widehat{Y}_i, \widehat{Z}_i)$ is given by $\widehat{\mathbf{Q}}_{i-1}$. The overall transformation from $(i + 1)$st frame with respect to *i*-th frame, $\widetilde{\mathbf{Q}}_\mathbf{i}$, is, thus, obtained as the product of $\mathbf{Q}_i$ and $\widehat{\mathbf{Q}}_{i-1}$, Eq. (3.1). The 3×3 rotation matrix, $\widehat{\mathbf{Q}}_{i-1}$, is due to the deformation of the *i*-th link, which is given by

$$\widehat{\mathbf{Q}}_{i-1} \equiv \begin{bmatrix} 1 & -\delta_{i-1}^z & \delta_{i-1}^y \\ \delta_{i-1}^z & 1 & -\delta_{i-1}^x \\ -\delta_{i-1}^y & \delta_{i-1}^x & 1 \end{bmatrix} \tag{3.2}$$

where $\boldsymbol{\delta}_{i-1} \equiv \left[\delta_{i-1}^{x}\ \delta_{i-1}^{y}\ \delta_{i-1}^{z}\right]^T$ is the vector of angular slopes of the (i-1)st link.

3.1.2 Kinematic Discretization

For the kinematic description of each flexible link, the assumed mode method (AMM) [18, 32] is used to describe its elastic deformation. For any element $\tilde{E}_i$, lying along $\widehat{X}_{i+1}$ axis, Fig. 3.1, its position vector from O_i' is $\overline{\mathbf{a}}_i$. Thus, $\overline{\mathbf{a}}_i = \overline{a}_i\hat{\mathbf{x}}_{i+1}$, where $\hat{\mathbf{x}}_{i+1}$ is the unit vector along $\widehat{X}_{i+1}$ axis and $\overline{a}_i$ is the axial distance of $\tilde{E}_i$ from O_i' along $\widehat{X}_{i+1}$ of the i-th link. Deformation of $\tilde{E}_i$ is given by the 3-dimensional vector, $\mathbf{u}_i$, i.e.,

$$\mathbf{u}_i(\overline{a}, t) \equiv \left[u_i^x(\overline{a}, t)\ u_i^y(\overline{a}, t)\ u_i^z(\overline{a}, t)\right]^T \tag{3.3a}$$

where u_i^x, u_i^y, and u_i^z are the projections of the deflection vector $\mathbf{u}_i$ on the $\widehat{X}_{i+1}$, $\widehat{Y}_{i+1}$ and $\widehat{Z}_{i+1}$ axes, respectively. Note that $\overline{a}_i$ varies from 0 to a_i—one of the DH parameters of the link. Moreover, $(\overline{a}, t)$ denotes the space and time dependence of the vector $\mathbf{u}_i$. In future texts, vector, $\mathbf{u}_i$ will be understood to have a dependence on the axial coordinate, $\overline{a}_i$ and time, t, and the argument $(\overline{a}_i, t)$ will be dropped. The dependence of u_i^x, u_i^y and u_i^z on the spatial coordinate $\overline{a}_i$ makes the system infinite-dimensional, leading to a coupled ordinary and partial differential equations of motion. For a revolute joint, the component u_i^x is a function of u_i^y and u_i^z, which is given by [7, 18],

$$u_i^x \approx \frac{1}{2}\int_0^{a_i}\left[\left(\frac{\partial u_i^y}{\partial \overline{a}_i}\right)^2 - \left(\frac{\partial u_i^z}{\partial \overline{a}_i}\right)^2\right]d\overline{a}_i \tag{3.3b}$$

The term u_i^x of Eq. (3.3b), results in axial deformation and foreshortening of the link [7], which are significant in the analysis of flexible multibody systems when the angular rates of the bodies are considerably greater than their first natural frequency. Kane et al. [14] have demonstrated the effect of centrifugal stiffening for a single beam rotating at a very large angular rate. For a flexible link used in industrial robots and in satellites whose speed of operation is generally slow, the effect of centrifugal stiffening is neglected. Moreover, the flexibility along the joint axis, Z_i, is ignored due to the assumption of i-th link rigid along with Z_i, i.e., O_iO_i' of Fig. 2.2. The above two assumptions are quite common in the literature, e.g., [7, 9, 32] and others. For a prismatic joint, the flexibility along the Z_i-axis is also ignored. Otherwise, the smooth translation along the axis is not possible. Now, using the assumed mode method, the vector, $\mathbf{u}_i$, can be expressed in terms of space-dependent eigenfunctions and time-dependent amplitudes, i.e.,

$$u_i^x = \begin{cases} 0 & : \text{For revolute joints} \\ \sum_{j=1}^{m_i} s_{i,j}^x d_{i,j}^x & : \text{For prismatic joints} \end{cases}$$

$$u_i^y = \sum_{j=1}^{m_i} s_{i,j}^y d_{i,j}^y$$

$$u_i^z = \begin{cases} \sum_{j=1}^{m_i} s_{i,j}^z d_{i,j}^z & : \text{For revolute joints} \\ 0 & : \text{For prismatic joints} \end{cases} \tag{3.3c}$$

Next, to express Eq. (3.3c) in matrix–vector form, the following three 3 × $m_i$matrices, namely $\mathbf{S}_i^x$, $\mathbf{S}_i^y$ and $\mathbf{S}_i^z$, are introduced:

$$\mathbf{S}_i^x \equiv \begin{bmatrix} s_{i,1}^x \cdots s_{i,m_i}^x \\ 0 \cdots 0 \\ 0 \cdots 0 \end{bmatrix}; \mathbf{S}_i^y \equiv \begin{bmatrix} 0 \cdots 0 \\ s_{i,1}^y \cdots s_{i,m_i}^y \\ 0 \cdots 0 \end{bmatrix}; \text{ and } \mathbf{S}_i^z \equiv \begin{bmatrix} 0 \cdots 0 \\ 0 \cdots 0 \\ s_{i,1}^z \cdots s_{i,m_i}^z \end{bmatrix} \tag{3.4a}$$

where $s_{i,1}^x \cdots s_{i,m_i}^x$ denote the shape functions along the $\widehat{X}_{i+1}$ axis corresponding to the m_i modes, as appeared in Eq. (3.3c). Similarly, $s_{i,1}^y \cdots s_{i,m_i}^y$ and $s_{i,1}^z \cdots s_{i,m_i}^z$ of Eq. (3.3c) or (3.4a) represent the shape functions of the link along $\widehat{Y}_{i+1}$ and $\widehat{Z}_{i+1}$ axes in m_i modes, respectively. Note that, although for simplification in representation, it is assumed that the flexible part of the links vibrates in an equal number of modes in three directions, it is considered to possess different shape functions along $\widehat{X}_{i+1}$, $\widehat{Y}_{i+1}$ and $\widehat{Z}_{i+1}$ axes. Thus, a more general case in which the beam possesses different stiffness properties in different directions is considered. In Eqs. (3.3c) and (3.4a), it is assumed that the flexible part of the link vibrates in m_i modes in X_i-, Y_i-, and Z_i-directions. If, however, the link vibrates in different number of modes in X_i-, Y_i-, and Z_i-directions, say, $m_i{}^x$, $m_{i,}{}^y$ and $m_i{}^{z,}$ respectively, then the dimensions of corresponding shape function matrices $\mathbf{S}_i^x$, $\mathbf{S}_i^y$ and $\mathbf{S}_i^z$, will change accordingly. Hence, the generality of the formulation will be maintained. The overall shape function matrix of the ith flexible beam vibrating in m_i modes is thus given by:

$$\mathbf{S}_i \equiv \left[\mathbf{S}_i^x \ \mathbf{S}_i^y \ \mathbf{S}_i^z \right] \tag{3.4b}$$

Moreover, three m_i-dimensional vectors, d_i^x, d_i^y and d_i^z, are introduced as

$$\mathbf{d}_i^x \equiv \left[d_{i,1}^x \cdots d_{i,m_i}^x\right]^T; \mathbf{d}_i^y \equiv \left[d_{i,1}^y \cdots d_{i,m_i}^y\right]^T \text{ and } \mathbf{d}_i^z \equiv \left[d_{i,1}^z \cdots d_{i,m_i}^z\right]^T \tag{3.4c}$$

where the vectors, d_i^x, d_i^y, and d_i^z, are, respectively, the vectors of time-dependent amplitudes corresponding to the shape function along $\widehat{X}_{i+1}$, $\widehat{Y}_{i+1}$, and $\widehat{Z}_{i+1}$ axes, Eq. (3.4a). The components of $\mathbf{d}_i^x$, $\mathbf{d}_i^y$, and $\mathbf{d}_i^z$ are treated here as the generalized coordinates to describe the deflection of i-th link, along with those associated with the i-th joint motion, namely, $\theta_i^{\prime}s$ of Eq. (3.3). Next, the $3m_i$-dimensional vector of time-dependent amplitudes, $\mathbf{d}_i$, is defined by

$$\mathbf{d}_i \equiv \begin{bmatrix} \mathbf{d}_i^x \\ \mathbf{d}_i^y \\ \mathbf{d}_i^z \end{bmatrix} \tag{3.4d}$$

Combining Eqs. (3.4b–d), Eq. (3.3c) is expressed as

$$\mathbf{u}_i = \mathbf{S}_i \mathbf{d}_i \tag{3.4e}$$

where the 3-dimensional vector, $\mathbf{u}_i$, is defined in Eq. (3.3a). Furthermore, since the flexible part of the link can undergo torsion about $\widehat{X}_{i+1}$ axis, it can be discretized similar to the bending deformation, Eqs. (3.3c) or (3.4e), as

$$\beta_i \equiv \bar{\mathbf{s}}_i^T \mathbf{c}_i \tag{3.5a}$$

where β_i is the scalar angular displacement of the cross section of the element, $\widetilde{E}_i$, and the $\overline{m}_i$-dimensional vectors, $\bar{\mathbf{s}}_i$ and $\mathbf{c}_i$, are the shape functions due to torsion and the time-dependent torsional amplitudes, respectively. Vectors $\bar{\mathbf{s}}_i$ and $\mathbf{c}_i$ are defined by

$$\bar{\mathbf{s}}_i \equiv \left[\bar{s}_{i,1} \cdots \bar{s}_{i,\overline{m}_i}\right]^T; \mathbf{c}_i \equiv \left[c_{i,1} \cdots c_{i,\overline{m}_i}\right]^T \tag{3.5b}$$

Note that the components of $\mathbf{c}_i$ will also be treated as the generalized coordinates for i-th link, along with $\boldsymbol{d}_i$ and θ_i.

3.1.3 Definitions for Flexible Systems

Referring to the serial rigid–flexible robot under study, Figs. 2.1 and 3.1, the following definitions are introduced in parallel to those appeared in Sect. 2.1.1 for rigid robots:

- $\mathbf{t}_i$ and $\mathbf{w}_i$: The $(6 + 3m_i + \overline{m}_i)$-dimensional twist and wrench of the i-th flexible link, respectively, i.e.,

$$\mathbf{t}_i \equiv \left[\mathbf{v}_i^T \ \boldsymbol{\omega}_i^T \ \dot{\mathbf{d}}_i^T \ \dot{\mathbf{c}}_i^T\right]^T; \mathbf{w}_i \equiv \left[\mathbf{f}_i^T \ \mathbf{n}_i^T \ \boldsymbol{\varepsilon}_i^T \ \boldsymbol{\vartheta}_i^T\right]^T \tag{3.6}$$

where $\mathbf{v}_i$ and $\boldsymbol{\omega}_i$, as defined after Eq. (2.1), are the 3-dimensional vectors of the velocity of point, O_i, of the i-th link, and its angular velocity, respectively. Moreover, the $3m_i$-dimensional vector, $\dot{\mathbf{d}}_i$, and the $\overline{m}_i$-dimensional vector, $\dot{\mathbf{c}}_i$, are the time derivatives of the time-dependent variables $\mathbf{d}_i$ and $\mathbf{c}_i$, defined in Eqs. (3.4c) and

(3.5b). The vectors, f_i and n_i, are the force at O_i and the moment about O_i of the i-th link, respectively, whereas $\boldsymbol{\varepsilon}_i$ is the $3m_i$-dimensional generalized force vector associated with the generalized coordinates $\mathbf{d}_i$ due to bending, and $\boldsymbol{\vartheta}_i$ is the $\overline{m}_i$-dimensional generalized force vector associated with the generalized coordinates $\mathbf{c}_i$ due to torsion. For a rigid body system, vectors $\mathbf{d}_i$ $\mathbf{c}_i$, $\boldsymbol{\varepsilon}_i$ and $\boldsymbol{\vartheta}_i$ vanish and the dimension of the vectors, $\mathbf{t}_i$ and $\mathbf{w}_i$, reduce to six only, as defined in Eq. (2.1).

- **t** and **w**: The $\hat{n} \equiv 6n_r + (6 + 3m_i + \overline{m}_i)n_f$-dimensional vector of generalized twist and wrench, respectively, which are defined similar to Eq. (2.2) as

$$\mathbf{t} \equiv \begin{bmatrix} \mathbf{t}_1^T & \mathbf{t}_2^T & \cdots & \mathbf{t}_n^T \end{bmatrix}^T; \mathbf{w} \equiv \begin{bmatrix} \mathbf{w}_1^T & \mathbf{w}_2^T & \cdots & \mathbf{w}_n^T \end{bmatrix}^T \tag{3.7}$$

where $\mathbf{t}_i$ and $\mathbf{w}_i$, for i=1, …, n, are given by Eq. (3.6).

- $\mathbf{q}_i$ and $\boldsymbol{\tau}_i$: The $(1 + 3m_i + \overline{m}_i)$-dimensional vectors of joint-and-amplitudes, and the corresponding generalized forces of the i-th flexible link, i.e.,

$$\mathbf{q}_i \equiv \begin{bmatrix} \theta_i & \mathbf{d}_i^T & \mathbf{c}_i^T \end{bmatrix}^T \text{ and } \boldsymbol{\tau}_i \equiv \begin{bmatrix} \tau_i & \boldsymbol{\varepsilon}_i^T & ,_i^T \end{bmatrix}^T \tag{3.8}$$

where θ_i is the rotational or translational displacement of the i-th joint depending on its type, i.e., revolute or prismatic, respectively, and vectors $\mathbf{d}_i$ and $\mathbf{c}_i$ are defined in Eqs. (3.4d) and (3.5b), respectively. Moreover, $\boldsymbol{\tau}_i$, and the vectors, $\boldsymbol{\varepsilon}_i$ and $\boldsymbol{\vartheta}_i$ are, respectively, the generalized forces corresponding to the joint coordinate, θ_i, and the amplitude vectors, $\mathbf{d}_i$, and $\mathbf{c}_i$. Note that for a rigid body, vectors, $\boldsymbol{\varepsilon}_i$ and $\boldsymbol{\vartheta}_i$ vanish, and $\boldsymbol{\tau}_i$ reduces to a scalar.

- $\dot{\mathbf{q}}$ and $\boldsymbol{\tau}$: The $\overline{n}$-dimensional vector of rates of joint-and-amplitude vector and the vector of corresponding generalized forces, i.e.,

$$\dot{\mathbf{q}} \equiv \begin{bmatrix} \dot{\mathbf{q}}_1 & \dot{\mathbf{q}}_2 & \cdots & \dot{\mathbf{q}}_n \end{bmatrix}^T \text{ and } \boldsymbol{\tau} \equiv \begin{bmatrix} \boldsymbol{\tau}_1 & \boldsymbol{\tau}_2 & \cdots & \boldsymbol{\tau}_n \end{bmatrix}^T \tag{3.9}$$

where $\dot{\mathbf{q}}_i$ is the time-rate of change of the joint-and-amplitude vector of the i-th flexible link, and $\boldsymbol{\tau}_i$ is the corresponding generalized forces of the i-th flexible link. For a rigid link, $\mathbf{q}_i \equiv \theta_i$.

3.1.4 The DeNOC Matrices for Rigid–Flexible Robots

Following the steps of Sect. 2.1.2 for rigid robots, the DeNOC matrices for serial rigid–flexible robots are derived below:

(1) The twist of the i-th link is expressed in terms of the $(i-1)$st link as

$$\mathbf{t}_i = \mathbf{A}_{i,i-1}\mathbf{t}_{i-1} + \mathbf{P}_i\dot{\mathbf{q}}_i \tag{3.10a}$$

where $\mathbf{t}_i$ is the $(6+3m_i+\overline{m}_i)$-dimensional twist vector of the ith flexible link defined in Eq. (3.6), and $\dot{\mathbf{q}}_i$ is the time-rate of change of vector $\mathbf{q}_i$ defined in Eq. (3.8), whereas the $(6+3m_i+\overline{m}_i)\times(6+3m_i+\overline{m}_i)$ twist propagation matrix, $\mathbf{A}_{i,i\text{-}1}$, and the $(6+3m_i+\overline{m}_i)\times(1+3m_i+\overline{m}_i)$ joint-and-amplitude propagation matrix, $\mathbf{P}_i$, for the flexible links are defined as

$$\mathbf{A}_{i,i-1} \equiv \begin{bmatrix} \mathbf{R}_{i,i-1} & \mathbf{F}_{i-1} \\ \tilde{\mathbf{O}} & \overline{\mathbf{O}} \end{bmatrix};\ \mathbf{P}_i \equiv \begin{bmatrix} \mathbf{p}_i & \tilde{\mathbf{0}}^T \\ \tilde{\mathbf{0}} & \overline{\mathbf{1}} \end{bmatrix} \tag{3.10b}$$

In Eq. (3.10b), the 6×6 matrix, $\mathbf{R}_{i,i-1}$, and the 6-dimensional vector, $\mathbf{p}_i$, are associated with the vectors $\mathbf{a}_{i,i-1}$, and $\mathbf{z}_i$, of the i-th link, respectively. Furthermore, the $6\times(3m_i+\overline{m}_i)$ lower block triangular matrix, $\mathbf{F}_{i-1}$, is given by

$$\mathbf{F}_{i-1} \equiv \begin{bmatrix} \mathbf{S}_{i-1} & \hat{\mathbf{O}}_{i-1} \\ \boldsymbol{\Delta}_{i-1} & \mathbf{C}_{i-1} \end{bmatrix} \tag{3.10c}$$

in which the $3\times 3m_i$ matrix, $\mathbf{S}_{i-1}$, contains the shape functions corresponding to the 3-dimensional bending deflections, as defined in Eqs. (3.4a–b), whereas the $3\times 3m_i$ matrix, $\boldsymbol{\Delta}_{i-1}$, contains the first derivatives of the bending shape functions, corresponding to the $(i\text{-}1)^{\text{st}}$ flexible link. For the i-th flexible link, if s_{ij} denotes the shape function corresponding to the j-th mode, its first derivative with respect to its length of the flexible part, $\overline{a}_i$, evaluated at the link end, a_i, is given by $\partial s_{ij}/\partial\overline{a}_i\big|_{a_i}$. Accordingly, the matrix Δ_i is represented as

$$\boldsymbol{\Delta}_i \equiv \begin{bmatrix} 0\cdots 0 & 0\cdots 0 & 0\cdots 0 \\ 0\cdots 0 & 0\cdots 0 & -\dfrac{\partial s^y_{i,1}}{\partial\overline{a}_i}\bigg|_{a_i} \cdots -\dfrac{\partial s^y_{i,m_i}}{\partial\overline{a}_i}\bigg|_{a_i} \\ 0\cdots 0 & \dfrac{\partial s^z_{i,1}}{\partial\overline{a}_i}\bigg|_{a_i} \cdots \dfrac{\partial s^z_{i,m_i}}{\partial\overline{a}_i}\bigg|_{a_i} & 0\cdots 0 \end{bmatrix} \tag{3.10d}$$

in which $s^y_{i,j}$ and $s^z_{i,j}$ for $j = 1,\ldots, m_i$ are those appeared in Eq. (3.4a). Moreover, the $3\times\overline{m}_i$ matrix $\mathbf{C}_{i-1}$, contains the shape functions, corresponding to the torsion of $(i\text{-}1)^{\text{st}}$ flexible link in $\overline{m}_i$ modes. Accordingly, the matrix $\mathbf{C}_i$ is represented as

$$\mathbf{C}_i \equiv \begin{bmatrix} \overline{s}_{i,1}\big|_{a_i} \cdots \overline{s}_{i,\overline{m}_i}\big|_{a_i} \\ 0\cdots 0 \\ 0\cdots 0 \end{bmatrix} \tag{3.10e}$$

where $\overline{s}_{i,h}$ for $h = 1,\ldots, \overline{m}_i$ are the shape functions of the beam in torsional vibration as defined in Eq. (3.5b). Note that $\widetilde{\mathbf{O}}$, $\overline{\mathbf{O}}$ and $\widehat{\mathbf{O}}$ in Eqs. (3.10b–c) are, respectively, the $(3m_i + \overline{m}_i) \times 6$, $(3m_i + \overline{m}_i) \times (3m_i + \overline{m}_i)$ and $3 \times \overline{m}_i$ zero matrices. Also, $\tilde{\mathbf{0}}$ and $\overline{\mathbf{1}}$ are, respectively, the $(3m_i + \overline{m}_i)$-dimensional vectors of zeros and the $(3m_i + \overline{m}_i) \times (3m_i + \overline{m}_i)$ identity matrix. Now, the interpretations similar to those given after Eq. (2.4), for the rigid body system, are given here from Eq. (3.10a), for the flexible body system. They are:

- If $\dot{\mathbf{q}}_i = 0$, i.e., when the $(i$-1)st and i-th flexible links are locked at the i-th joint with no vibration in the i-th link, they behave like a single flexible link. The term, $\mathbf{A}_{i,i-1}\mathbf{t}_{i-1}$, propagates the twist, $\mathbf{t}_{i-1}$, to $\mathbf{t}_i$ without any external excitation due to the i-th joint or the vibrations in the i-th link. Hence, in line with the rigid link system, the $(6 + 3m_i + \overline{m}_i) \times (6 + 3m_i + \overline{m}_i)$ matrix, $\mathbf{A}_{i,i-1}$, is termed here as the twist propagation matrix for the flexible link which satisfies the property, $\mathbf{A}_{i,j}\mathbf{A}_{j,k} = \mathbf{A}_{i,k}$. However, $\mathbf{A}_{i,i-1}$ does not satisfy the beautiful properties applicable to serial rigid links systems, namely, Eq. (2.6). Hence, for a flexible link system,

$$\mathbf{A}_{i,j}^{-1} \neq \mathbf{A}_{j,i} \text{ and } \mathbf{A}_{i,i} \neq 1 \tag{3.11}$$

The above conditions cause one not to have full linear order forward dynamics algorithm based on the analytical expressions of the associated inertia matrix, as evident from Sect. 3.4 of this chapter.

- If $\dot{\mathbf{q}}_i \neq 0$, i.e., when both the i-th joint motion and the vibrations are allowed, $\mathbf{P}_i\dot{\mathbf{q}}_i$, adds to the components of $\mathbf{A}_{i,i-1}\mathbf{t}_{i-1}$, thus, giving rise to the actual twist of i-th flexible link, $\mathbf{t}_i$. Hence, the $(6 + 3m_i + \overline{m}_i) \times (1 + 3m_i + \overline{m}_i)$ matrix, $\mathbf{P}_i$, is termed as the joint-and-amplitude propagation matrix.

(2) Now, for the n-link serial rigid–flexible robot, as in Figs. 2.1 and 3.1, the generalized twist, $\mathbf{t}$ of Eq. (3.7), can be expressed using Eq. (3.10a) as

$$\mathbf{t} = \mathbf{A}\mathbf{t} + \mathbf{N}_d\dot{\mathbf{q}} \tag{3.12}$$

where the $\hat{n}$-dimensional vector, $\mathbf{t}$, and the $\overline{n}$-dimensional vector, $\mathbf{q}$, are defined in Eqs. (3.7) and (3.8), respectively. In Eq. (3.10a), the $\hat{n} \times \hat{n}$ matrix, $\mathbf{A}$, and the $\hat{n} \times \overline{n}$ matrix, $\mathbf{N}_d$, are given by

$$\mathbf{A} \equiv \begin{bmatrix} \mathrm{O} & \cdots & \cdots & \mathrm{O} \\ \mathbf{A}_{21} & \mathrm{O} & \cdots & \mathrm{O} \\ \vdots & \vdots & \ddots & \vdots \\ \mathrm{O} & \cdots & \mathbf{A}_{n,n-1} & \mathrm{O} \end{bmatrix}; \mathbf{N}_d \equiv \begin{bmatrix} \mathbf{P}_1 & \mathbf{0} & \cdots & \mathbf{0} \\ \mathbf{0} & \mathbf{P}_2 & \cdots & \mathbf{0} \\ \vdots & \vdots & \ddots & \\ \mathbf{0} & \mathbf{0} & \cdots & \mathbf{P}_n \end{bmatrix} \tag{3.13}$$

where $\mathbf{O}$ and $\mathbf{0}$ are the $(6 + 3m_i + \overline{m}_i) \times (6 + 3m_i + \overline{m}_i)$ matrix and the $(6 + 3m_i + \overline{m}_i)$-dimensional vector of zeros, respectively. Henceforth, $\mathbf{O}$ and $\mathbf{0}$ should be understood as of compatible dimensions based on the expressions where they appear.

(3) Since Eqs. (3.12) and (3.13) have the same structure as those for the rigid system, Eq. (2.7), and $\mathbf{A}_{i,j}\mathbf{A}_{j,k} = \mathbf{A}_{i,k}$, the generalized twist, $\mathbf{t}$, can be written similar to Eq. (2.9) as

$$\mathbf{t} = \mathbf{N}\dot{\boldsymbol{\theta}}, \text{ where } \mathbf{N} \equiv \mathbf{N}_l\mathbf{N}_d \tag{3.14}$$

In Eq. (3.14), the $\hat{n} \times \hat{n}$ matrix, $\mathbf{N}_l$, is given by:

$$\mathbf{N}_1 \equiv \begin{bmatrix} \mathbf{1} & \mathrm{O} & \cdots & \mathrm{O} \\ \mathbf{A}_{21} & \mathbf{1} & \cdots & \mathrm{O} \\ \vdots & \vdots & \ddots & \\ \mathbf{A}_{n1} & \mathbf{A}_{n2} & \cdots & \mathbf{1} \end{bmatrix} \tag{3.15}$$

where, $\mathbf{1}$ denotes the $(6 + 3m_i + \overline{m}_i) \times (6 + 3m_i + \overline{m}_i)$ identity matrix. Like $\mathbf{O}$ and $\mathbf{0}$, henceforth, $\mathbf{1}$ should be understood as of compatible size based on where it appears. The matrix, $\mathbf{N}$, in a coupled form is the natural orthogonal complement (NOC) matrix for the serial rigid–flexible robot, as reported in Cyril [7]. In this book, the decoupled form, namely $\mathbf{N}_l$ and $\mathbf{N}_d$ matrices, are derived for the flexible systems for the first time, which are referred as the decoupled natural orthogonal complement (DeNOC) matrices for flexible robots. For the rigid–flexible robots understudy, matrix $\mathbf{A}_{ij}$ will take the form of $\mathbf{R}_{ij}$, Eq. (2.5), while a rigid link is present. The decoupled form, Eq. (3.14), allows one to write the matrix and vector elements associated with the dynamic equations of motion in analytical form, leading to a recursive forward dynamics algorithm.

3.2 Dynamic Modeling of Rigid–Flexible Robots

The dynamic model of serial robots having both rigid and flexible links, i.e., rigid–flexible robots, is now derived by extending the methodology for the rigid robots proposed in Sect. 2.2. These are explained in the following steps:

(1) Referring to Fig. 3.1, the position vectors of the elements, $\overline{E}_i$, $\tilde{E}_i$ and payload of mass m_{pi}, on the i-th link, namely, $\mathbf{r}_i$, $\tilde{\mathbf{r}}_i$ and $\mathbf{r}_{pi}$ are, respectively, given by,

$$\mathbf{r}_i = \mathbf{o}_i + \overline{\mathbf{b}}_i, \text{ where } \overline{\mathbf{b}}_i = \overline{b}_i\mathbf{z}_i$$

$$\tilde{\mathbf{r}}_i = \mathbf{o}_i + \overline{\mathbf{r}}_i, \text{ where } \overline{\mathbf{r}}_i = b_i\mathbf{z}_i + \overline{a}_i\hat{\mathbf{x}}_{i+1} + \mathbf{u}_i, \text{ and}$$
$$\mathbf{r}_{pi} = \mathbf{o}_i + \overline{\mathbf{r}}_{pi}, \text{ where } \overline{\mathbf{r}}_{pi} = b_i\mathbf{z}_i + a_i\hat{\mathbf{x}}_{i+1} + \mathbf{u}_{pi} \quad (3.16)$$

where a_i and b_i, are the DH parameters of the link, as defined in Appendix A, $\overline{b}_i$ is the axial distance of element $\overline{E}_i$ along Z_i from O_i, and $\overline{a}_i$ is the axial distance of element $\tilde{E}_i$ along $\widehat{X}_{i+1}$ from O_i', as shown in Fig. 3.1. Note that $\overline{\mathbf{b}}_i$ is the position vector of element $\overline{E}_i$ along Z_i from O_i, whose magnitude is $\overline{b}_i$. The term, $\overline{b}_i$, varies from 0 to b_i, and $\overline{\mathbf{a}}_i$ varies from 0 to a_i. Moreover, the unit vectors along Z_i and $\widehat{X}_{i+1}$-axes are denoted with $\mathbf{z}_i$ and $\hat{\mathbf{x}}_{i+1}$, respectively, and vector $\mathbf{o}_i$ denotes the position vector of the point, O_i, of the ith frame with respect to the origin of the fixed first frame. Furthermore, vectors $\mathbf{u}_i$ and $\mathbf{u}_{pi}$ are, respectively, the positions of the element, $\tilde{E}_i$, and the payload m_{pi}, on the deformed flexible link from its undeformed state, which are given by $\mathbf{u}_i = u_i\mathbf{e}_u$ and $\mathbf{u}_{pi} = u_{pi}\mathbf{e}_u$, respectively, where $\mathbf{e}_u$ is the unit vector along the direction of displacement. Vector $\mathbf{u}_i$ is shown in Fig. 3.1. Note that, the payload is considered as a concentrated point mass at the tip of the link that accounts for any assembly with sensors attached to the ith link, and the real load to be carried by the last link, etc.

(2) The kinetic energy, T_i, for the i-th flexible link is then given by

$$T_i = \frac{1}{2}\int_0^{b_i} \rho_i \dot{\mathbf{r}}_i^T \dot{\mathbf{r}}_i d\overline{b}_i + \frac{1}{2}\int_0^{a_i} \rho_i \dot{\tilde{\mathbf{r}}}_i^T \dot{\tilde{\mathbf{r}}}_i d\overline{a}_i + \frac{1}{2} m_{pi} \dot{\mathbf{r}}_{pi}^T \dot{\mathbf{r}}_{pi} + \frac{1}{2}\int_0^{a_i} \rho_i I_{pi} \dot{\beta}_i^2 dx_{i+1} + T_{hi} \quad (3.17)$$

where ρ_i is the mass per unit length of the i-th link, and the vectors, $\dot{\mathbf{r}}_i$, $\dot{\tilde{\mathbf{r}}}_i$ and $\dot{\mathbf{r}}_{pi}$, are the velocities of the elements, $\overline{E}_i$, $\tilde{E}_i$ and payload, respectively, which can be written from Eq. (3.16) as

$$\dot{\mathbf{r}}_i = \mathbf{v}_i + \boldsymbol{\omega}_i \times \overline{\mathbf{b}}_i;$$
$$\dot{\tilde{\mathbf{r}}}_i = \mathbf{v}_i + \boldsymbol{\omega}_i \times \overline{\mathbf{r}}_i + \dot{b}_i\mathbf{z}_i + \dot{u}_i\mathbf{e}_u; \text{ and}$$
$$\dot{\mathbf{r}}_{pi} = \mathbf{v}_i + \boldsymbol{\omega}_i \times \overline{\mathbf{r}}_{pi} + \dot{b}_i\mathbf{z}_i + \dot{u}_{pi}\mathbf{e}_u \quad (3.18a)$$

where $\mathbf{v}_i$ is substituted for $\dot{\mathbf{o}}_i$, i.e., $\mathbf{v}_i \equiv \dot{\mathbf{o}}_i$. Moreover, $\dot{\overline{a}}_i = \dot{a}_i = 0$ is used in Eq. (3.18a) due to the assumption of no extension along $\widehat{X}_{i+1}$ axis of the i-th flexible link shown in Fig. 3.1. Similarly, $\dot{\overline{b}}_i = 0$, as the link is rigid along Z_i-axis. Furthermore, $\dot{b}_i$ represents the linear joint rate in the presence of prismatic joint. If the i-th joint is revolute then $\dot{b}_i = 0$. For that, Eq. (3.18a) is modified as

$$\dot{\mathbf{r}}_i = \mathbf{v}_i + \boldsymbol{\omega}_i \times \overline{\mathbf{b}}_i;$$
$$\dot{\tilde{\mathbf{r}}}_i = \mathbf{v}_i + \boldsymbol{\omega}_i \times \overline{\mathbf{r}}_i + \dot{u}_i\mathbf{e}_u; \text{ and}$$
$$\dot{\mathbf{r}}_{pi} = \mathbf{v}_i + \boldsymbol{\omega}_i \times \overline{\mathbf{r}}_{pi} + \dot{u}_{pi}\mathbf{e}_u \quad (3.18b)$$

in which $\boldsymbol{\omega}_i$ is the angular velocity of the i-th flexible link and the vectors, $\dot{\mathbf{u}}_i$ and $\dot{\mathbf{u}}_{pi}$, are, respectively, the time-rate of change of vector $\mathbf{u}_i$ and $\mathbf{u}_{pi}$ with respect to their undeformed position, Fig. 3.1. Also, the scalar, I_{pi}, denotes the polar moment of inertia of the cross section of the element, $\tilde{E}_i$ belonging to the i-th flexible link, whereas β_i is the scalar angular deformation of the cross section of the element $\tilde{E}_i$, as defined after Eq. (3.5a). Finally, the term, T_{hi}, represents the kinetic energy due to the hub inertia at the joint. A hub can include the effect of a motor at joints and the associated gear assembly that is given by,

$$T_{hi} = \frac{1}{2}\boldsymbol{\omega}_i^T \mathbf{I}_{hi} \boldsymbol{\omega}_i \tag{3.19}$$

where $\mathbf{I}_{hi}$ is the 3×3 inertia tensor for the hub.

(3) The Euler–Lagrange (EL) equations of motion for the whole system are then given by [18]:

$$\frac{d}{dt}\left(\frac{\partial T}{\partial \dot{\mathbf{q}}_i}\right) - \frac{\partial T}{\partial \mathbf{q}_i} = \tau_i, \text{ for } i = 1, \cdots, n \tag{3.20a}$$

where n is the total number of rigid and flexible bodies, whereas $\mathbf{q}_i$ is the $(1+3m_i+\overline{m}_i)$-dimensional vector of independent generalized coordinates defined in Eq. (3.8). Accordingly, vector $\boldsymbol{\tau}_i$ is the associated generalized forces given by $\boldsymbol{\tau}_i \equiv \boldsymbol{\tau}_i^E + \boldsymbol{\tau}_i^s$, in which $\boldsymbol{\tau}_i^E$ is the generalized forces due to external forces and moments on the whole system, and $\boldsymbol{\tau}_i^s$ is the generalized forces due to strains in the i-th link. Vector $\boldsymbol{\tau}_i^s$ has the following form:

$$\boldsymbol{\tau}_i^s \equiv \frac{\partial \mathbf{V}_s}{\partial \mathbf{q}_i} \equiv \mathbf{K}_i \mathbf{q}_i \tag{3.20b}$$

where $\mathbf{V}_s$ is the potential energy of the system at hand due to strains and $\mathbf{K}_i$ is the stiffness matrix of the i-th link. Note here that the effect of potential energy due to gravity is included by adding negative of the acceleration due to gravity to the linear acceleration of the first link, $\dot{\mathbf{v}}_1$, as proposed by Walker and Orin [35], and also done for the rigid bodies in Chap. 2. The potential energy of the system at hand due to strain energy, V_s, is evaluated as:

$V_s = \sum_{i=1}^{n_f} V_{si}$, where

$$V_{si} \equiv \frac{1}{2}\int_0^{a_i} E_i I_i^y \left(\frac{\partial^2 u_i^y}{\partial \overline{a}_i^2}\right)^2 d\overline{a}_i + \frac{1}{2}\int_0^{a_i} E_i I_i^z \left(\frac{\partial^2 u_i^z}{\partial \overline{a}_i^2}\right)^2 d\overline{a}_i + \frac{1}{2}\int_0^{a_i} G_i I_{pi} \left(\frac{\partial^2 \beta_i}{\partial \overline{a}_i^2}\right)^2 d\overline{a}_i \tag{3.20c}$$

where $E_i I_i^y$, $E_i I_i^z$ are the flexure rigidity and $G_i I_{pi}$ is the torsional rigidity of the i-th link, in which E_i is Young's modulus of elasticity, I_i^y and I_i^z are the moment of inertia of the link about its $\widehat{Y}_{i+1}$ and $\widehat{Z}_{i+1}$-axes, respectively, and I_{pi} is the polar moment of inertia of the cross section of the link. The deflection of the link, u_i^y, u_i^z, β_i, is given by Eqs. (3.3c) and (3.5a), whereas the deflection along X_{i+1}-axis is ignored. Now, the partial differentiation of V_s with respect to $\mathbf{q}_i$, as required in Eq. (3.20b), is evaluated. Note that the vector of generalized coordinates, $\mathbf{q}_i$, is the array of θ_i, d_i^y, d_i^z and $\mathbf{c}_i$, but V_s is a function of only d_i^y, d_i^z and $\boldsymbol{c}_i$. As a result, $\boldsymbol{\tau}_i^s \equiv \frac{\partial V_s}{\partial \mathbf{q}_i}$ is obtained as

$$\boldsymbol{\tau}_i^s = \begin{bmatrix} 0 \\ \mathbf{0} \\ \boldsymbol{\tau}_i^{sy} \\ \boldsymbol{\tau}_i^{sz} \\ \boldsymbol{\tau}_i^{sx} \end{bmatrix} \tag{3.20d}$$

Since strain energy due to axial centrifugal stiffening is neglected, the m_i-dimensional zero vector $\mathbf{0}$ appears in $\boldsymbol{\tau}_i^s$ after the scalar zero of Eq. (3.20d). In Eq. (3.20d) the m_i-dimensional vectors $\boldsymbol{\tau}_i^{sy}$ and $\boldsymbol{\tau}_i^{sz}$, and the $\overline{m}_i$-dimensional vector, $\boldsymbol{\tau}_i^{sx}$, are given as,

$$\boldsymbol{\tau}_i^{sy} \equiv \left[\tau_{i1}^{sy} \cdots \tau_{im_i}^{sy} \right]^T; \boldsymbol{\tau}_i^{sz} \equiv \left[\tau_{i1}^{sz} \cdots \tau_{im_i}^{sz} \right]^T \text{ and } \boldsymbol{\tau}_i^{sx} \equiv \left[\tau_{i1}^{sx} \cdots \tau_{i\overline{m}_i}^{sx} \right]^T \tag{3.20e}$$

where τ_{ij}^{sy} and τ_{ij}^{sz} are defined, for $j = 1,\ldots, m_i$, as

$$\tau_{ij}^{sy} \equiv E_i I_i^y \int_0^{a_i} \left(\tilde{k}_{ij}^y \sum_{h=1}^{m_i} \tilde{k}_{ih}^y d_{ih}^y \right) d\overline{a}_i \text{ and } \tau_{ij}^{sz} \equiv E_i I_i^z \int_0^{a_i} \left(\tilde{k}_{ij}^z \sum_{h=1}^{m_i} \tilde{k}_{ih}^z d_{ih}^z \right) d\overline{a}_i \tag{3.20f}$$

Similarly, τ_{il}^{sx} are given as, for $l = 1,\ldots, \overline{m}_i$, as

$$\tau_{il}^{sx} = G_i I_{pi}^x \int_0^{a_i} \left(\overline{k}_{il}^x \sum_{h=1}^{\overline{m}_i} \overline{k}_{ih}^x c_{ih}^x \right) d\overline{a}_i \tag{3.20g}$$

where for $j = 1, \ldots, m_i$: $\tilde{k}_{ij}^y \equiv \frac{\partial^2 s_{ij}^y}{\partial \overline{a}_i^2}$, $\tilde{k}_{ij}^z \equiv \frac{\partial^2 s_{ij}^z}{\partial \overline{a}_i^2}$ and, for $l = 1,\ldots, \overline{m}_i$, $\overline{k}_{il}^x \equiv \frac{\partial^2 \overline{s}_{il}^x}{\partial \overline{a}_i^2}$ in which s_{ij}^y and s_{ij}^z are the shape functions of the link in its j-th mode of vibration in $\widehat{Y}_{i+1}$ and $\widehat{Z}_{i+1}$ axis, respectively, as defined after Eq. (3.4a), whereas $\overline{s}_{il}^x$ is the shape function of the link in its l-th mode in torsion, as given by Eq. (3.5b). Note that $d_{i,h}^y$, $d_{i,h}^z$, and $c_{i,l}^x$ are the generalized coordinates of the system. Moreover, for a revolute joint, the generalized coordinate, θ_i, represents joint rotation, and for a prismatic joint, it should be understood as joint displacement. The total kinetic energy of the system is then, $T = \sum_{i=1}^{n} T_i$ – n being the total number of rigid and flexible links, n_r and n_f, respectively, i.e., $n \equiv n_r + n_f$. The dynamic equations of motion, as derived in Appendix D, are accordingly expressed as

$$\left[\left(\frac{\partial \mathbf{t}_1}{\partial \dot{\mathbf{q}}_i}\right)^T \cdots \left(\frac{\partial \mathbf{t}_n}{\partial \dot{\mathbf{q}}_i}\right)^T\right]\begin{bmatrix}\mathbf{w}_1^* \\ \vdots \\ \mathbf{w}_n^*\end{bmatrix} = \boldsymbol{\tau}_i, \text{ for } i = 1, \ldots, n \tag{3.21a}$$

where the $(6 + 3m_i + \overline{m}_i) \times (1 + 3m_i + \overline{m}_i)$ matrix, $\frac{\partial \mathbf{t}_i}{\partial \dot{\mathbf{q}}_j}$, and the $(6 + 3m_i + \overline{m}_i)$-dimensional vector, $\mathbf{w}_i^*$, for $i = 1, \cdots, n$, are reproduced from Eq. (D.10) as

$$\frac{\partial \mathbf{t}_i}{\partial \dot{\mathbf{q}}_j} \equiv \begin{bmatrix} \frac{\partial \mathbf{v}_i}{\partial \dot{\mathbf{q}}_j} \\ \frac{\partial \boldsymbol{\omega}_i}{\partial \dot{\mathbf{q}}_j} \\ \frac{\partial \mathbf{d}_i}{\partial \dot{\mathbf{q}}_j} \\ \frac{\partial \dot{\mathbf{c}}_i}{\partial \dot{\mathbf{q}}_j} \end{bmatrix}; \mathbf{w}_i^* \equiv \begin{bmatrix} \int_0^{b_i} \rho_i \ddot{\mathbf{r}}_i d\overline{b}_i + \int_0^{a_i} \rho_i \ddot{\overline{\mathbf{r}}}_i d\overline{a}_i + m_{pi}\ddot{\mathbf{r}}_{pi} \\ \left[\int_0^{b_i} \rho_i \overline{b}_i (\mathbf{z}_i \times \ddot{\mathbf{r}}_i) d\overline{b}_i + \int_0^{a_i} \rho_i (\overline{\mathbf{r}}_i \times \ddot{\overline{\mathbf{r}}}_i) d\overline{a}_i + m_{pi}(\overline{\mathbf{r}}_{pi} \times \ddot{\mathbf{r}}_{pi}) \right. \\ \left. +(\mathbf{I}_{hi}\dot{\boldsymbol{\omega}}_i + \boldsymbol{\omega}_i \times \mathbf{I}_{hi}\boldsymbol{\omega}_i)\right] \\ \int_0^{a_i} \rho_i \mathbf{S}_i^T \ddot{\overline{\mathbf{r}}}_i d\overline{a}_i + m_{pi}\, \mathbf{S}_i|_{a_i}^T \ddot{\mathbf{r}}_{pi} \\ \int_0^{a_i} \rho_i I_{pi} \overline{\mathbf{s}}_i \ddot{\mathbf{c}}_i^T \overline{\mathbf{s}}_i d\overline{a}_i \end{bmatrix} \tag{3.21b}$$

where $\mathbf{S}_i|_{a_i}$ implies the value of the shape function matrix $\mathbf{S}_i$, Eq. (3.4b), evaluated at $\overline{a}_i = a_i$.

(4) The vector, $\mathbf{w}_i^*$ of Eq. (3.21b), can be physically interpreted as the inertia wrench of the flexible link and can be written similar to Eq. (2.18) as

$$\mathbf{w}_i^* = \mathbf{M}_i \dot{\mathbf{t}}_i + \boldsymbol{\gamma}_i \tag{3.22}$$

where $\mathbf{M}_i$ is the $(6 + 3m_i + \overline{m}_i) \times (6 + 3m_i + \overline{m}_i)$ mass matrix, and $\boldsymbol{\gamma}_i$ is the $(6 + 3m_i + \overline{m}_i)$-dimensional vector. Next $\ddot{\mathbf{r}}_i$, $\ddot{\overline{\mathbf{r}}}_i$, and $\ddot{\mathbf{r}}_{pi}$ of the link coupled to a revolute joint are obtained from the time derivative of the expressions given in Eq. (3.18b), i.e.,

$$\begin{aligned} \ddot{\mathbf{r}}_i &= \dot{\mathbf{v}}_i + \dot{\boldsymbol{\omega}}_i \times \overline{\mathbf{b}}_i + \boldsymbol{\omega}_i \times (\boldsymbol{\omega}_i \times \overline{\mathbf{b}}_i) \\ \ddot{\overline{\mathbf{r}}}_i &= \dot{\mathbf{v}}_i + \dot{\boldsymbol{\omega}}_i \times \overline{\mathbf{r}}_i + \boldsymbol{\omega}_i \times (\boldsymbol{\omega}_i \times \overline{\mathbf{r}}_i) + \ddot{\mathbf{u}}_i + 2\boldsymbol{\omega} \times \dot{u}_i \mathbf{e}_u \\ \ddot{\overline{\mathbf{r}}}_{pi} &= \dot{\mathbf{v}}_i + \dot{\boldsymbol{\omega}}_i \times \overline{\mathbf{r}}_{pi} + \boldsymbol{\omega}_i \times \left(\boldsymbol{\omega}_i \times \overline{\mathbf{r}}_{pi}\right) + \ddot{\mathbf{u}}_{pi} + 2\boldsymbol{\omega} \times \dot{u}_{pi} \mathbf{e}_u \end{aligned} \tag{3.23}$$

Substituting $\ddot{\mathbf{r}}_i$, $\ddot{\bar{\mathbf{r}}}_i$ and $\ddot{\bar{\mathbf{r}}}_{pi}$ from Eqs. (3.23) into Eq. (3.21b), the mass matrix, $\mathbf{M}_i$, and vector $\boldsymbol{\gamma}_i$ are obtained as:

$$\boldsymbol{M}_i \equiv \int_0^{b_i} \rho_i \begin{bmatrix} \mathbf{1} & -\bar{\mathbf{b}}_i \times \mathbf{1} & \mathbf{O} & \mathbf{0} \\ & -\bar{\mathbf{b}}_i \times (\bar{\mathbf{b}}_i \times \mathbf{1}) & \mathbf{O} & \mathbf{0} \\ & & \mathbf{O} & \mathbf{0} \\ sym & & & \mathbf{0} \end{bmatrix} d\bar{b}_i$$

$$+ \int_0^{a_i} \rho_i \begin{bmatrix} \mathbf{1} & -\bar{\mathbf{r}}_i \times \mathbf{1} & \mathbf{S}_i & \mathbf{0} \\ & -\bar{\mathbf{r}}_i \times (\bar{\mathbf{r}}_i \times \mathbf{1}) & \bar{\mathbf{r}}_i \times \mathbf{S}_i & \mathbf{0} \\ & & \mathbf{S}_i^T \mathbf{S}_i & \mathbf{0} \\ sym & & & I_{pi} \bar{\mathbf{s}}_i \bar{\mathbf{s}}_i^T \end{bmatrix} d\bar{a}_i$$

$$+ m_{pi} \begin{bmatrix} \mathbf{1} & -\bar{\mathbf{r}}_{pi} \times \mathbf{1} & \mathbf{S}_i|_{a_i} & \mathbf{0} \\ & -\bar{\mathbf{r}}_{pi} \times (\bar{\mathbf{r}}_{pi} \times \mathbf{1}) & \bar{\mathbf{r}}_{pi} \times \mathbf{S}_i|_{a_i} & \mathbf{0} \\ & & \mathbf{S}_i|_{ai}^T \mathbf{S}_i|_{a_i} & \mathbf{0} \\ sym & & & \mathbf{0} \end{bmatrix} + \begin{bmatrix} \mathbf{O} & \mathbf{O} & \mathbf{O} & \mathbf{0} \\ & \mathbf{I}_{hi} & \mathbf{O} & \mathbf{0} \\ & & \mathbf{O} & \mathbf{0} \\ sym & & & \mathbf{0} \end{bmatrix} \tag{3.24a}$$

$$\boldsymbol{\gamma}_i \equiv \int_0^{b_i} \rho_i \begin{bmatrix} \bar{\bar{\boldsymbol{\omega}}}_i \\ \bar{\mathbf{b}}_i \times \bar{\bar{\boldsymbol{\omega}}}_i \\ \mathbf{0} \\ \mathbf{0} \end{bmatrix} d\bar{b}_i + \int_0^{a_i} \rho_i \begin{bmatrix} \bar{\boldsymbol{\omega}}_i \\ \bar{\mathbf{r}}_i \times \bar{\boldsymbol{\omega}}_i \\ \mathbf{S}_i^T \bar{\boldsymbol{\omega}}_i \\ \mathbf{0} \end{bmatrix} d\bar{a}_i + m_{pi} \begin{bmatrix} \bar{\boldsymbol{\omega}}_{pi} \\ \bar{\mathbf{r}}_{pi} \times \bar{\boldsymbol{\omega}}_{pi} \\ \mathbf{S}_i|_{ai}^T \bar{\boldsymbol{\omega}}_{pi} \\ \mathbf{0} \end{bmatrix}$$

$$+ \begin{bmatrix} \mathbf{0} \\ \boldsymbol{\omega}_i \times \mathbf{I}_{hi} \boldsymbol{\omega}_i \\ \mathbf{0} \\ \mathbf{0} \end{bmatrix} \tag{3.24b}$$

where $\bar{\bar{\boldsymbol{\omega}}}_i \equiv \boldsymbol{\omega}_i \times (\boldsymbol{\omega}_i \times \bar{\mathbf{b}}_i)$, $\bar{\boldsymbol{\omega}}_i \equiv \boldsymbol{\omega}_i \times [(\boldsymbol{\omega}_i \times \bar{\mathbf{r}}_i) + \dot{\mathbf{u}}_i]$, $\bar{\boldsymbol{\omega}}_{pi} \equiv \boldsymbol{\omega}_i \times [(\boldsymbol{\omega}_i \times \bar{\mathbf{r}}_{pi}) + \dot{\mathbf{u}}_{pi}]$ and the word '*sym*' denotes the symmetric elements of the matrix, $\mathbf{M}_i$.

(5) Combining Eq. (3.22) for all *n*-links, i.e., $i = 1,\ldots, n$, one can now express them exactly in the same form as for the rigid body system, namely Eq. (2.21), which is

$$\left(\frac{\partial \mathbf{t}}{\partial \dot{\mathbf{q}}}\right)^T (\mathbf{M}\dot{\mathbf{t}} + \boldsymbol{\gamma}) = \boldsymbol{\tau} \tag{3.25}$$

where the $\hat{n} \times \hat{n}$ matrix, $\mathbf{M}$, the $\hat{n}$-dimensional vector, $\boldsymbol{\gamma}$, and the $\bar{n}$-dimensional vector, $\boldsymbol{\tau}$, are given by

$$\mathbf{M} \equiv diag[\mathbf{M}_1 \cdots \mathbf{M}_n]; \boldsymbol{\gamma} \equiv [\boldsymbol{\gamma}_1^T \cdots \boldsymbol{\gamma}_n^T]^T; \text{ and } \boldsymbol{\tau} \equiv [\tau_1 \cdots \tau_n]^T \tag{3.26}$$

The vectors, $\boldsymbol{t}$ and $\boldsymbol{w}$, are defined in Eq. (3.7), whereas the $\hat{n} \times \hat{n}$ matrix, $\frac{\partial \mathbf{t}}{\partial \dot{\mathbf{q}}}$, is defined as

$$\frac{\partial \mathbf{t}}{\partial \dot{\mathbf{q}}} \equiv \begin{bmatrix} \frac{\partial \mathbf{t}_1}{\partial \dot{\mathbf{q}}_1} & \cdots & \frac{\partial \mathbf{t}_1}{\partial \dot{\mathbf{q}}_n} \\ \vdots & & \vdots \\ \frac{\partial \mathbf{t}_n}{\partial \dot{\mathbf{q}}_1} & \cdots & \frac{\partial \mathbf{t}_n}{\partial \dot{\mathbf{q}}_n} \end{bmatrix} \tag{3.27}$$

(6) Similar to rigid body system, Eq. (3.24), it is clear from Eq. (3.14) that

$$\frac{\partial \mathbf{t}}{\partial \dot{\mathbf{q}}} = \mathbf{N}_l \mathbf{N}_d \tag{3.28}$$

Hence, Eq. (3.25) yields

$$\mathbf{N}_d^T \mathbf{N}_l^T \left(\mathbf{M}\dot{\mathbf{t}} + \boldsymbol{\gamma}\right) = \boldsymbol{\tau} \tag{3.29}$$

(7) Now, differentiating Eq. (3.14) with respect to time similar to Eq. (2.26), one obtains

$$\dot{\mathbf{t}} = \mathbf{N}_l \mathbf{N}_d \ddot{\mathbf{q}} + \mathbf{N}_l \dot{\mathbf{N}}_d \dot{\mathbf{q}} + \dot{\mathbf{N}}_l \mathbf{N}_d \dot{\mathbf{q}} \tag{3.30}$$

Then, substituting Eq. (3.30) into Eq. (3.29) the independent set of constraint dynamic equations of motion for the rigid–flexible robot is obtained as

$$\mathbf{I}\ddot{\mathbf{q}} + \mathbf{h} = \boldsymbol{\tau} \tag{3.31}$$

where the $\overline{n} \times \overline{n}$ generalized inertia matrix (GIM), $\mathbf{I}$, is also appeared in Eq. (3.27), but with different dimensions. It is pointed out here that the (i, j) elements of the matrix, $\mathbf{I}$ of Eq. (3.31), for purely flexible link system, are the $(1 + 3m_i + \overline{m}_i) \times (1 + 3m_i + \overline{m}_i)$ block matrices, in contrast to the scalars for purely rigid link system given by Eq. (2.31). The $(1 + 3m_i + \overline{m}_i) \times (1 + 3m_i + \overline{m}_i)$ block matrices of the GIM denoted with $\mathbf{I}_{ij}$ are given by

$$\mathbf{I}_{ij} = \mathbf{I}_{ji}^T = \mathbf{P}_i^T \widetilde{\mathbf{M}}_i \mathbf{A}_{ij} \mathbf{P}_j, \text{ for } i = 1, \ldots n;\ j = 1, \ldots, i. \tag{3.32}$$

where the $(1 + 3m_i + \overline{m}_i) \times (1 + 3m_i + \overline{m}_i)$ matrix, $\widetilde{\mathbf{M}}_i$, is obtained recursively using an expression similar to Eq. (2.30), i.e.,

$$\widetilde{\mathbf{M}}_i \equiv \mathbf{M}_i + \mathbf{A}_{i+1,i}^T \widetilde{\mathbf{M}}_{i+1} \mathbf{A}_{i,i+1}$$

where $\widetilde{\mathbf{M}}_{n+1} \equiv \mathbf{O}$, as there is no $(n+1)^{st}$ link in the chain. Hence, $\widetilde{\mathbf{M}}_n \equiv \mathbf{M}_n$. Moreover, the $n(6+3m_i)$-dimensional vector $\mathbf{h}$ is given by

$$\mathbf{h} = \mathbf{N}_d^T \mathbf{N}_l^T \left[\mathbf{M}(\mathbf{N}_l \dot{\mathbf{N}}_d + \dot{\mathbf{N}}_l \mathbf{N}_d)\dot{\mathbf{q}} + \gamma \right] \tag{3.33}$$

which is again similar to Eq. (2.29) except that, $\dot{\boldsymbol{\theta}}$, is replaced with $\dot{\boldsymbol{q}}$.

Equations (3.31)–(3.32) represent the analytical and recursive expressions for the matrix elements of the GIM. Like the rigid body system, $\widetilde{\boldsymbol{M}}_i$ is interpreted here as the mass matrix for the 'composite flexible body' comprising of the rigidly attached flexible bodies #*i*, …, #*n*. Note that the expression of the DeNOC matrices and the recursive expressions associated with the dynamic models for the rigid and flexible systems are exactly the same except their dimensions. This feature is exploited to build up a single forward dynamic algorithm for the robots with both rigid and flexible links. The algorithm and the associated computational complexity are presented in the following sections.

3.3 Geometric Stiffness

Usually, in the analysis of flexible multibody systems, the elastic deformations were assumed to be small. These deformations were calculated by the product of vibration modes and the modal coordinates. The vibration modes are assumed to have derived from linear elastic theory by neglecting the effect of the geometric nonlinearity, i.e., the coupling between the axial and the transverse displacements. This assumption is valid for problems where the bodies are rotating at lower angular rates. However, if the angular rates of the bodies are considerably greater than their first natural frequency, then the solution of the system causes the erroneous dynamic softening effect [2, 14, 17].

To solve the flexible system undergoing the overall large rotation, the above-said premature linearization is compensated by introducing the geometric stiffness in the equations of motion of the system. The typical examples of such systems are helicopter rotor blades, turbine blades, etc. The geometric stiffness matrix is derived using the nonlinear stress–strain relationship.

The nonlinear stress–strain relationship for a beam element is given by

$$\varepsilon_{xx} \equiv \frac{\partial u_i^x}{\partial \overline{a}_i} - \frac{\partial^2 u_i^y}{\partial \overline{a}_i^2} y_i + \frac{1}{2}\left(\frac{\partial u_i^y}{\partial \overline{a}_i}\right)^2 \tag{3.34}$$

Then, the potential energy of the system due to strain energy, $\mathbf{V}_{si}$ of Eq. (3.20c), is rewritten as

$$\mathbf{V}_{si} \equiv \frac{E_i}{2} \int_{v} \varepsilon_{xx}^2 \mathrm{d}V_i$$

$$\mathbf{V}_{si} \equiv \frac{E_i}{2}\int_0^{a_i}\int_{A_i}\left[\left(\frac{\partial u_i^x}{\partial \overline{a}_i}\right)^2 + \left(\frac{\partial^2 u_i^y}{\partial \overline{a}_i^2}y_i\right)^2 + \frac{1}{4}\left(\frac{\partial u_i^y}{\partial \overline{a}_i}\right)^4 - 2\frac{\partial u_i^x}{\partial \overline{a}_i}\frac{\partial^2 u_i^y}{\partial \overline{a}_i^2}y_i \right.$$
$$\left. -\frac{\partial^2 u_i^y}{\partial \overline{a}_i^2}\left(\frac{\partial u_i^y}{\partial \overline{a}_i}\right)^2 y + \frac{\partial u_i^x}{\partial \overline{a}_i}\left(\frac{\partial u_i^y}{\partial \overline{a}_i}\right)^2\right] d\overline{a}_i dA_i \tag{3.35}$$

The higher-order term in the above equation is neglected, and since y_i is measured from the neutral axis, the $\int y_i dA$ must vanish. Therefore, Eq. (3.35) becomes

$$\mathbf{V}_{si} \equiv \frac{E_i A_i}{2}\int_0^{a_i}\left(\frac{\partial u_i^x}{\partial \overline{a}_i}\right)^2 d\overline{a} + \frac{E_i I_i}{2}\int_0^{a_i}\left(\frac{\partial^2 u_i^y}{\partial \overline{a}_i^2}\right)^2 d\overline{a}_i + \frac{E_i A_i}{2}\int_0^{a_i}\frac{\partial u_i^x}{\partial \overline{a}_i}\left(\frac{\partial u_i^y}{\partial \overline{a}_i}\right)^2 d\overline{a}_i \tag{3.36}$$

where A_i is the cross-sectional area of the i-th link, a_i is its length, and $I_i = \int y_i^2 dA$ is the second moment of area. In Eq. (3.36), the first two integrals represent the linear strain energy $\boldsymbol{\tau}_i^{sl}$, whereas the third integral is the nonlinear component of the strain $\boldsymbol{\tau}_i^{snl}$. Then, the vector $\boldsymbol{\tau}_i^s$ of Eq. (3.20b) for the flexible link is given by $\boldsymbol{\tau}_i^s = \boldsymbol{\tau}_i^{sl} + \boldsymbol{\tau}_i^{snl}$ and is evaluated as

$$\boldsymbol{\tau}_i^s \equiv \frac{\partial \mathbf{V}_s}{\partial \mathbf{q}_i} \equiv (\mathbf{K}_i + \mathbf{K}_i^g)\mathbf{q}_i \tag{3.37}$$

where $\mathbf{K}_i$ is the standard elastic stiffness matrix, and $\mathbf{K}_i^g$ is the geometric stiffness matrix for the i-th link. The geometric stiffening is also referred to as stress stiffening or centrifugal stiffening effect. The effect of including the geometric stiffness in the formulation for high-speed, large displacement system is demonstrated by solving the benchmark problem for the flexible multibody system, i.e., a spinning cantilever beam. The geometric stiffness can be added to the DeNOC-based formulation for flexible systems to solve the problems of large overall motion.

3.4 Forward Dynamics for Rigid–Flexible Robots

This section presents a recursive, computationally efficient, and numerically stable algorithm for the forward dynamics of serial rigid–flexible robotic systems. First, the recursive algorithm is presented in Sect. 3.4.1. For clarity of representation and to show the computer implementation, a flowchart of the algorithm is presented in Sect. 3.4.2.

3.4.1 Recursive Algorithm

It is shown in Chap. 2 and Sect. 3.2 that the analytical expressions of the matrices/vectors associated with the dynamic equations of motion for the rigid and the flexible systems are exactly the same except their dimensions. Using the similarity in the analytical expressions, a single, unified forward dynamics algorithm can be built up for the robots with both types of links. The forward dynamics algorithm for the serial rigid robotic systems was presented in Sect. 2.3.1. Moreover, here the forward dynamics algorithm for a serial n-link robot with flexible links is provided. For this, the GIM, $\boldsymbol{I}$ of Eq. (3.31), is decomposed using the reverse Gaussian elimination used for the rigid link systems, namely

$$\mathbf{I} = \mathbf{U}\mathbf{D}\mathbf{U}^T \tag{3.38}$$

where $\mathbf{U}$ and $\mathbf{D}$ are, respectively, the $\overline{n} \times \overline{n}$ upper block triangular and block diagonal matrices, and $\overline{n}$ is the degree of freedom of the system given by, $\overline{n} \equiv n + \sum_{i=1}^{n_f} (3m_i + \overline{m}_i)$. The term, $n \equiv n_r + n_f$, is the total number of links–n_r and n_f being the number of rigid and flexible links, respectively, and m_i is the number of modes of bending vibrations for each flexible link. Furthermore, $\overline{m}_i$ is the number of modes of torsional vibration. Matrices $\mathbf{U}$ and $\mathbf{D}$ are given by

$$\mathbf{U} \equiv \begin{bmatrix} \mathbf{1} & \mathbf{U}_{12} & \cdots & \mathbf{U}_{1n} \\ \mathrm{O} & \mathbf{1} & \cdots & \mathbf{U}_{2n} \\ \vdots & \vdots & \ddots & \vdots \\ \mathrm{O} & \mathrm{O} & \cdots & \mathbf{1} \end{bmatrix}; \mathbf{D} \equiv \begin{bmatrix} \hat{\mathbf{I}}_1 & \mathrm{O} & \cdots & \mathrm{O} \\ \mathrm{O} & \hat{\mathbf{I}}_2 & \cdots & \mathrm{O} \\ \vdots & \vdots & \ddots & \vdots \\ \mathrm{O} & \cdots & \cdots & \hat{\mathbf{I}}_n \end{bmatrix} \tag{3.39}$$

where the $(1 + 3m_i + \overline{m}_i) \times (1 + 3m_1 + \overline{m}_i)$ matrices, $\mathbf{U}_{ij}$ and $\hat{\mathbf{I}}_i$, for $i,j = 1, \ldots, n$, are obtained from the application of the Gaussian Elimination rules in reverse order [24–26]. For the rigid bodies, matrices $\mathbf{U}_{ij}$ and $\hat{\mathbf{I}}_i$ are scalars and $\mathbf{U}$ and $\mathbf{D}$ reduce to $n \times n$ upper triangular and diagonal matrices. The expressions for $\mathbf{U}_{ij}$ and $\hat{\mathbf{I}}_i$, are as follows: For $i = 1,\ldots, n$ and $j = i + 1, \ldots, n$,

$$\mathbf{U}_{ij} \equiv \mathbf{P}_i^T \boldsymbol{\Psi}_{ij}, \text{ and } \hat{\mathbf{I}}_i \equiv \mathbf{P}_i^T \widehat{\Psi}_i \tag{3.40}$$

where $\boldsymbol{P}_i$, is the $(6 + 3m_i + \overline{m}_i) \times (1 + 3m_i + \overline{m}_i)$-dimensional joint-and-amplitude propagation matrix, as defined in Eq. (3.10b), and the $(6+3m_i+\overline{m}_i)\times(1+3m_i+\overline{m}_i)$ matrices, $\widehat{\Psi}_i$ and $\boldsymbol{\Psi}_{ij}$ are given by

$$\widehat{\Psi}_i \equiv \widehat{\mathbf{M}}_i \mathbf{P}_i; \boldsymbol{\Psi}_i \equiv \widehat{\Psi}_i \hat{\mathbf{I}}_i^{-1}; \boldsymbol{\Psi}_{ij} \equiv \mathbf{A}_{ji}^T \boldsymbol{\Psi}_i. \tag{3.41}$$

Again, for the rigid bodies, matrices $\widehat{\Psi}_i$ and $\boldsymbol{\Psi}_{ij}$ reduce to 6-dimensional vectors, whereas the $(6 + 3m_i + \overline{m}_i) \times (1 + 3m_i + \overline{m}_i)$ dimensional joint-and-amplitude propagation matrix, $\mathbf{P}_i$, reduces to the 6-dimensional vector, $\mathbf{p}_i$, defined in Eq. (2.4). Also, the $(6 + 3m_i+\overline{m}_i) \times (6 + 3m_i+\overline{m}_i)$ flexible twist propagation matrix, $\mathbf{A}_{ij}$, defined in Eq. (3.10b), reduces to the 6×6 matrix, $\mathbf{R}_{ij}$ of Eq. (2.6). Note that the matrix, $\widehat{\mathbf{M}}_i$, represents the mass and inertia properties of the articulated flexible body i, defined similar to the articulated inertia matrix for the rigid body system, consisting of flexible links, #i, …, #n, which are coupled by joints $i + 1$, …, n [10, 25, 26]. The $(6 + 3m_i+\overline{m}_i) \times (6 + 3m_i+\overline{m}_i)$ matrix, $\widehat{\mathbf{M}}_i$, is obtained recursively as

$$\widehat{\mathbf{M}}_i \equiv \mathbf{M}_i + \mathbf{A}_{i,i+1}^T \overline{\mathbf{M}}_{i+1} \mathbf{A}_{i,i+1}, \text{ where } \overline{\mathbf{M}}_i \equiv \widehat{\mathbf{M}}_i - \widehat{\Psi}_i \mathbf{\Psi}_i^T \tag{3.42}$$

for $i = n - 1$, …, 1, and $\widehat{\mathbf{M}}_n \equiv \mathbf{M}_n$. For the recursive algorithm, the $\overline{n}$-dimensional vector of the generalized forces due to external moments, forces, gravity, strain associated with the link deformations, and the centrifugal and Coriolis accelerations, etc., $\mathbf{0} = \mathbf{\tau} - \mathbf{h}$, are assumed to be evaluated recursively. The solution for $\ddot{\mathbf{q}}$ can now be obtained from Eq. (3.31) in the following three steps:

Step A. Solution for $\hat{\boldsymbol{\varphi}}$: The solution, $\hat{\boldsymbol{\varphi}} = \mathbf{U}^{-1}\boldsymbol{\varphi}$, is evaluated for, $i = n-1$, …,1, as

$$\hat{\boldsymbol{\varphi}}_i = \boldsymbol{\varphi}_i - \mathbf{P}_i^T \boldsymbol{\eta}_i \tag{3.43}$$

where $\hat{\boldsymbol{\varphi}}_n \equiv \boldsymbol{\varphi}_n$. Note in Eq. (3.43) that unlike the rigid body system, where $\hat{\boldsymbol{\varphi}}_i$ is a scalar, here it is $(1 + 3m_i + \overline{m}_i)$-dimensional vector,$\hat{\boldsymbol{\varphi}}_i$. Moreover, the $(6 + 3m_i + \overline{m}_i)$-dimensional vector, $\boldsymbol{\eta}_{i,}$, is obtained as,

$$\boldsymbol{\eta}_i = \mathbf{\Psi}_i \hat{\boldsymbol{\varphi}}_i + \overline{\boldsymbol{\eta}}_i;\ \overline{\boldsymbol{\eta}}_i \equiv \mathbf{A}_{i+1,i}^T \boldsymbol{\eta}_{i+1}, \text{ and } \overline{\boldsymbol{\eta}}_n = 0$$

Step B. Solution for $\tilde{\boldsymbol{\varphi}}$: $\mathbf{D}\tilde{\boldsymbol{\varphi}} = \hat{\boldsymbol{\varphi}}$, for $i = 1$, …, n,

$$\tilde{\varphi}_i = \hat{\mathbf{I}}_i^{-1} \varphi_i \tag{3.44}$$

in which the $(1 + 3m_i + \overline{m}_i) \times (1 + 3m_i + \overline{m}_i)$ matrix $\hat{\mathbf{I}}_i$ is defined in Eq. (3.40). For the rigid body system, $\hat{\mathbf{I}}_i$ becomes scalar and $\hat{\mathbf{I}}_i^{-1}$ simply requires one division.

Step C. Solution for $\ddot{\mathbf{q}}_i$: $\ddot{\mathbf{q}}_i \equiv \mathbf{U}^{-T}\tilde{\boldsymbol{\varphi}}$, for $i = 2$, …, n,

$$\ddot{\mathbf{q}}_i = \tilde{\boldsymbol{\varphi}}_i - \mathbf{\Psi}_i^T \bar{\boldsymbol{\mu}}_i \tag{3.45}$$

where the $(1 + 3m_i + \overline{m}_i)$-dimensional vector, $\ddot{\mathbf{q}}_1 \equiv \tilde{\boldsymbol{\varphi}}_1$, and the $(6 + 3m_i + \overline{m}_i)$ -dimensional vector, $\overline{\boldsymbol{\mu}}_i$, is obtained from

$$\overline{\boldsymbol{\mu}}_i \equiv \mathbf{A}_{i,i-1}\boldsymbol{\mu}_{i-1} \text{ and } \boldsymbol{\mu}_i \equiv \mathbf{P}_i \ddot{\mathbf{q}}_i + \overline{\boldsymbol{\mu}}_i;\ \overline{\boldsymbol{\mu}}_1 = 0 \tag{3.46}$$

For the rigid bodies,$\mathbf{A}_{i,i-1} \equiv \mathbf{R}_{i,i-1}$, $\mathbf{P}_i \equiv \mathbf{p}_i$, and $\overline{\boldsymbol{\mu}}_i$ is a 6-dimensional vector.

3.4.2 Computational Complexity

The proposed recursive forward dynamics algorithm presented in Sect. 3.4.1 is summarized in the flowchart in Fig. 3.2. The computational counts in terms of the number of multiplications/divisions (M), additions/subtractions (A), and trigonometric operations (T) associated with each step are also shown. The corresponding detailed steps are shown in Appendix E and referred in the flowchart. For the purpose of evaluation of computational complexity, it has been assumed that all the links are flexible and vibrating in an equal number of modes, namely m in bending and $\overline{m}$ in torsion. In cases when some of the links are rigid or vibrate in less than m or $\overline{m}$ number of modes in bending and torsion, the computational complexity of the algorithm will reduce, thereby increasing its efficiency. Moreover, the effect of the payload and hub is not considered while computing the computational complexity so as to be able to compare the final complexity with those reported in the literature, where these effects are not reported. The total computational counts are given as

$$\begin{aligned}
&\left[(2m^2+\bar{m}^2+180m+4\bar{m}+299+M_{m1})n-(124m+\bar{m}+229+M_{m2})\right]M;\\
&\left[(6m^2+2\bar{m}^2+4m\bar{m}+122m+8\bar{m}+220+A_{m1})n\right.\\
&\quad\left.-(94m+\bar{m}+192+A_{m2})\right]\mathrm{A};\ \text{and } 2n\mathrm{T}.
\end{aligned}\tag{3.47}$$

where M_{m1}, M_{m2}, A_{m1} and A_{m2} are the functions of $\hat{m}$ and depend on the number of modes considered for vibrations. They are given as:

$$\begin{aligned}
M_{m1} &\equiv \frac{\hat{m}^3}{6}+\left(2m+\overline{m}+\frac{15}{2}\right)\hat{m}^2\\
&\quad+\left(\frac{4m^2+\overline{m}^2+4m\overline{m}+30m+15\overline{m}}{2}+\frac{100}{3}\right)\hat{m};\\
M_{m2} &\equiv \left(27+\frac{4m^2+\overline{m}^2+4m\overline{m}+26m+13\overline{m}}{2}\right)\hat{m};\\
A_{m1} &\equiv \frac{\hat{m}^3}{6}+\frac{\hat{m}^2}{2}+\left(6m^2+\frac{3}{2}\overline{m}^2+6m\overline{m}+27m+\frac{27}{2}\overline{m}+\frac{82}{3}\right)\hat{m};\\
A_{m2} &\equiv \left(2m^2+\frac{\overline{m}^2}{2}+2m\overline{m}+13m+\frac{13}{2}\overline{m}+27\right)\hat{m}
\end{aligned}$$

where $\hat{m}=1+2m+\overline{m}$.

From Eq. (3.47), it is clear that the proposed recursive forward dynamics algorithm for flexible link robots is of $O(n)+O(m^3)$.

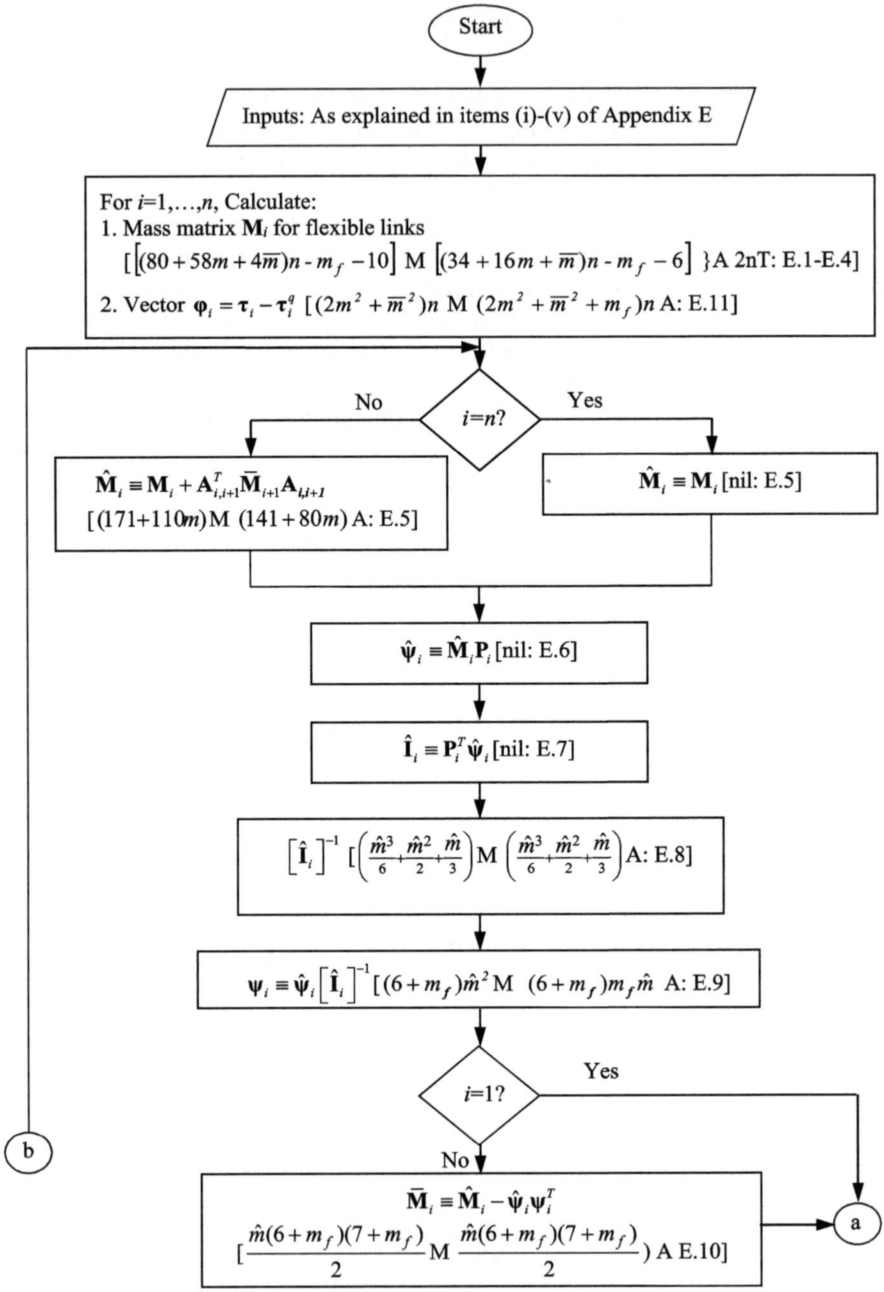

Fig. 3.2 Flowchart for the proposed recursive algorithm

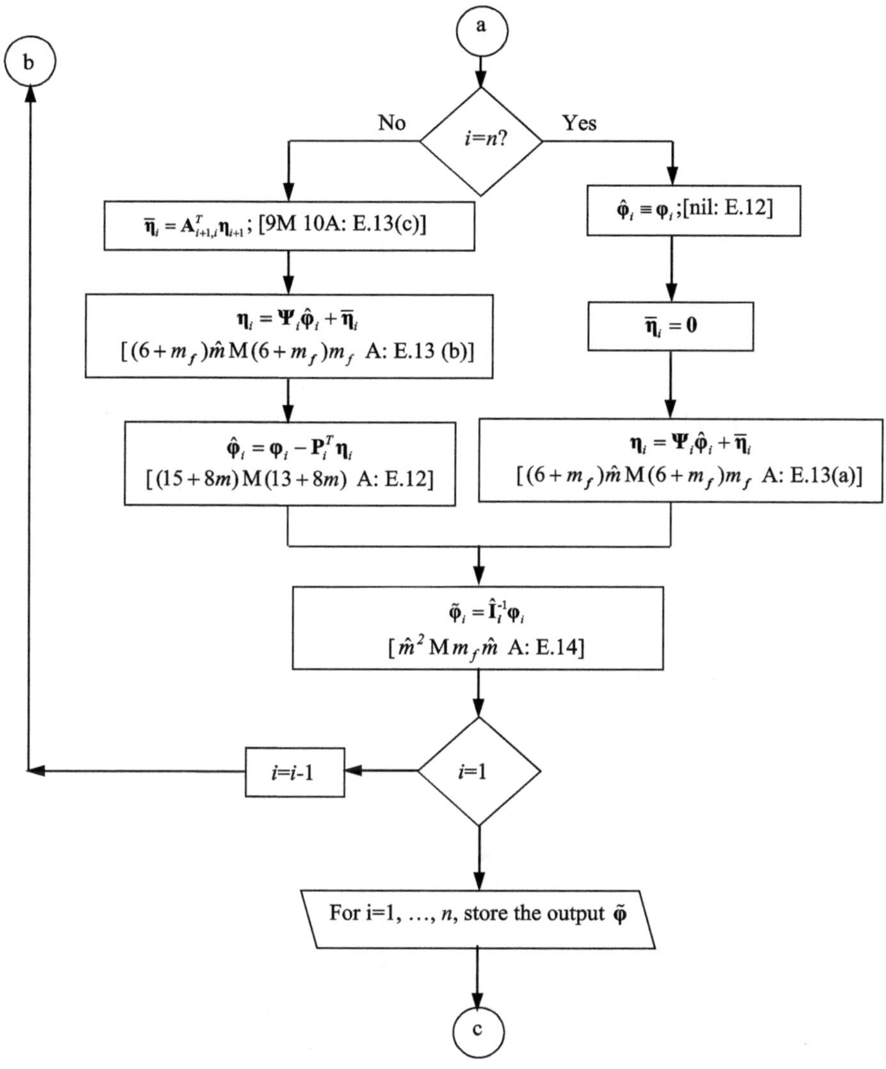

Fig. 3.2 (continued)

3.4.3 Comparison

The computational complexity of the proposed algorithm is compared in Table 3.1 with the two other algorithms available in the literature, namely [7, 13]. No other recent literature reported such complexity count except for the RR—R stands for revolute joints—and RRR robots [32], where only the computational complexity for the generalized inertia matrix (GIM) is reported, not the forward dynamics algorithm. This may be apparently due to the complexity in counting the number of operations

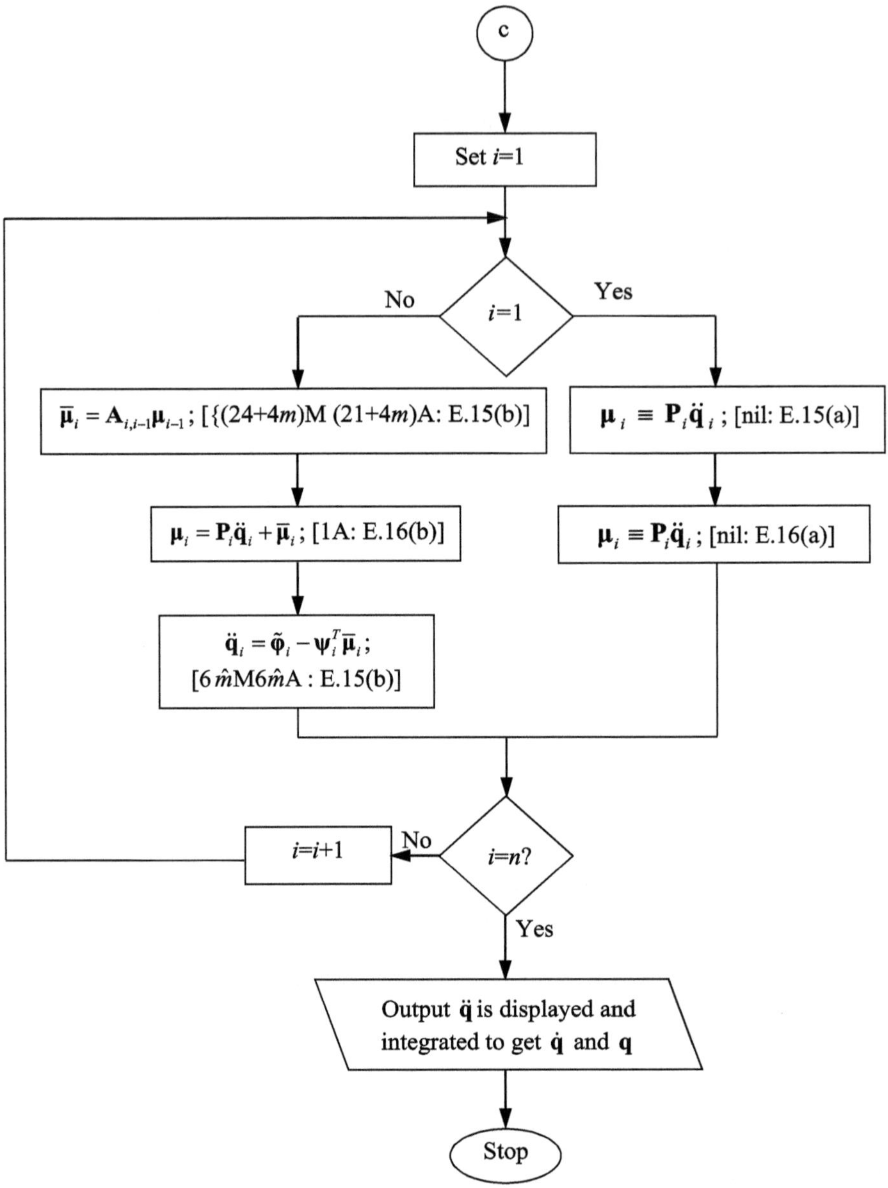

Fig. 3.2 (continued)

Table 3.1 Comparison of computational complexity

(**a**) Considering vibrations in bending ($m = 2$) and torsion ($\overline{m} = 1$)

Algorithm	Multiplications (M)	Additions (A)	M	A
Proposed ($n^* = 6$)	$M_{1pt}n + M_{2pt}$	$A_{1pt}n + A_{2pt}$	**9038**	**7311**
Cyril [7] ($n^* = 6$)	$M_{1c}n^3 + M_{2c}n^2 + M_{3c}n + M_{4c}$	$A_{1c}n^3 + A_{2c}n^2 + A_{3c}n + A_{4c}$	10,812	8538

(**b**) Considering vibrations in bending only ($m = 2$)

Algorithm	Multiplications (M)	Additions (A)	M	A
Proposed ($n^* = 8$)	$M_{1p}n + M_{2p}$	$A_{1p}n + A_{2p}$	**9874**	**7699**
Jain and Rodriguez$^{\Delta}$ [10] ($n^* = 9$)	$M_{1j}n$	$A_{1j}n$	11,889	9387
Jain and Rodriguez† [10]	$M_{1j}^{nr}n^3 + M_{2j}^{nr}n^2 + M_{3j}^{nr}n$	$A_{1j}^{nr}n^3 + A_{2j}^{nr}n^2 + A_{3j}^{nr}n$	9920 ($n^* = 8$) 12,360 ($n^* = 9$)	8605 ($n^* = 8$) 10,660 ($n^* = 9$)

$^{\Delta}$ Recursive algorithm; † Non-recursive algorithm
n^*: Number of links for which the recursive algorithm benefits over the non-recursive algorithm.

in terms of multiplications, additions, etc., and the availability of fast modern-day computers. However, it is emphasized here that even a few savings in multiplications or additions may have significant impact on efficiency when these computations are performed over thousands or millions of discretized steps. As a result, one can apply the algorithm in real-time applications or use less advanced, less costly computing system to perform the same work satisfactorily.

Note that, the algorithm and the computational count given by Cyril [7] include the effect of torsional vibrations, whereas the algorithm and the computational count given by [13] do not consider them. Hence, the proposed algorithm is compared with the above two algorithms separately, namely with [7] in Table 3.1a and with [13] in Table 3.1b. From Table 3.1(a–b), it is clear that the recursive algorithm proposed in this book outperforms the other two.

The coefficients in Table 3.1a are:

$$
\begin{aligned}
M_{1pt} &\equiv 2m^2 + \overline{m}^2 + 180m + 4\overline{m} + 299 + M_{m1};\\
M_{2pt} &\equiv -(124m + \overline{m} + 229 + M_{m2});\\
A_{1pt} &\equiv 6m^2 + 2\overline{m}^2 + 4m\overline{m} + 122m + 8\overline{m} + 220 + A_{m1};\\
A_{2pt} &\equiv -(94m + \overline{m} + 192 + A_{m2});
\end{aligned}
$$

where M_{m1}, M_{m2}, A_{m1} and A_{m2} are given after Eq. (3.43), and

$$
M_{1c} \equiv \frac{\hat{m}^3}{6} + \hat{m}^2\left(-\frac{m}{2} + 1\right) + \hat{m}\left(\frac{m^2}{2} - 2m\right) - \frac{m^3}{6} + m^2;
$$

$$M_{2c} \equiv -\frac{3\hat{m}^2}{2} + \hat{m}\left(3m + \frac{33}{2}\right) - \frac{3m^2}{2} - \frac{15m}{2} + \frac{15\overline{m}}{2} + 12;$$

$$M_{3c} \equiv 5\hat{m}^2 - \hat{m}\left(10m + \frac{109}{6}\right) + \frac{m^3}{6} + \frac{53m^2}{6} + \frac{233m}{3} - \frac{7\overline{m}}{2} + 246;$$

$$M_{4c} \equiv -123$$

$$A_{1c} \equiv \frac{\hat{m}^3}{6} + \hat{m}^2\left(-\frac{m}{2} + 1\right) + \hat{m}\left(\frac{m^2}{2} - 2m\right) - \frac{m^3}{6} + m^2;$$

$$A_{2c} \equiv -\frac{3\hat{m}^2}{2} + \hat{m}(3m + 13) - \frac{3m^2}{2} - \frac{17m}{2} + 3\overline{m} + 9;$$

$$A_{3c} \equiv 5\hat{m}^2 - \hat{m}\left(10m + \frac{85}{6}\right) + \frac{m^3}{6} + \frac{53m^2}{6} - \frac{251m}{6} + 228;$$

$$A_{4c} \equiv -102$$

The coefficients in Table 3.1b are:

$$M_{1p} \equiv 2m^2 + 180m + 299 + M_{m3};$$
$$M_{2p} \equiv -(124m + 229 + M_{m4});$$
$$A_{1p} \equiv 6m^2 + 122m + 220 + A_{m3};$$
$$A_{2p} \equiv -(94m + 192 + A_{m4});$$

where

$$M_{m3} \equiv \frac{\hat{m}^3}{6} + \left(2m + \frac{15}{2}\right)\hat{m}^2 + \left(2m^2 + 15m + \frac{100}{3}\right)\hat{m};$$

$$M_{m4} \equiv \left(2m^2 + 13m + 27\right)\hat{m};$$

$$A_{m3} \equiv \frac{\hat{m}^3}{6} + \frac{\hat{m}^2}{2} + \left(6m^2 + 27m + \frac{82}{3}\right)\hat{m};\ A_{m4} \equiv \left(2m^2 + 13m + 27\right)\hat{m};$$

Moreover,

$$M_{1j} \equiv \begin{cases} \frac{5m^3}{6} + \frac{27m^2}{2} + \frac{893m}{3} + 281; & \text{for } n \le 7 \\ 13m^2 + 298m + 673; & \text{for } n \ge 7 \end{cases}$$

$$A_{1j} \equiv \begin{cases} \frac{5m^3}{6} + \frac{23m^2}{2} + \frac{677m}{3} + 256; & \text{for } n \le 7 \\ 114m^2 + 225m + 537; & \text{for } n \ge 7 \end{cases}$$

$$M_{1j}^{nr} \equiv \frac{1}{6};$$

$$M_{2j}^{nr} \equiv 12m^2 + 34m + \frac{29}{2};$$

$$M_{3j}^{nr} \equiv 48m + \frac{268}{3};$$

$$A_{1j}^{nr} \equiv \frac{1}{6};$$

$$A_{2j}^{nr} \equiv 11m^2 + 24m + 14;$$

$$A_{3j}^{nr} \equiv m^2 + \frac{97}{2}m + 116;$$

3.5 Simulation Results

Simulation of a robotic system is defined as the determination of its configuration, i.e., the position and orientation of the links, rigid or flexible, constituting the robotic system at hand, for a given set of external forces and moments, be it at the joints or elsewhere. Thus, the overall process of the simulation consists of two steps, namely forward dynamics, i.e., the determination of the generalized accelerations, $\ddot{\boldsymbol{q}}$, from the dynamic equations of motion, Eq. (3.31), and its subsequent integrations to obtain $\dot{\boldsymbol{q}}$ and $\boldsymbol{q}$. The information on $\boldsymbol{q}_i$ is then used to find the link configurations mentioned above. In this chapter, simulation results mean the determination of $\ddot{\boldsymbol{q}}$, followed by $\dot{\boldsymbol{q}}$ and $\boldsymbol{q}$. Several popular industrial and space robots are considered for simulation based on the algorithms proposed in this book. To start with, evaluation of shape functions, as required in the modeling of flexible links, is given in Sect. 3.5, followed by the simulation of a single-link, a planar two-link robotic arm, and a planar three-link robotic arm. In all three cases, only rigid links, both rigid and flexible links, i.e., as rigid and rigid–flexible systems, respectively, are considered. The 3-degree-of-freedom (DOF) planar Canadarm is then presented in Sect. 3.9, followed by the simulation of the three-link spatial arms having only rigid and rigid–flexible links. The reason for considering both rigid and rigid–flexible links in the robots is to compare the effect of flexible links on the system behavior which would help designers to select the links or a controller accordingly. Figure 3.3 shows the scheme for studying the flexibility effect, where a set of input joint torques/forces to follow a specified joint trajectory was calculated using the inverse dynamics algorithm of a rigid robot. These torques and forces were then given as inputs to two forward dynamics algorithms, one for rigid and another for rigid–flexible manipulator. The simulated joint trajectories were next compared with the specified input joint trajectories, which would show the effect of flexibility in the links, provided the numerical integration errors are negligible. All the numerical integrations carried out in this book use the MATLAB functions, ‘ode45’ and ‘ode23s’ [21] that are based on the Runge–Kutta method [19], where the step size and tolerance were taken as 0.001 and 10^{-8}, respectively, unless otherwise stated. For the purpose of simulations, it has always been assumed that a robot arm starts from rest, i.e., its velocity or the joint and other rates, are zeros, whereas the configurations are stated in the sections and subsections where the results are reported.

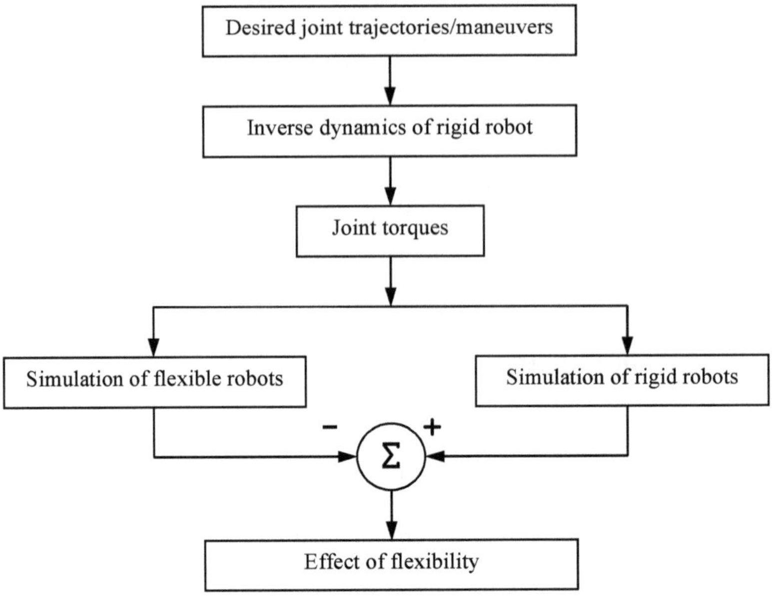

Fig. 3.3 Scheme for studying flexibility

3.6 Shape Functions Evaluation

For the dynamic modeling presented in this chapter, the deflection due to the flexibility in a link is discretized using the Rayleigh–Ritz discretization and the assumed modes method (AMM) [18]. In AMM, any deflection in a link is assumed to be a sum of the set of appropriately chosen shape functions and corresponding time-dependent amplitudes of vibrations, e.g., in Eq. (3.3c). Note here that any admissible function satisfying the geometric boundary conditions of the flexible link can be a candidate for the shape function. These admissible functions could be polynomials, harmonic functions, splines, trigonometric, or eigenfunctions. In this book, the eigenfunctions are selected to represent the shape functions since they have orthogonal properties and lead to computational efficiency [31]. Moreover, using a specific number of eigenfunctions, one is assured of reproducing exactly the same natural frequency. The eigenfunction used in this book to represent the shape function of the i-th link in its j-th mode of bending is given by [31, 33]:

$$s_{ij} = c_j\left[(\sin \varsigma_j + \sinh \varsigma_j) - \sigma(\cos \varsigma_j + \cosh \varsigma_j)\right] \tag{3.48a}$$

where

$$\varsigma_j \equiv \tilde{m}_j \bar{a}_i \big/ a_i; \text{ and } \sigma = (\sin \tilde{m}_j + \sinh \tilde{m}_j)\big/(\cos \tilde{m}_j + \cosh \tilde{m}_j) \tag{3.48b}$$

In Eq. (3.48b), the modal constant, $\tilde{m}_j$, for different modes and different boundary conditions of the i-th link is given in Table 3.2 [23], whereas $\overline{a}_i$ is the distance along $\widehat{X}_{i+1}$ axis and a_i is one of the DH parameter of the link defined in Sect. 3.1.2, and c_j depends on the boundary conditions.

Note that, the selection of the proper mode shapes is based upon the boundary conditions of the link. The flexible behavior of a link is dependent on its shape functions, which depend upon its boundary conditions. An appropriate selection of boundary conditions is important because this choice affects the resulting vibration modes, and a flexible link can only deform in the space spanned by the selected modes (Table 3.2).

Moreover, in a real robotic system, the boundary conditions of the links used to calculate the deformation mode shapes do not usually fit in a standard description, such as simply supported, fixed–fixed, or cantilevered. Furthermore, the description of the link and its deformation is configuration-dependent also. In addition, the choice of the deformation modes also depends on the choice of the definition of the coordinate frames—fixed or moving body axes. Several researchers have addressed the issue of the selection of the deformation modes and their relation to the boundary conditions. The suitable boundary conditions for the modeling of flexible links of robots are analyzed in Rao [22]. Aggarwal and Shabana [1] have analyzed the flexible beams using body mean axis formulation, in which the local frame attached to the link is a floating frame that follows the mean displacement of the flexible body and thus does not necessarily coincide with any specific material point, which may introduce artificial asymmetries in the response. Craig [6], Kim and Haug [15], Shabana [27], Cardona [4] have presented various criteria for the selection of proper boundary conditions. For serial robots with flexible links, mostly the cantilever end conditions have been preferred by the researchers [7, 9, 16, 28, 29, 31, 32].

The shape functions of the link in torsion $\overline{s}_{ij}$, as defined in Eq. (3.5b), can also be represented by similar trigonometric functions [18, 22]. However, it should be noted that most of the controlling schemes for the control of the flexible links consider only bending vibrations and neglect the torsional vibrations. This is because the effect of torsional vibrations is very small in comparison with the bending vibrations, while its consideration imposes severe constraints on the design and performance of the controller. Consequently, researchers tend to lay more emphasis on simulation results for vibrations in bending [3, 7–9, 12, 13, 16, 20, 31, 36]. Cetinkunt et al. [5],

Table 3.2 Modal constant $\tilde{m}_j$ for different modes

Boundary conditions	Mode Number, j				
	1	2	3	4	5
Both ends simply supported	3.142	6.283	9.425	12.566	15.708
Cantilever	**1.875**	**4.694**	**7.855**	**10.996**	**14.137**
Both ends free/ fixed	4.730	7.853	10.996	14.137	17.279
One end fixed other end free	3.927	7.069	10.210	13.352	16.493

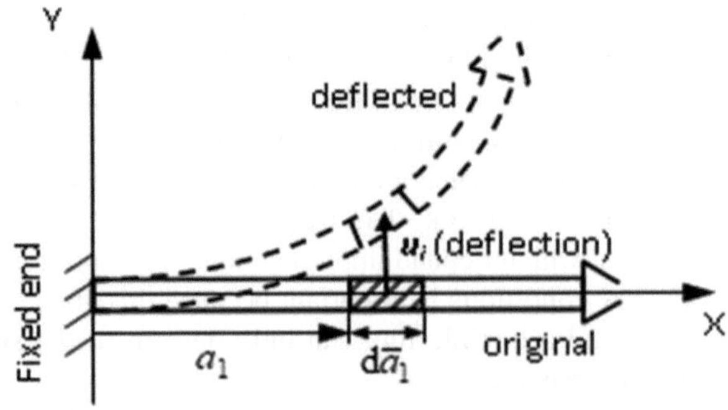

Fig. 3.4 A cantilever beam, i

Hastings and Book [11], and others have shown through experiments and analytical expressions that the shape functions of a flexible link serial robotic arm can be best approximated by the eigenfunctions of a cantilever type, i.e., clamped-free beam as shown in Fig. 3.4. Hence, the eigenfunctions corresponding to a cantilever beam are used here to represent the shape functions of the flexible robot links. The shape functions of the beam with respect to $\widehat{Y}_{i+1}$ and $\widehat{Z}_{i+1}$ axes are denoted by, S_{ij}^y and S_{ij}^z, respectively, which are defined in Eq. (3.4a). The shape functions with respect to $\widehat{Y}_{i+1}$, i.e., along the beam in Y_{i+1}-axis, S_{ij}^y are evaluated at $\bar{a}_i = a_i$ for $j = 1, \ldots, m_i$. This is shown in Step E.2 of Appendix E. The evaluated expressions are denoted as,

$$K_{1i}^y \equiv \int_0^{a_i} \begin{bmatrix} 0 \cdots 0 \\ k_{i,1}^y \cdots k_{i,m}^y \\ 0 \cdots 0 \end{bmatrix} d\bar{a}_i \tag{3.48c}$$

In Eq. (3.48c), $k_{ij}^y = \int_0^{a_i} s_{ij}^y d\bar{a}_i$ for, $j = 1, \ldots, m_i$. The function k_{ij}^y for $m = 4$, i.e., $j = 1, \ldots, 4$ is plotted in Fig. 3.5 for the cantilever beam shown in Fig. 3.4, whose physical properties are given in Table 3.3. The function k_{ij}^z is obtained similarly, as outlined in Step E.2 of Appendix E.

3.7 Single-Link Arm

In this section, the simulation of a single-link arm is performed, where the link is considered rigid and flexible.

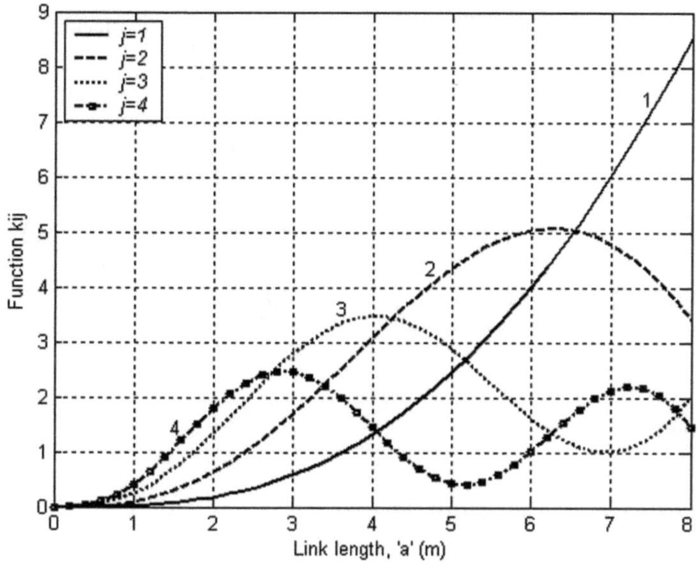

Fig. 3.5 First four mode shapes of a cantilever beam ($m_i = 4$)

Table 3.3 Physical properties of a cantilever beam

Length, a_i(m)	Mass, m_i(kg)	Flexure stiffness, E_iI_i(Nm^2)
8.0	1.616	336

3.7.1 Simulation Results

Simulation results of a single rigid and single flexible link rotating about an axis perpendicular to the plane of motion, as shown in Fig. 3.6a, b, respectively, are presented. To distinguish between the rigid and flexible links, rigid link is shown by rectangle, whereas the flexible link is by a thick line. However, in future texts, no such distinction in the representation of links will be made, and only a kinematic line sketch of the robotic arm will be given. The link parameters are taken as the same as shown in Fig. 3.4 and Table 3.3. Its rotational degree of freedom is denoted by θ_1, Fig. 3.6. Moreover, a step torque input of 5 Nm under no gravity is applied at the joint. Corresponding simulation results are plotted in Fig. 3.7 (a-c). The link is assumed to vibrate in its first two modes, as considered by several researchers, namely [7, 9, 31], and others. The justification is given in Sect. 3.7.2. Its simulation results, as shown in Fig. 3.7a–c, match exactly with those reported in Cyril [7]. In Fig. 3.7a–b, the joint angle and its rate corresponding to the rigid and flexible links are compared. Deflection of the tip of the flexible link, i.e., the end-effector, is shown in Fig. 3.7c. It can be seen that the joint rate for the flexible link has undergone considerable fluctuations, Fig. 3.7b. This fluctuation is due to the strong coupling between rotation and vibration (Fig. 3.6).

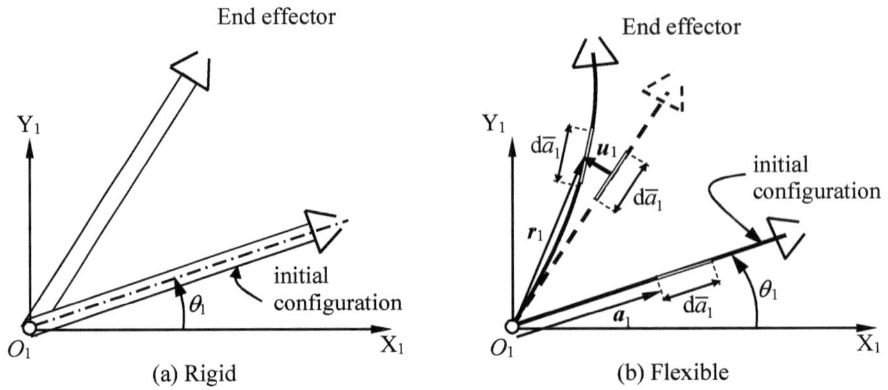

Fig. 3.6 Single-link arm

3.7.2 Number of Vibration Modes

In the literature, e.g., Cyril [7], DeLuca and Siciliano [14], and others, it is stated that the first two modes of vibration are sufficient for the study of flexible link robotic arms. In this subsection, a justification is provided for such consideration. To do this, the tip deflections of the single flexible link, Fig. 3.4 and Table 3.3, are obtained considering a different number of modes. Here, a free-fall simulation, i.e., the link is assumed to fall under gravity from its horizontal position, i.e.,$\theta_1 = 0°$, is performed to capture the natural behavior of the beam, in contrast to the forced simulation as shown in Fig. 3.7. Its initial conditions for the numerical integration are:$\theta_1 = d_j = 0$ and $\dot{\theta}_1 = \dot{d}_j = 0$ where d_j, for $j = 1,\ldots, 4$, is the generalized coordinates associated with the vibration amplitudes. The tip deflections are then compared considering single ($j = 1$), first two ($j = 1, 2$), first three ($j = 1, 2, 3$), and first four ($j = 1,\ldots,4$) modes. Assuming the first four modes would give the best results, their differences from those obtained using one, two, and three modes are plotted in Fig. 3.8.

It is seen from the differences of Fig. 3.8 that the results of the first two and three modes are very close. Hence, first two modes are sufficient to study the link flexibility, which would provide a computationally more efficient algorithm than the three or more mode assumptions. This is also in agreement with the comments of Cyril [7] and DeLuca and Siciliano [14] that the consideration of more than two modes of vibrations in simulation does not affect the accuracy of the results significantly. Henceforth, in all the simulations, only the first two modes will be considered, and the effect of higher modes will be neglected. It is, however, emphasized here that the proposed model and algorithm are valid for any number of modes of vibrations, and the advantage associated with the consideration of only the first two modes is just the computational efficiency leading to a reduction in the simulation time. Similar conclusions can also be obtained using other simulations, for example, the one given in Sect. 3.7.1.

Fig. 3.7 Simulation results for single-link arm

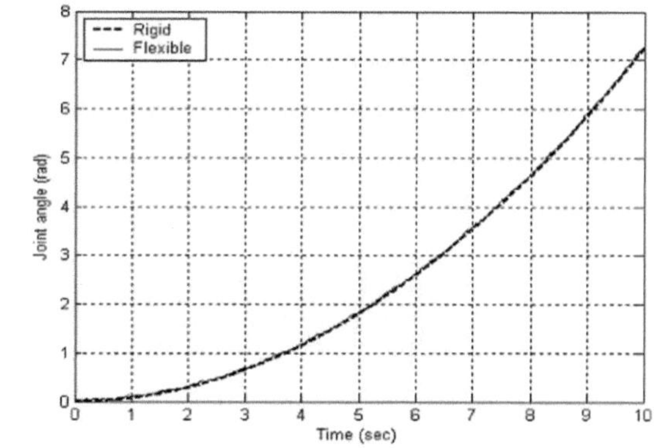

(a) Joint angle

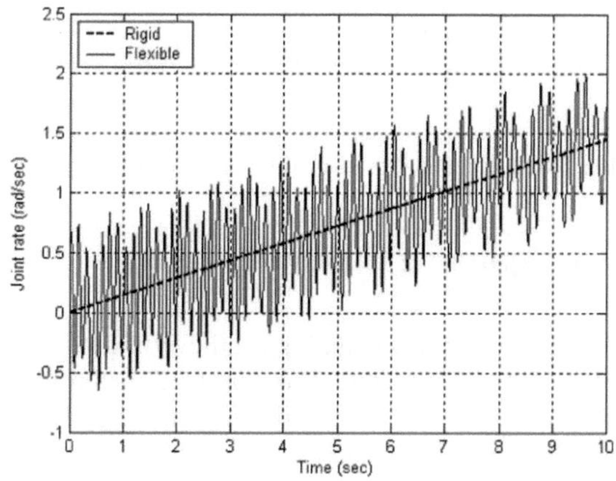

(b) Joint rate

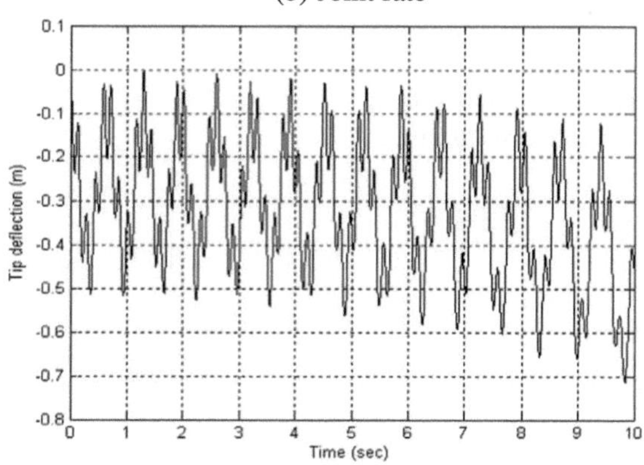

(c) Tip deflection of the link

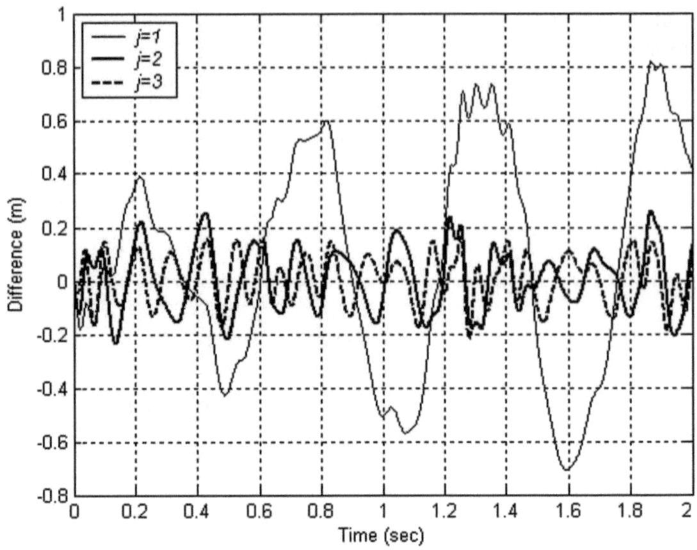

Fig. 3.8 Tip deflections using different number of modes

3.8 Spinning Cantilever Beam

An illustration of a spinning cantilever beam is considered to demonstrate the effect of geometric stiffness. The cantilever beam C is attached to rotating base B, as shown in Fig. 3.9. In this illustration, the deflection due to bending, i.e., transverse displacement (d_y) with 2-modes, is considered for the flexible link neglecting the axial deflection and the torsional effect. The GIM **I** of Eq. (3.31) for the flexible link is given by

Fig. 3.9 Cantilever beam attached to rotating base [2]

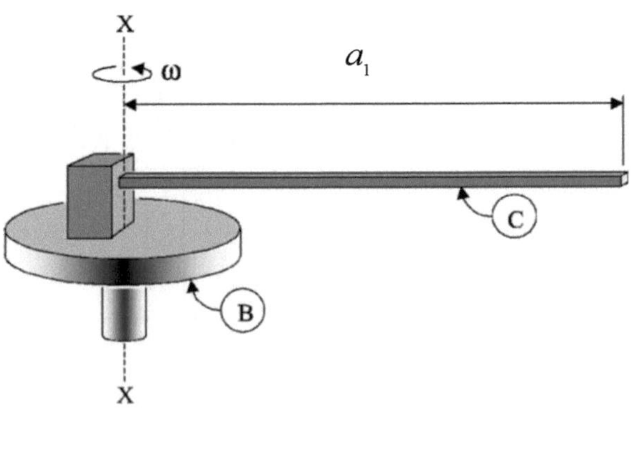

$$\mathbf{I} \equiv \rho_1 \begin{bmatrix} i_{11} & k3_1^{y1} & k3_2^{y1} \\ & k2_{11}^{y1} & k2_{12}^{y1} \\ sym & & k2_{22}^{y1} \end{bmatrix} \tag{3.49}$$

where $i_{11} \equiv \frac{a_1^3}{3} + k2_{11}^{y1} dy_{11}^2 + 2 \cdot k2_{12}^{y1} dy_{11} dy_{12} + k2_{22}^{y1} dy_{12}^2$ in which a_1 is the length of the output link. The constants $k2^{y1}$ and $k3^{y1}$ are the modal integrals which are given by

$$k2_{11}^{y1} \equiv \int_0^{a_1} s_1^{y1} s_1^{y1} d\bar{a},\ k2_{12}^{y1} \equiv \int_0^{a_1} s_1^{y1} s_2^{y1} d\bar{a},\ k2_{22}^{y1} \equiv \int_0^{a_1} s_2^{y1} s_2^{y1} d\bar{a}$$

$$k3_1^{y1} \equiv \int_0^{a_1} a_1 s_1^{y1} d\bar{a}, \text{ and } k3_2^{y1} \equiv \int_0^{a_1} a_1 s_2^{y1} d\bar{a} \tag{3.50}$$

where s_1^{y1} and s_2^{y1} are the shape functions in the transverse direction, i.e., along Y-direction of the link for 1st and 2nd modes, respectively. Further, dy_{11} and dy_{12} are the time-dependent modal coordinates for 1st and 2nd modes, respectively. The vector of convective inertia terms $\mathbf{h}$ is given by

$$\mathbf{h} \equiv \rho_1 \dot{\theta}_1 \begin{bmatrix} 2\left(k2_{11}^{y1} dy_{11} \dot{d}y_{11} + k2_{12}^{y1} dy_{12} \dot{d}y_{11} + k2_{12}^{y1} dy_{11} \dot{d}y_{12} + k2_{22}^{y1} dy_{12} \dot{d}y_{12}\right) \\ -k2_{11}^{y1} dy_{11} \dot{\theta}_1 - k2_{12}^{y1} dy_{12} \dot{\theta}_1 \\ -k2_{12}^{y1} dy_{11} \dot{\theta}_1 - k2_{22}^{y1} dy_{12} \dot{\theta}_1 \end{bmatrix} \tag{3.51}$$

Then, the generalized force vector $\boldsymbol{\tau}$ for the spinning beam is given by $\boldsymbol{\tau} = \boldsymbol{\tau}^E + \boldsymbol{\tau}^g + \boldsymbol{\tau}^s$, where $\boldsymbol{\tau}^E$ is the vector of external joint torques, $\boldsymbol{\tau}^g$ is the vector of generalized force due to gravity, and $\boldsymbol{\tau}^s$ is the vector of internal forces due to strain. The vector $\boldsymbol{\tau}^s$ is taken as the sum of generalized force due to linear strain $\boldsymbol{\tau}^{sl}$ and nonlinear strain $\boldsymbol{\tau}^{snl}$ to take into account the geometric stiffening effect. The vectors $\boldsymbol{\tau}^g$, $\boldsymbol{\tau}^{sl}$ and $\boldsymbol{\tau}^{snl}$ are given by

$$\boldsymbol{\tau}^g \equiv \rho_1 g \begin{bmatrix} \frac{1}{2}\left(a_1^2 c_1\right) - a_1 c_1 - \left(k1_1^{y1} dy_{11} + k1_2^{y1} dy_{12}\right) s_1 \\ k1_1^{y1} c_1 \\ k1_2^{y1} c_1 \end{bmatrix} \tag{3.52}$$

$$\boldsymbol{\tau}^{sl} \equiv \mathbf{K}\mathbf{q} = \frac{EI^y}{2} \begin{bmatrix} 0 \\ k8_1^{y1} dy_{11} \\ k8_2^{y1} dy_{12} \end{bmatrix}, \text{ and } \boldsymbol{\tau}^{snl} \equiv \mathbf{K}^g \mathbf{q} = \frac{F^a}{2} \begin{bmatrix} 0 \\ k9_1^{y1} dy_{11} \\ k9_2^{y1} dy_{12} \end{bmatrix} \tag{3.53}$$

where g is the acceleration due to gravity, E is Young's modulus of elasticity, I^y is the area moment of inertia, F^a is the axial force on the rotating beam, $c_1 \equiv \cos\theta_1$

and $s_1 \equiv \sin\theta_1$. The constants $k1_1^{y1}$, $k1_2^{y1}$, $k8_1^{y1}$, $k8_2^{y1}$, $k9_1^{y1}$, and $k9_2^{y1}$ are given by

$$k1_1^{y1} \equiv \int_0^{a_1} s_1^{y1} d\bar{a}, k2_2^{y1} \equiv \int_0^{a_1} s_2^{y1} d\bar{a}$$

$$k8_1^{y1} \equiv \int_0^{a_1} \left(\frac{\partial^2 s_1^{y1}}{\partial^2 \bar{a}}\right)^2 d\bar{a}, k8_2^{y1} \equiv \int_0^{a_1} \left(\frac{\partial^2 s_2^{y1}}{\partial^2 \bar{a}}\right)^2 d\bar{a}$$

$$k9_1^{y1} \equiv \int_0^{a_1} \left(\frac{\partial s_1^{y1}}{\partial \bar{a}}\right)^2 d\bar{a}, \text{ and } k9_2^{y1} \equiv \int_0^{a_1} \left(\frac{\partial s_2^{y1}}{\partial \bar{a}}\right)^2 d\bar{a} \tag{3.54}$$

Using the above equations of motion, one can evaluate the dynamic behavior of the rotating beam.

3.8.1 Numerical Simulation

For the numerical simulation, the parameters taken were similar to [2] so that the obtained results can be compared. The base is rotating with an angular velocity $\boldsymbol{\omega}$ which is given by

$$\boldsymbol{\omega} = \begin{cases} \frac{2}{5}\left[t - \left(\frac{15}{2\pi}\right)\sin\left(\frac{2\pi}{15}t\right)\right] & 0 \le t \le 15 \\ 6 & 15 \le t \le 30 \end{cases} \tag{3.55}$$

The link length is considered as 10 m, Young's modulus $E = 7 \times 10^{10}\text{N/m}^2$, mass per unit length $\rho = 1.2$ kg/m, and area moment of inertia $I^y = 2 \times 10^{-7}$ m^4. The simulation is run for 30 s. The transverse displacements of spinning cantilever beam without and with geometric stiffening effects are plotted in Figs. 3.10 and 3.11, respectively. Figure 3.10 shows the unbounded solution, which is the result of not considering the geometric coupling between the axial and transverse displacement causing the geometric softening effect. The consideration of the geometric stiffness corrects the error in the formulation, as shown in Fig. 3.11. Both results obtained here are comparable with those given in [2]. This validates the proposed formulation for the geometric stiffening.

3.9 Two-Link Planar Arm

In this section, a two-link robot arm in a plane motion is considered. Three cases are considered, namely (1) both links are rigid; (2) one link flexible and the other one rigid; (3) both links are flexible.

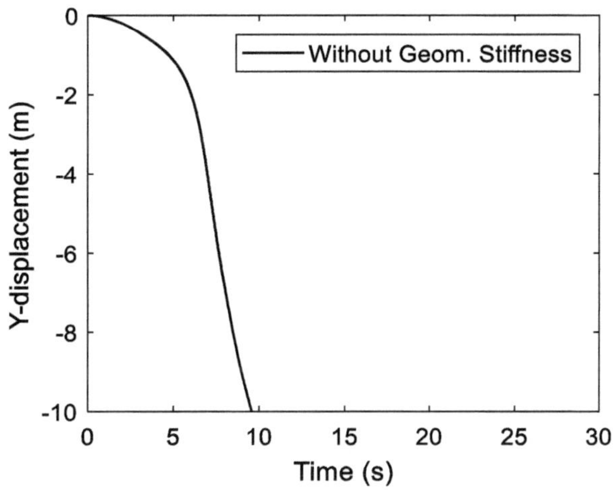

Fig. 3.10 Transverse displacement without geometric stiffness

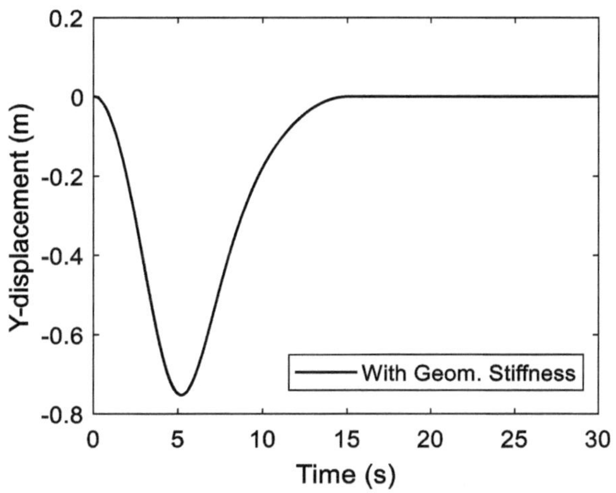

Fig. 3.11 Transverse displacement with geometric stiffness

It should be noted that while the first case represents a simple system, the second one is more realistic, particularly, in industrial and space robotic applications. Consideration of all the links as flexible, as in the third case, is rare in terms of practical applications because its analyses and control are complex and difficult. Hence, most of the industrial and space robots have rigid as well as flexible links, as in Case 2. Consequently, the position of the flexible link in the robot's kinematic chain becomes vital to the motion accuracy and vibrations, which is studied in Sect. 3.8.2.

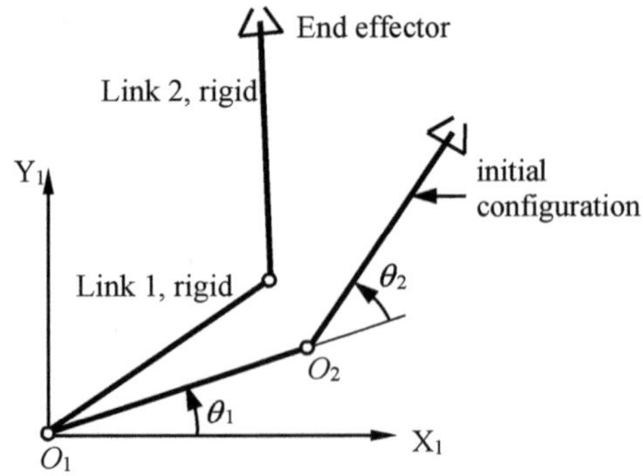

Fig. 3.12 A two-link arm with both links rigid (*RR*)

Table 3.4 Physical parameters for two-link rigid and rigid–flexible arms

Link, i	Length, a_i (m)	Mass, m_i (kg)	Flexure stiffness, E_iI_i (Nm^2)
(a) Rigid–Rigid (*RR*)			
1	1	5	–
2	1	5	–
(b) Rigid–Flexible (*RF*)			
1	1	5	–
2	1	5	1000
(c) Flexible–Rigid (*FR*)			
1	1	5	1000
2	1	5	–
(d) Flexible–Flexible (*FF*)			
1	1	5	1000
2	1	5	1000

3.9.1 Rigid Links

In this subsection, simulation results of a two-link planar arm with both links rigid, as shown in Fig. 3.12, are presented. This configuration is referred here as rigid–rigid (*RR* [1]). Physical parameters of the link are given in Table 3.4a. The rotational degrees of freedom of the robot are denoted by θ_1 and θ_2, as shown in Fig. 3.12. The free-fall simulation results with initial conditions, $\theta_1 = -90°, \theta_2 = 5°$, and $\dot{\theta}_1 = \dot{\theta}_2 = 0$ are shown in Fig. 3.13a–d. The results could be easily verified with the formulas available in any robotic textbook, e.g., Yoshikawa (1998).

[1] In this book, upper-case letters R and F with italic font represent the rigid and flexible links, whereas the non-italic letters R and P represent revolute and prismatic joints.

Fig. 3.13 Simulation results for two-link rigid and rigid–flexible arm

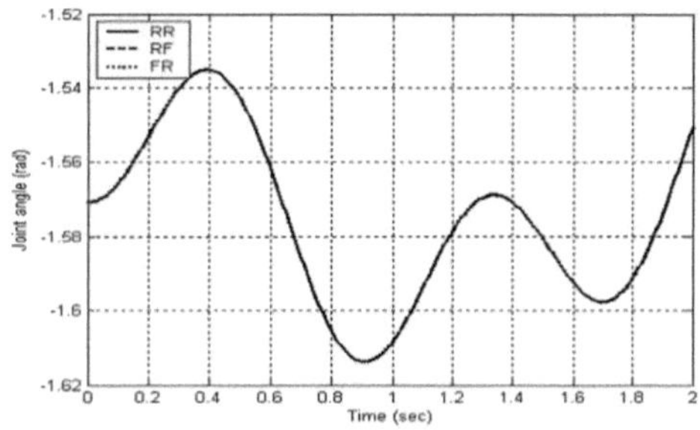

(a) Joint angle 1

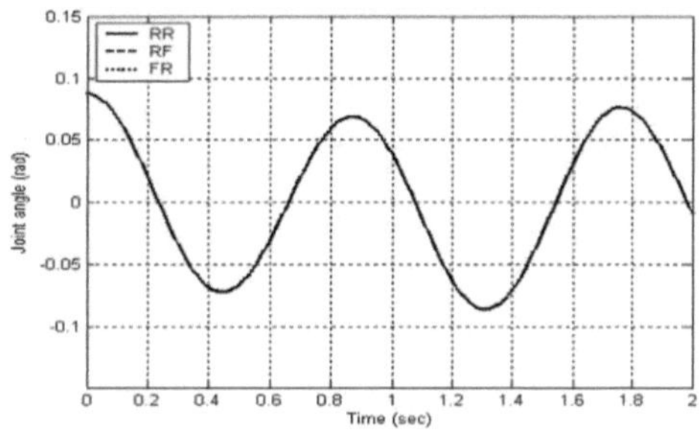

(b) Joint angle 2

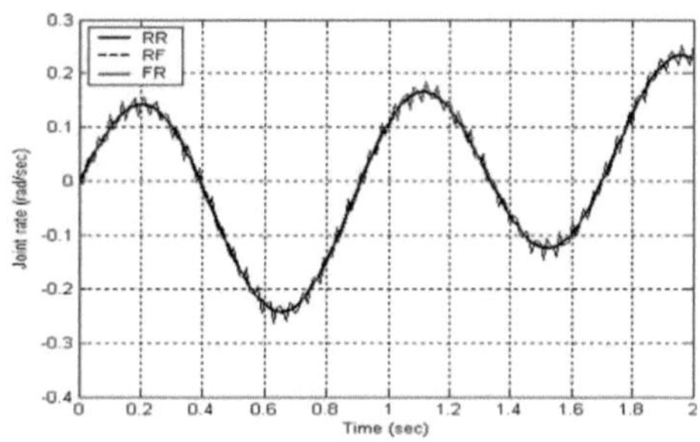

(c) Joint rate 1

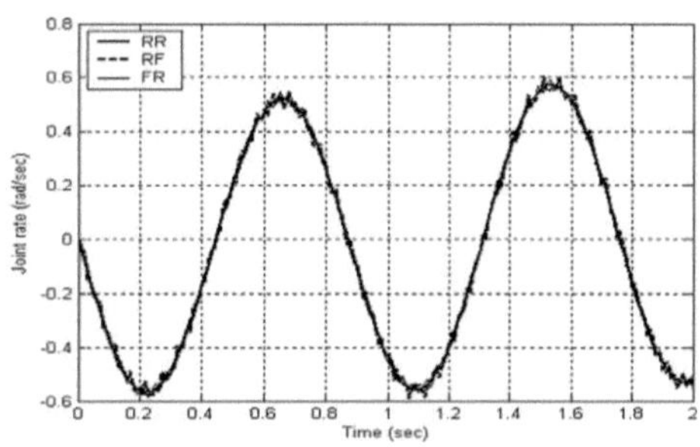

(d) Joint rate 2 response

Fig. 3.13 (continued)

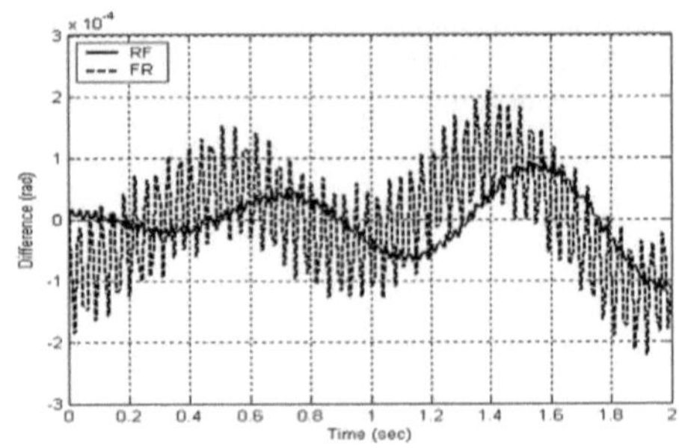

e) Difference in joint 1 response

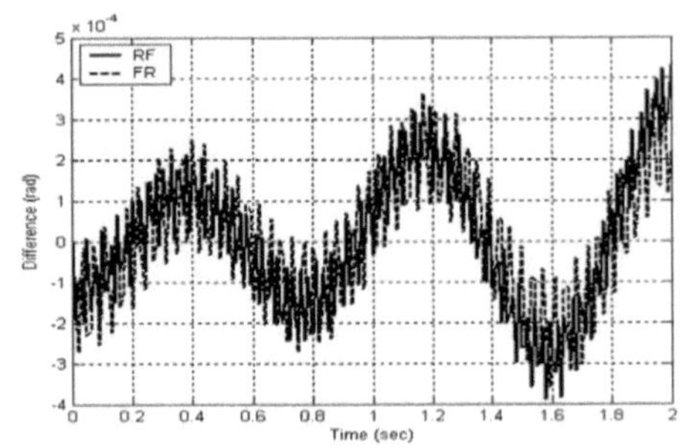

f) Difference in joint 2 response

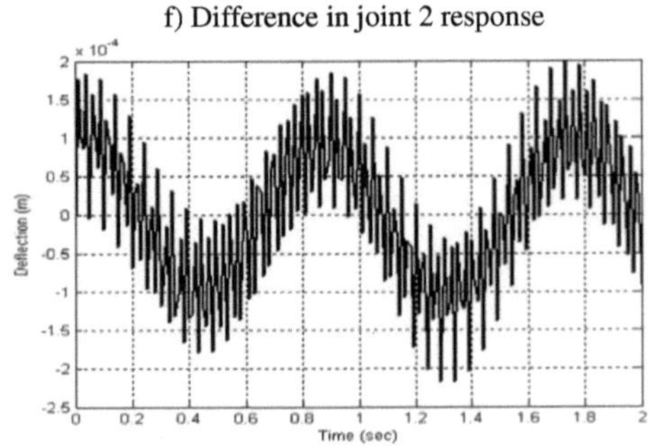

(g) Tip deflection of link 2 (RF)

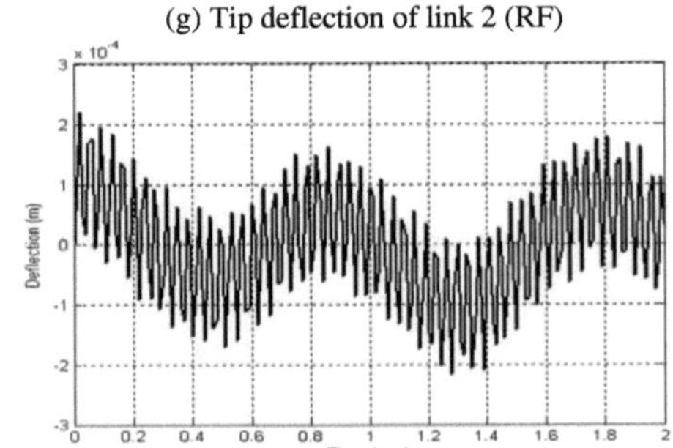

(h) Tip deflection of link 1 (*FR*)

3.9.2 *Rigid and Flexible Links*

Most of the real industrial and space robots have rigid as well as flexible links. In this subsection, two-link planar arm with one link rigid and another one flexible is undertaken. Since two configurations are possible, namely, rigid link first followed by the flexible one and vice-versa, as shown in Fig. 3.14a, b, respectively, and they are analyzed separately.

3.9.2.1 Rigid–Flexible (*RF*)

In this case, the first or the inner link is considered rigid, followed by the flexible link, i.e., the second or outer link is flexible, as shown in Fig. 3.14a. This configuration is referred here as rigid–flexible (*RF*). Using the modeling methodology presented in this chapter, simulation results are obtained that are shown in Fig. 3.13a–g, whereas the input parameters are shown in Table 3.4b. The joint angles, θ_1 and θ_2, are shown in Fig. 3.14a, b, whereas d_{21} and d_{22} are the generalized coordinates corresponding to the time amplitudes of vibration of link 2 in 1st and 2nd modes. For this system, free-fall simulation was performed while the initial conditions were, $\theta_1 = -90°, \theta_2 = 5°, d_{2j} = 0$ and $\dot{\theta}_1 = \dot{\theta}_2 = \dot{d}_{2j} = 0$, for $j = 1, 2$. Simulation results for the joint angles and the joint rates are shown in Figs 3.13a–d, where differences in the results with respect to the *RR* arm are invisible. As a result, the differences in the joint angle response for the *RR* and *RF* arms are separately plotted in Fig. 3.13e–f, which are very small, in the range of 10^{-4}. Since link 2 is flexible, its tip deflection with respect to O_2 is shown in Fig. 3.13g.

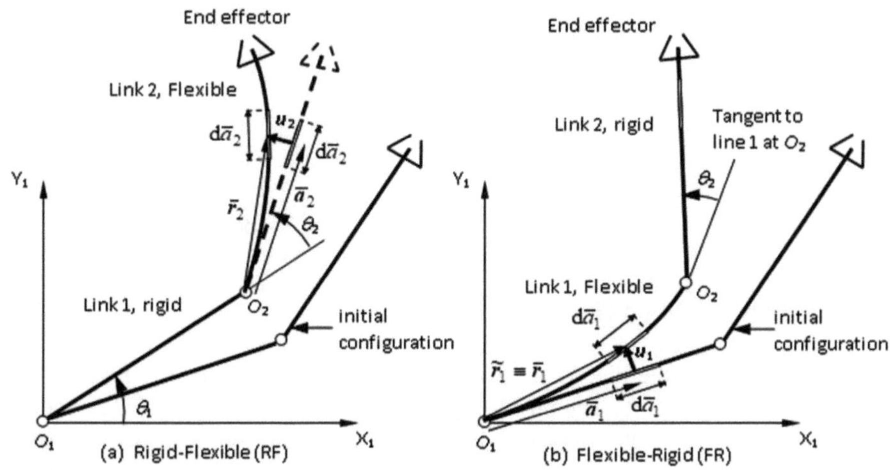

Fig. 3.14 Two-link arm with rigid–flexible links

Fig. 3.15 End-effector coordinates of the two-link arm

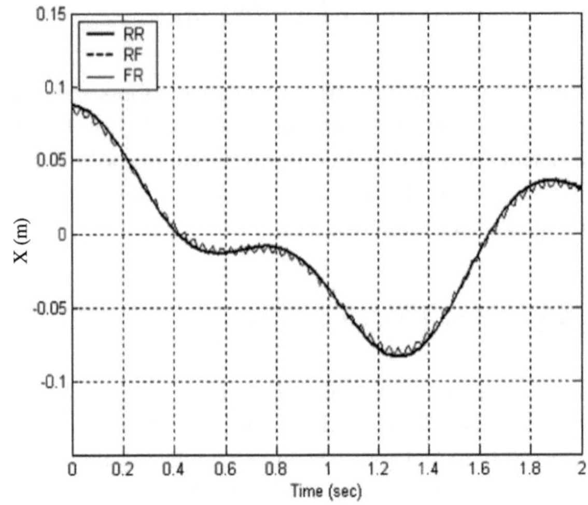

(a) X-coordinate

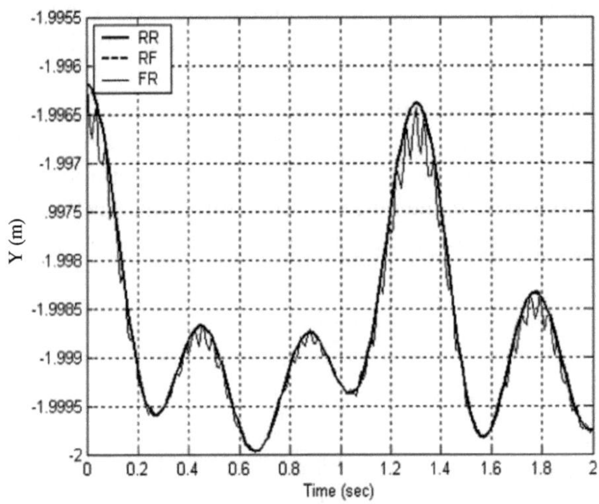

(b) Y-coordinate

3.9.2.2 Flexible–Rigid (*FR*)

In this arm, the flexible link is put first as link 1, followed by the rigid one as link 2, as shown in Fig. 3.14b. The physical properties are given in Table 3.4c, whereas the simulation results are shown in Fig. 3.13a–f, h. The initial conditions were, $\theta_1 = -90°, d_{1j} = 0, \theta_2 = 5°$ and $\dot{\theta}_1 = \dot{d}_{1j} = \dot{\theta}_2 = 0$, for $j = 1, 2$. The joint angles θ_1 and θ_2 are shown in Fig. 3.14b, and d_{11} and d_{12} are the generalized coordinates corresponding to the time amplitudes of vibration of link 1 in 1st and 2nd modes. Figure 3.13a, b shows the simulation results for the joint angles and the simulation results for joint rates are shown in Fig. 3.13c, d, which have invisible differences compared to *RR* and *RF* arms. Hence, the differences in the joint angles with respect to *RR* arm are shown in Fig. 3.13e, f. Note here that the joint 2 behavior for *RF* and *FR* systems is very similar, as seen in Fig. 3.13h. Finally, the tip deflection of the flexible link, i.e., point O_2 on link 1 with respect to O_1, is shown in Fig. 3.13h.

Comparing the results of Fig. 3.13e, f, which were calculated as the differences between the simulated joint angles for the rigid–flexible arms and the rigid only arm, it is very clear that the *RF* configuration has less differences compared to the *FR* configuration, particularly with respect to joint 1, Fig. 3.13e. Hence, if a choice is available *RF* should be preferred over *FR* for any two-link rigid–flexible arm. In order to check if the *X*- and *Y*-coordinates of the end-effector give the same conclusion or not, they are plotted in Fig. 3.15a, b. while the *RF* almost coincides with the *RR*, the *FR* has significant variations.

3.9.3 Both Links Flexible

A two-link arm with both links flexible, as shown in Fig. 3.16, is considered next, whose simulation results while falling freely under gravity are shown in Fig. 3.17. The architecture and the inertial parameters of the arm are given in Table 3.4d. The response of the system is obtained with the following initial conditions: $\theta_1 = -90°$, $\theta_2 = 5°$, $\dot{\theta}_1 = 0$ and $\dot{\theta}_2 = 0$, and $d_{i,j} = 0$ and $\dot{d}_{i,j} = 0$, for $i, j = 1, 2$. Note that, the results reported in Fig. 3.17 match exactly with those given by [7, 34]. It is pointed

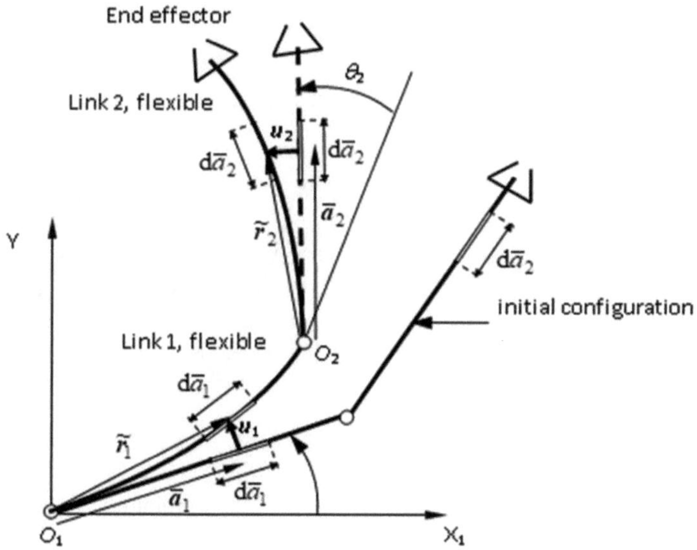

Fig. 3.16 A two-link flexible arm

Fig. 3.17 Simulation results for two-link flexible arm

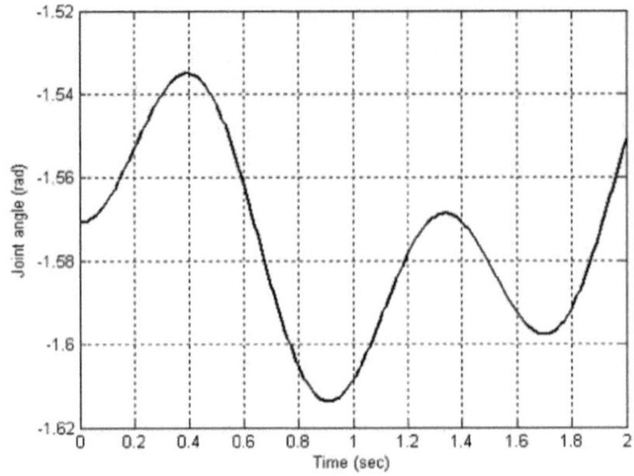

(a) Joint angle 1

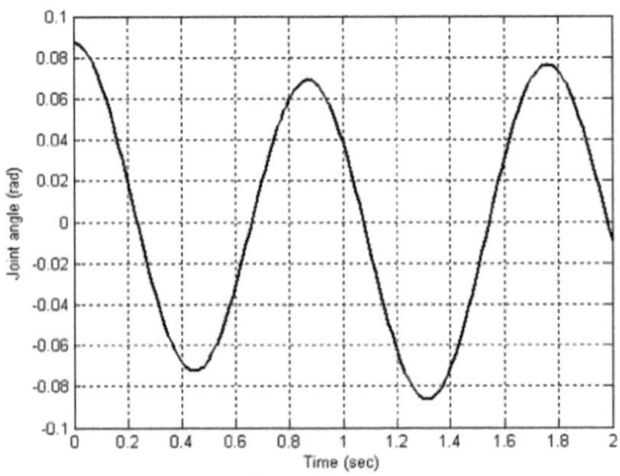

b) Joint angle 2

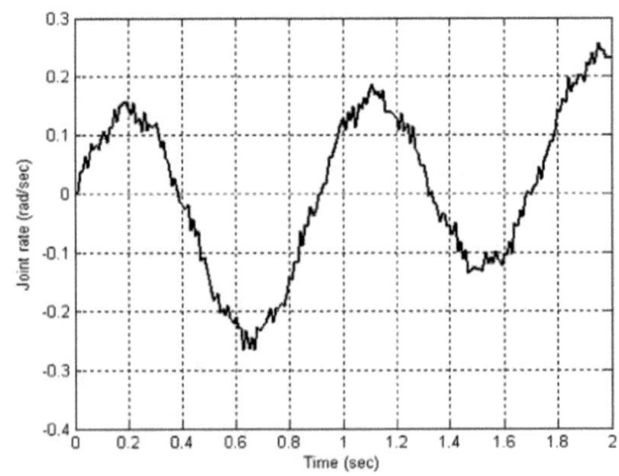

(c) Joint rate 1

Fig. 3.17 (continued)

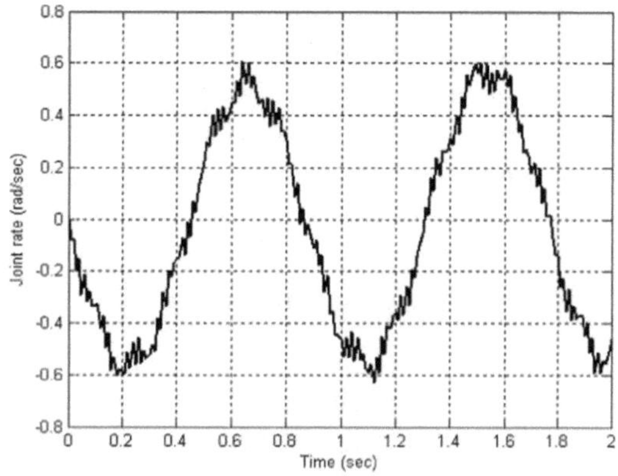

(d) Joint rate 2

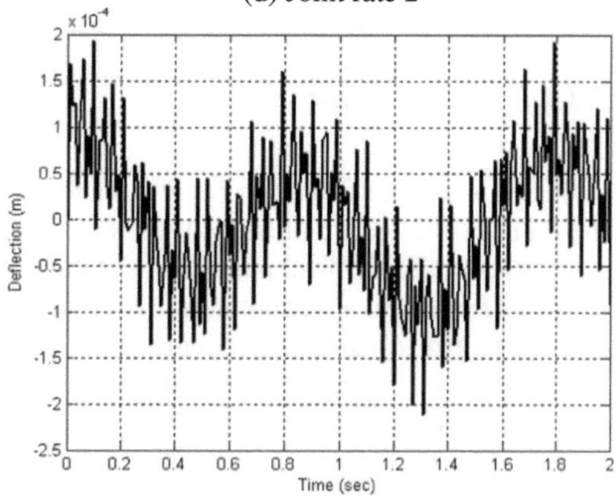

(e) Tip deflection of link 1

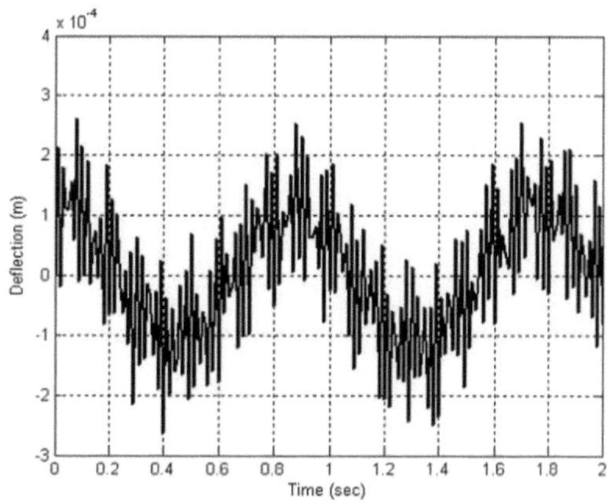

(f) Tip deflection of link 2

out here that [7] had observed artificial damping in his simulation results, which are absent using the present formulation. This can be attributed to the numerical stiffness present in the original NOC-based non-recursive formulation proposed by Cyril [7]. The present DeNOC-based recursive algorithm avoids such artificial stiffness. These aspects are also elaborated in Chap. 7.

3.10 Three-Link Planar Arm

In this section, simulations of a three-link planar arm are presented. First, all links were treated as rigid, followed by the first two links flexible and the last one as rigid, as in the case of Canadarm [7].

3.10.1 Rigid Links

In this subsection, simulation results of a three-link planar arm with all links rigid, as shown in Fig. 3.18, are presented. This configuration is referred here as rigid–rigid–rigid (*RRR*). The physical parameters of the link are given in Table 3.5. Its rotational degrees of freedom are denoted by θ_1, θ_2, and θ_3, Fig. 3.18. Forced simulation of the arm is performed using the scheme outlined in Fig. 3.3. The joint trajectories were prescribed for a representative maneuver of the arm [30] for which the joint torques are computed using the inverse dynamics algorithm, developed separately. All links

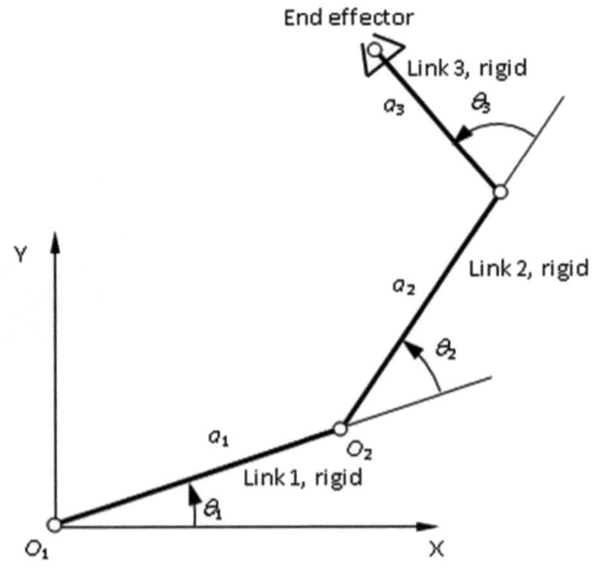

Fig. 3.18 A three-link planar arm with rigid links

Table 3.5 DH and inertial parameters for the three-link arms

Link, i	a_i (m)	b_i (m)	α_i (rad)	θ_i(rad)	m_i (kg)	E_iI_i (Nm^2)	
						RRR	*FFR* (Canadarm)
1	6	0	0	θ_1 [0]*	140	–	1×10^5
2	7	0	0	θ_2 [0]	85	–	2×10^5
3	2	0	0	θ_3 [0]	95	–	–

*The values in square brackets show initial configuration of the arm

are assumed rigid, and the prescribed trajectories are:

$$\theta_i = k_i\left[t - \frac{T_i}{\pi}\sin\left(\frac{\pi}{T_i}t\right)\right] \text{ for } i = 1, 2, 3 \tag{3.56}$$

where $k_1 = 0.075\ \text{s}^{-1}$, $k_2 = 0.05\ \text{s}^{-1}$, $k_3 = 0.1\ \text{s}^{-1}$, $T_1 = 10$ s, $T_2 = T_3 = 5$ s. The joint torques thus obtained are shown in Figs 3.19a–c, which has been verified by Cyril [7]. These torques are then used as the inputs to the dynamic simulation of the three-link planar arm shown in Fig. 3.18. Figures 3.20 and 3.21 show the forced simulation results, where the simulated joint angles and rates are compared with those obtained using Eq. (3.48). The differences are almost insignificant. Hence, the simulated motion almost reproduces the desired one, thus indicating negligible numerical integration errors of the order of 10^{-8}, introduced during simulation.

3.10.2 Canadarm with Two Flexible Links

In this system, the first two links of the three-link arm are considered flexible, as shown in Fig. 3.22. Such architecture is referred as *FFR* that is used in Canadarm [7, 30], whose (DH)-parameters, along with the inertial parameters, are shown in Table 3.5. The numerical values in Table 3.5 are taken from Sharf and D'Eleuterio [30] and Cyril [7] in order to be able to compare the simulation results with their results. The simulated results are shown in Figs. 3.20, 3.21, and 3.23 where the effect of flexibility is quite significant, particularly in Figs. 3.20 and 3.21. The deflections of each flexible link from its rigid position are shown in Fig. 3.23, which has also been verified with those reported in Cyril [7]. Note that the axial stiffening effect resulting from the term, $u_{i,1}^x$—the projection of the displacement of the i-th link, on the $\widehat{X}_{i+1}$-axis, as in Eq. (3.3b),—becomes significant only when the angular rates of the bodies are considerably greater than their first natural frequency. Unlike the case of a single beam rotating at a very large angular rate, as in the case of spinning cantilever beam in this chapter, the flexible links of the Canadarm do not rotate at a large angular rate. Hence, the component, $u_{i,1}^x$, is ignored here, as done in Cyril [7] and DeLuca and Siciliano [14].

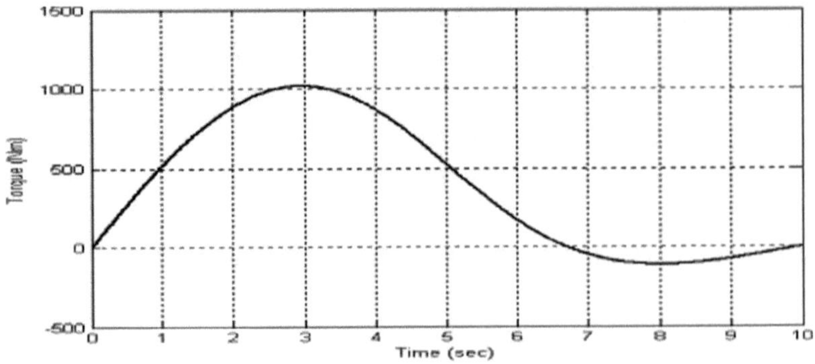

(a) Joint 1

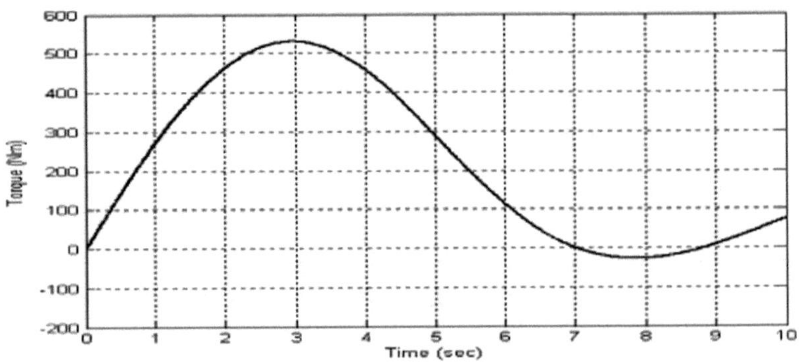

(b) Joint 2

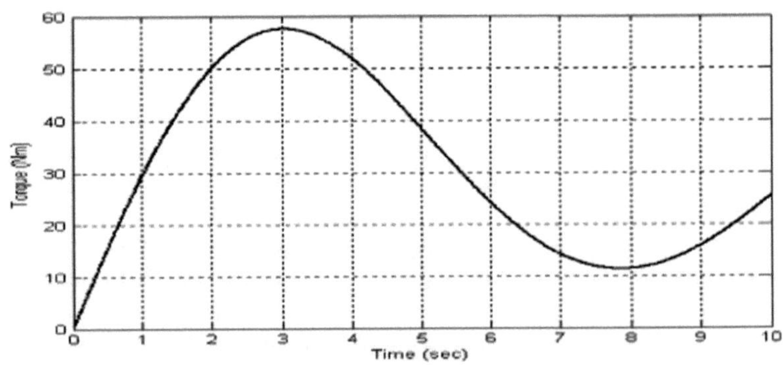

(c) Joint 3

Fig. 3.19 Joint torques for *RRR* arm

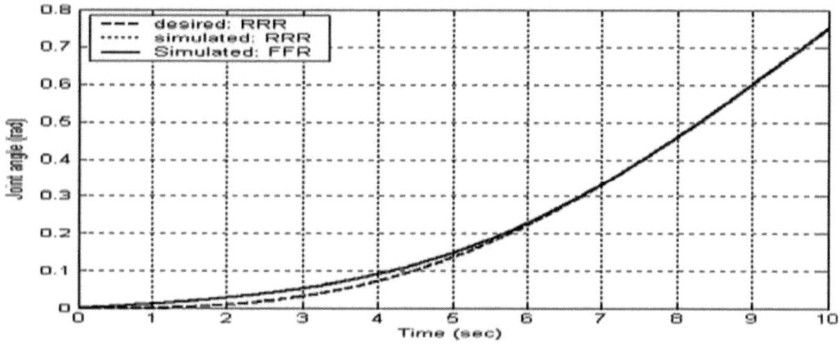

(a) Joint 1

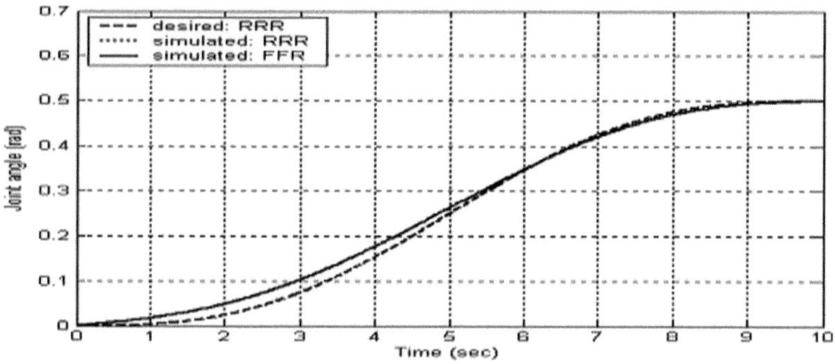

(b) Joint 2

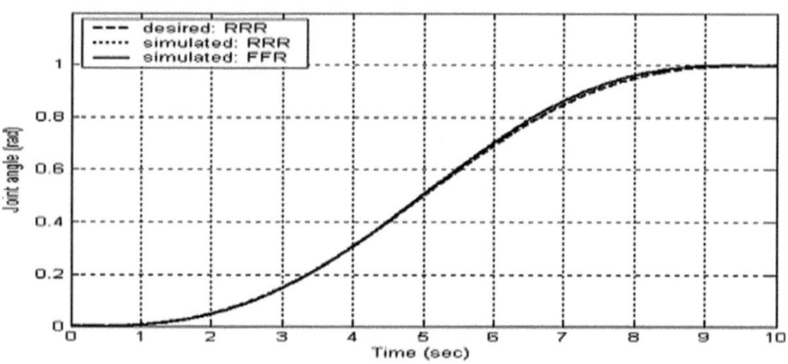

(c) Joint 3

Fig. 3.20 Simulated joint angles for the *RRR* and *FFR* arms

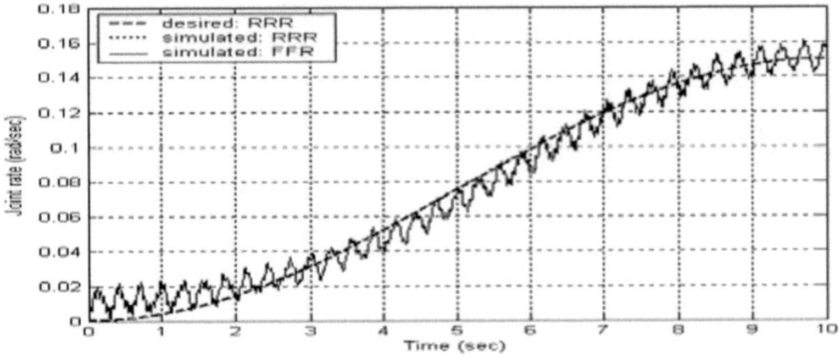

(a) Joint 1

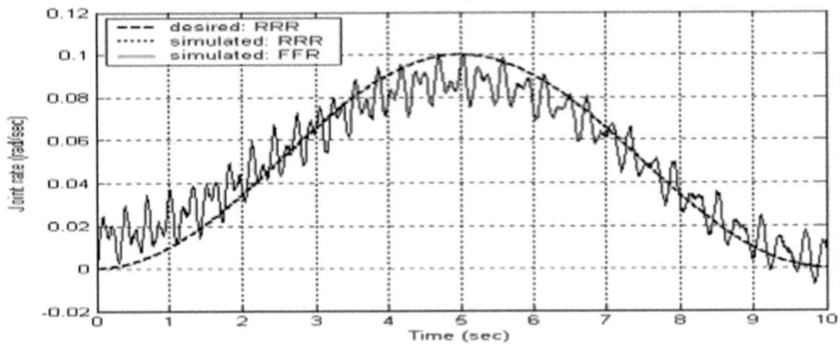

(b) Joint 2

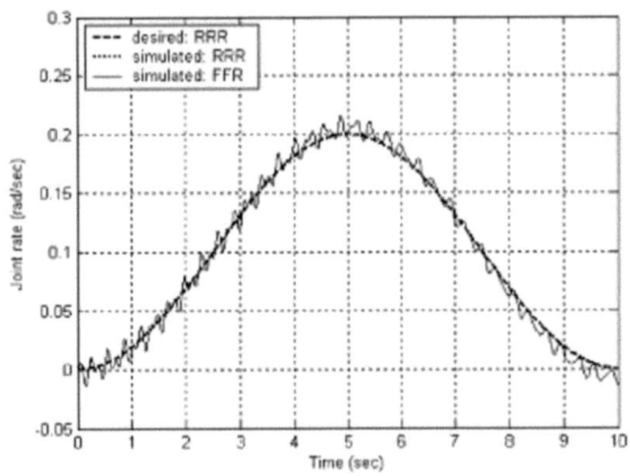

(c) Joint 3

Fig. 3.21 Simulated joint rates for the RRR and FFR arms

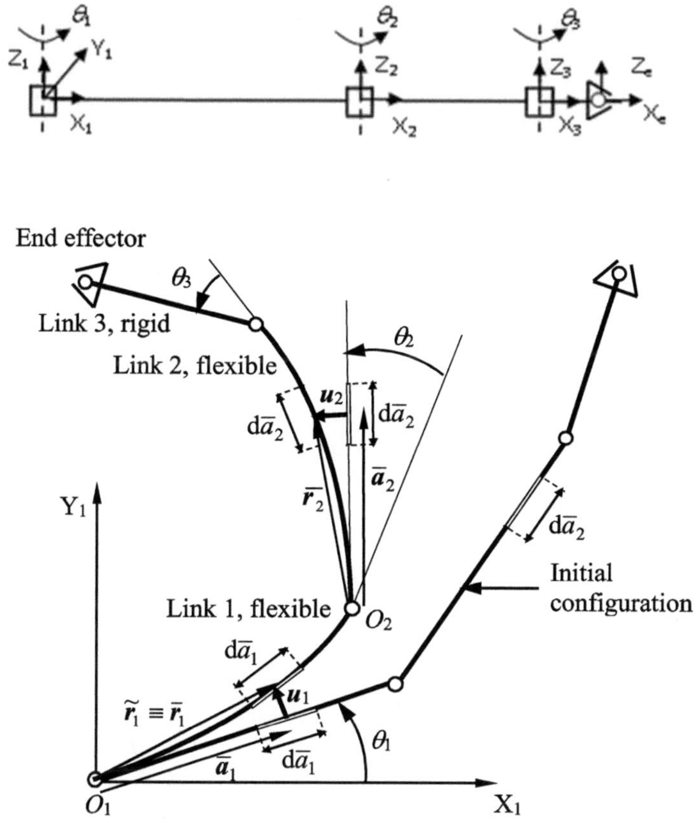

(b) Top view

Fig. 3.22 Canadarm with 1st and 2nd links flexible

The components, $u^y_{i,1}$ and $u^z_{i,1}$–the projections of displacement of the i-th link on the $\widehat{Y}_{i+1}$- and $\widehat{Z}_{i+1}$-axes, respectively, are however, modeled using the assumed mode method, as stated above in Sect. 3.6.

3.11 Summary

The dynamic modeling of open-loop serial-chain, rigid–flexible robots comprising both rigid and flexible links is presented in this chapter. The dynamic equations of motion of the systems are derived using the decoupled natural orthogonal complement (DeNOC) matrices. The analytical expressions for matrices and vectors associated with the dynamic equations of motion of the system are also obtained. The effect of geometric stiffness is also incorporated in the DeNOC-based formulation to

Fig. 3.23 Tip deflection of the flexible links

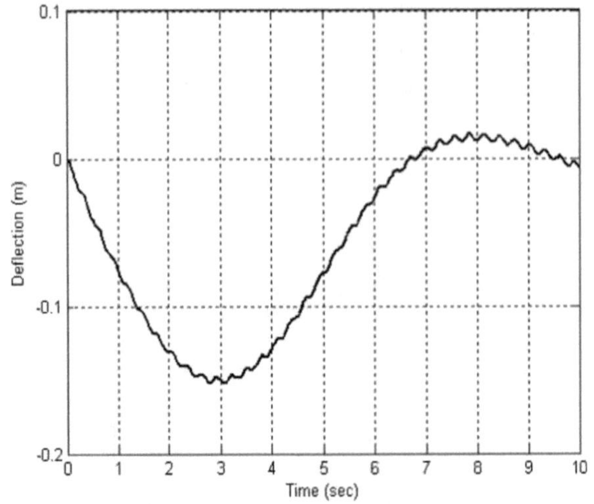

(a) Link 1

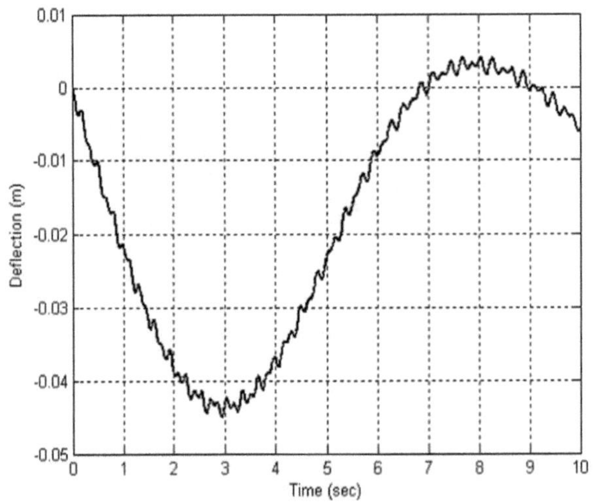

(b) Link 2

deal with the high-speed, large displacement problem. A recursive forward dynamics algorithm for rigid–flexible robots is presented. The implementation scheme and the computational complexity of the algorithm are also given in a systematic manner. It can be seen from Eq. (3.47) that the order of the algorithm is linear in n—the number of links, and cubic in m and $\overline{m}$—the number of modes of vibration of the robot links. The present formulation leads to an efficient forward dynamics algorithm, as shown in Table 3.1. The simulation of a single link, a spinning cantilever beam with the inclusion of geometric stiffness, a planar two-link robotic arm, and a planar three-link robotic arm is presented. In all three cases, only rigid links, both rigid and flexible links, i.e., as rigid and rigid–flexible systems, respectively, are considered. The 3-degree-of-freedom (DOF) planar Canadarm is then presented in Sect. 3.9, followed by the simulation of the three-link spatial arms with only rigid and rigid–flexible links.

References

1. Agrawal OP, Shabana AA (1985) Dynamic analysis of multibody systems using component modes. Comput Struct 21(6):1303–1312. https://doi.org/10.1016/0045-7949(85)90184-1
2. Banerjee AK (2003) Contributions of multibody dynamics to space flight: a brief review. J Guid Control Dyn 26(3):385–394. https://doi.org/10.2514/2.5069
3. Book WJ (2004) Flexible robot arms. In: Kurfess T (ed) Robotics handbook. CRC Press
4. Cardona A (2000) Superelements modelling in flexible multibody dynamics. Multibody Syst Dyn 4:245–266
5. Cetinkunt S, Siciliano B, Book W (1986) Symbolic modeling and dynamic analysis of flexible manipulators. In: Proceedings of the IEEE conference on systems, man and cybernetics, pp 798–803
6. Craig RR (1981) Structural dynamics. Wiley, Singapore
7. Cyril X (1988) Dynamics of flexible-link manipulators. McGill University, Canada
8. De Luca A (2003) Robots with flexible links: modeling and control. Scuola di Dottorato CIRA Controllo di Sistemi Robotici per la Manipolazione e la Cooperazione. http://www.prisma.unina.it/cira03/TA_DeLuca2.pdf
9. De Luca A, Siciliano B (1991) Closed-form dynamic model of planar multilink lightweight robots. IEEE Trans Syst Man Cybern 21(4):826–839. https://doi.org/10.1109/21.108300
10. Featherstone R (1983) Calculation of robot dynamics using articulated-body inertias. Int J Robot Res 2(1):13–30
11. Hastings G, Book W (1986) Verification of a linear dynamic model for flexible robotic manipulators. In: Proceedings of IEEE international conference on robotics and automation, pp 1024–1029. https://doi.org/10.1109/ROBOT.1986.1087617
12. Hwang YI (2005) A new approach for dynamic analysis of flexible manipulator systems. Int J Non-Linear Mech 40:925–938
13. Jain A, Rodriguez G (1992) Recursive flexible multibody system dynamics using spatial operators. J Guid Control Dyn 15(6):1453–1466
14. Kane TR, Ryan RR, Banerjee AK (1987) Dynamics of a cantilever beam attached to a moving base. J Guid Control Dyn 10(2):139–151
15. Kim S-S, Haug EJ (1988) A recursive formulation for flexible multibody dynamics part I: open-loop systems. Comput Methods Appl Mech Eng 71(3):293–314. https://doi.org/10.1016/0045-7825(88)90037-0
16. Martins J, Ayala Botto M, Sá da Costa J (2002) Modeling of flexible beams for robotic manipulators. Multibody Syst Dyn 7:79–100. https://doi.org/10.1023/A:1015239604152
17. Mayo JM, García-Vallejo D, Domínguez J (2004) Study of the geometric stiffening effect: comparison of different formulations. Multibody SysDyn 11(4):321–341. https://doi.org/10.1023/B:MUBO.0000040799.63053.d9
18. Meirovitch L (1967) Analytical methods in vibration. The Mcmillan Company, New York
19. Nikravesh PE (1988) Computer-aided analysis of mechanical systems. Prentice-Hall Englewood Cliffs, New Jersey
20. Pascal M (2006) Dynamics and stability of a two degree of freedom oscillator with an elastic stop. J Comput Nonlinear Dyn 1(1):94–102. https://doi.org/10.1115/1.1961873

21. Pratap R (2005) Getting started with MATLAB: a quick introduction for scientists and engineers. Oxford University Press, New York
22. Rao SS (1995) Mechanical vibrations. Addison-Wesley Publications, New York
23. Rao JS, Gupta K (1999) Introductory course on theory and practice of mechanical vibrations. New Age International
24. Saha SK (1997) A decomposition of the manipulator inertia matrix. IEEE Trans Robot Autom 13(2):301–304
25. Saha SK (1999a) Dynamics of serial multibody systems using the decoupled natural orthogonal complement matrices. J Appl Mech Trans ASME 66(4):986–995
26. Saha SK (1999b) Analytical expression for the inverted inertia matrix of serial robots. Int J Robot Res 18(1):116–124
27. Shabana AA (1997) Flexible multibody dynamics: review of past and recent developments. Multibody Syst Dyn 1:189–222
28. Shabana AA (2005) Dynamics of multibody systems. Cambridge University Press, Cambridge
29. Sharf I (1999) Nonlinear strain measures, shape functions and beam elements for dynamics of flexible beams. Multibody SysDyn 3(2):189–205
30. Sharf I, D'Eleuterio GMT (1988) Computer simulation of elastic chains using a recursive formulation. In: Proceedings of IEEE conference on robotics and automation, Pensylvania, pp 1539–1545
31. Shyu YJ, Gill KF (1997) Dynamic modelling of planar flexible manipulators: computational and algorithmic efficiency. Proc Inst Mech Eng C J Mech Eng Sci 211(2):119–133. https://doi.org/10.1243/0954406971521700
32. Theodore RJ, Ghosal A (1995) Comparison of the assumed modes and finite element models for flexible multilink manipulators. Int J Robot Res 14(2):91–111
33. Thompson WT (1988) Theory of vibration with applications. Prentice-Hall, London
34. Usoro PB, Nadira R, Mahil SS (1986) Finite element/Lagrange approach to modeling lightweight flexible manipulators. J Dyn Syst Measurement Control Trans ASME 108(3):198–205
35. Walker MW, Orin DE (1982) Efficient dynamic computer simulation of robotic mechanisms. J Dyn Syst Measurement Control Trans ASME V 104(N3):205–211
36. Yang B (2005) Stress, strain, and structural dynamics: an interactive handbook of formulas, solutions, and MATLAB toolboxes. Academic Press

Chapter 4
Dynamics of Six-Link Spatial Robot Arms

Simulation results of the six-link spatial industrial and space robots, namely PUMA robot, Space Shuttle Remote Manipulator System (SSRMS), and Stanford arm, with only rigid and rigid–flexible links, are presented in this chapter. A typical six-link robot has a three-link wrist attached at the end of a three-link spatial arm. Examples of PUMA robot, Space Shuttle Remote Manipulator System (SSRMS), which have revolute joints only, and Stanford arm with five revolute joints and one prismatic joint are considered here.

4.1 PUMA Robot

Figure 4.1 shows the configuration of the six-link PUMA robot [5]. The DH parameters of the robot, along with its mass and hub inertia, etc., are given in Table 4.1. It can be seen that, unlike the case of Canadarm, Fig. 3.22 and Table 3.5, all the links of a PUMA arm are short and stub, which can be dynamically modeled as rigid links.

However, when some of the link lengths are long, as in the case of SSRMS, Fig. 4.6 and Table 4.2, the system would be treated as rigid–flexible, whose results are discussed in Sect. 4.2. Free-fall simulation of the PUMA robot with the following initial conditions was performed: $\theta_i = 0$ and $\dot{\theta}_i = 0$, for $i = 1, \ldots, 6$. The results are shown in Fig. 4.2a–f, which were verified using another algorithm proposed by Bhangale [2]. This step confirms the correctness of the proposed algorithm implemented in MATLAB. Simulation for a time period of only 1.6 s is shown because the results show instability beyond this point. This aspect will be discussed later in Chap. 7. Next, following the scheme of Fig. 3.3, the joint torques required by the robot for the desired trajectory are determined using the inverse dynamics algorithm developed for the purpose. The joint trajectory is taken as, for $i = 1, \ldots, 6$:

P. V. Nandihal et al., *Dynamics of Rigid-Flexible Robots and Multibody Systems*, Intelligent Systems, Control and Automation: Science and Engineering 100,
https://doi.org/10.1007/978-981-16-2798-9_4

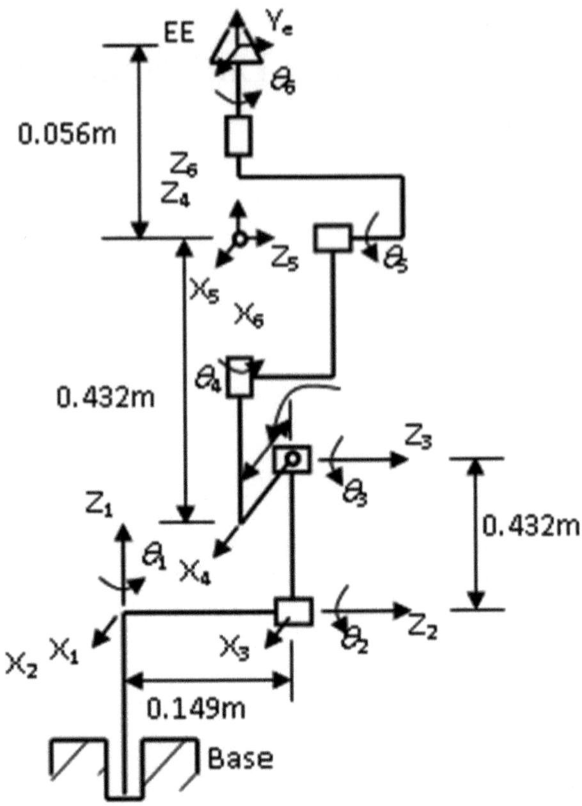

Fig. 4.1 Six-DOF PUMA robot

Table 4.1 DH and other parameters of the six-link PUMA robot

i	a_i (m)	b_i (m)	α_i (rad)	θ_i (rad)	m_i (Kg)	I_{zz}^{hub} (Kgm^2)
1	0	0.20	$-\pi/2$	$\theta_1[0]$[a]	10.521	0
2	0.432	0.149	0	$\theta_2[0]$	15.7612	0
3	0.02	0	$\pi/2$	$\theta_3[0]$	8.767	0
4	0	0.432	$-\pi/2$	$\theta_4[0]$	0	0.181
5	0	0	$\pi/2$	$\theta_5[0]$	0	0.0735
6	0	0.056	0	$\theta_6[0]$	0	0.0141

[a] The values in square brackets show the initial configuration of the robot

$$\theta_i = \left[\frac{\pi}{T}t - \frac{1}{2}\sin\left(\frac{2\pi}{T}t\right)\right];\ \theta_i(0) = 0\,\text{and}\,\theta_i(T) = \pi \tag{4.1}$$

where$T = 10$. The joint torques are shown in Fig.4.3a–f, which were then given as input to the PUMA robot. The forced simulation results are then shown in Figs. 4.4

Table 4.2 DH and other parameters of the SSRMS robot

Link no. (i)	a_i (m)	b_i (m)	α_i (rad)	θ_i (rad)	m_i (Kg)	E_iI_i (Nm^2)
1	1	0	$-\pi/2$	$\theta_1[0]$[a]	47.5	–
2	6	0	0	$\theta_2[0]$	140	1×10^5
3	7	0	$\pi/2$	$\theta_3[0]$	85	1×10^5
4	1	0	$-\pi/2$	$\theta_4[0]$	47.5	–
5	1	0	$\pi/2$	$\theta_5[0]$	47.5	–
6	1	0	0	$\theta_6[0]$	47.5	–

[a]The values in square brackets shows the initial configuration of the arm

and 4.5. The results match exactly with those obtained using a separately developed non-recursive algorithm based on the explicit Cholesky decomposition, similar to the one reported in Angeles [1].

4.2 Space Shuttle Remote Manipulator System Robot (SSRMS)

In this section, the six-link Space Shuttle Remote Manipulator System (SSRMS) [3, 4] is considered, whose architecture is very similar to the PUMA robot as discussed in Sect. 4.1, but the lengths of links 2 and 3 are very long, which are treated as flexible. Figure 4.6 shows the manipulator, and Table 4.2 gives its DH and other inertial parameters.

The joint trajectory is taken as, for $i = 1, \ldots, 6$:

$$\theta_i = 0.05\left[t - \frac{5}{\pi}\sin\left(\frac{\pi}{5}t\right)\right] \tag{4.2}$$

The joint torques required by the manipulator to prescribe the desired trajectory given by Eq. (4.2) are first determined using the inverse dynamics algorithm. The calculated joint torques shown in Fig. 4.7 are now given as input to the SSRMS. The simulation results are shown in Figs. 4.8 and 4.9. The results match with those presented in Cyril [3].

The tip deflection of the flexible links 2 and 3 are shown in Fig. 4.10. The simulation results obtained for the robot are stable up to about 6 s. After this, numerical error start building up, resulting in unstable results. These aspects are separately discussed in Chap. 7. However, up to 6 s whatever variations are noted, they are due to the flexibility of the links.

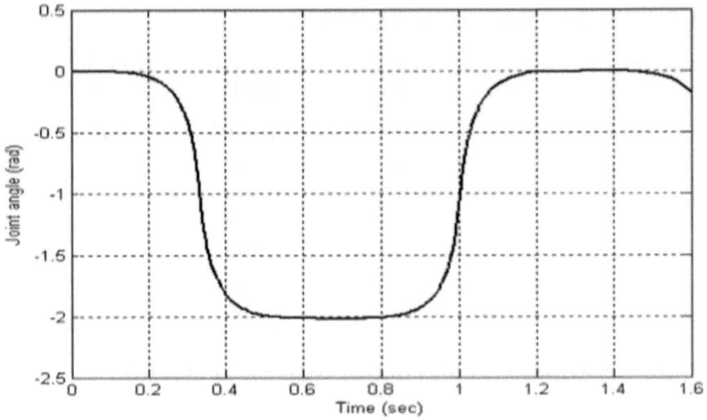

(a) Joint-1

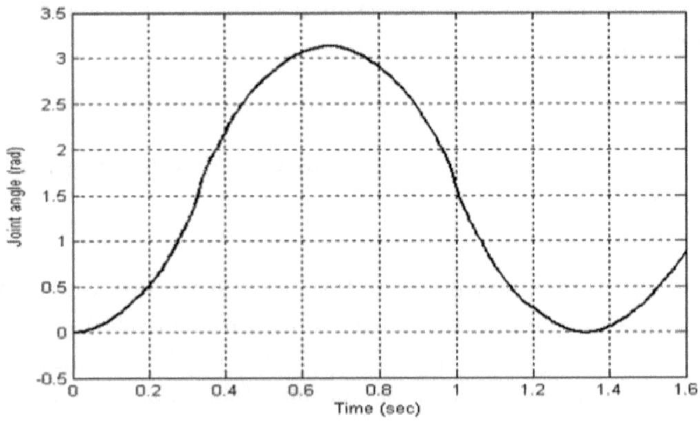

(b) Joint 2

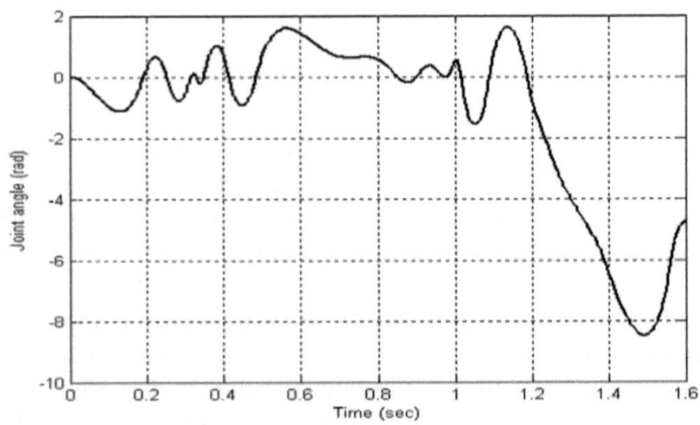

(c) Joint 3

Fig. 4.2 Free-fall simulation of the PUMA robot

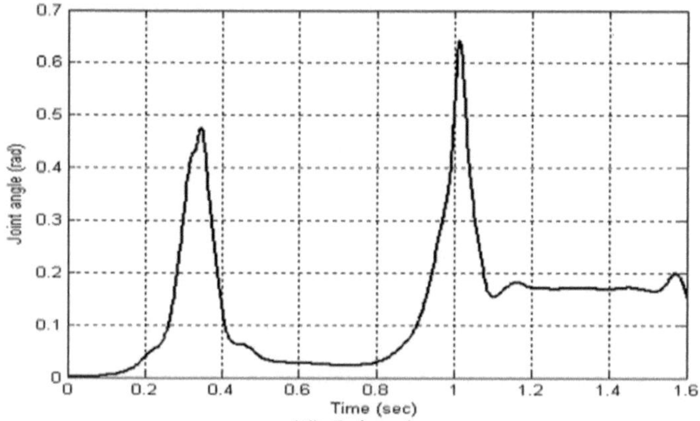

(d) Joint 4

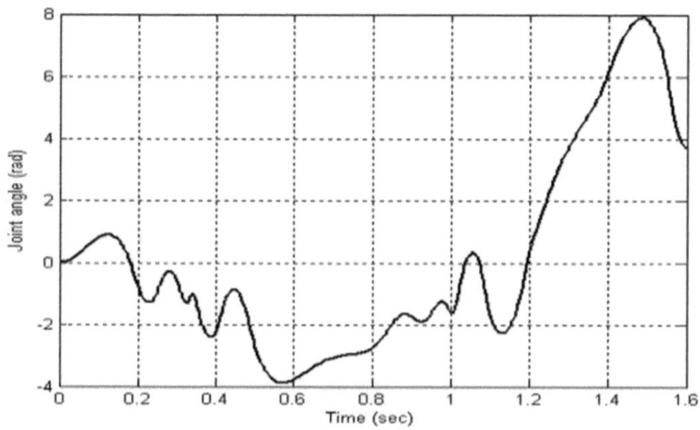

(e) Joint 5

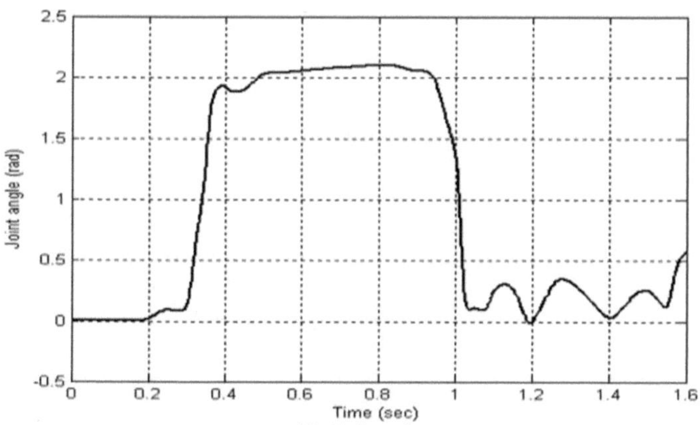

(f) Joint 6

Fig. 4.2 (continued)

Fig. 4.3 Joint torques for the PUMA robot

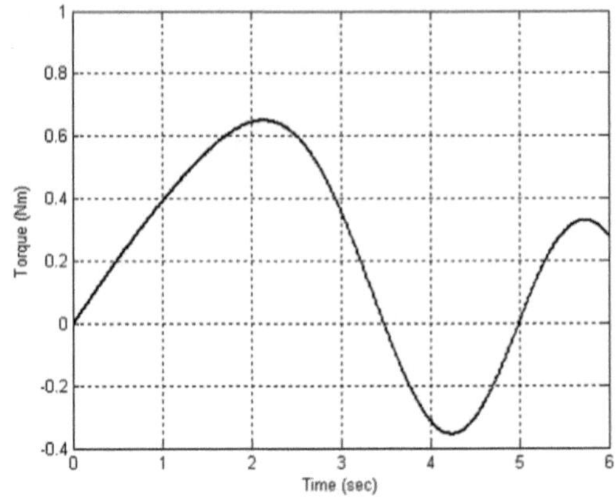

(a) Joint 1

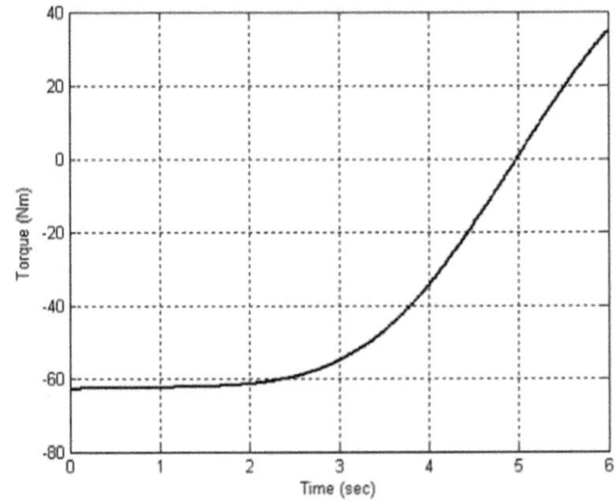

(b) Joint 2

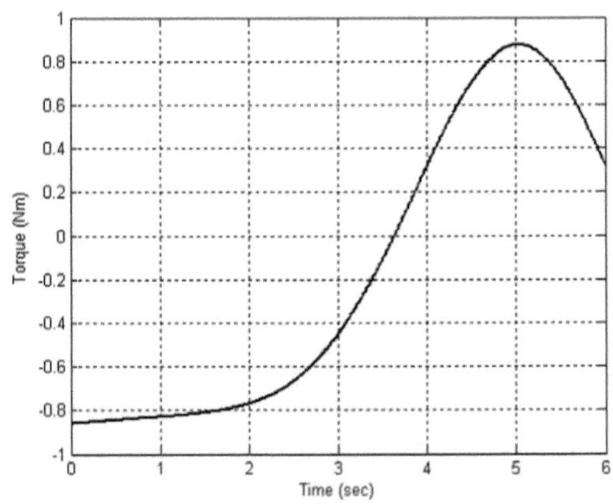

(c) Joint 3

Fig. 4.3 (continued)

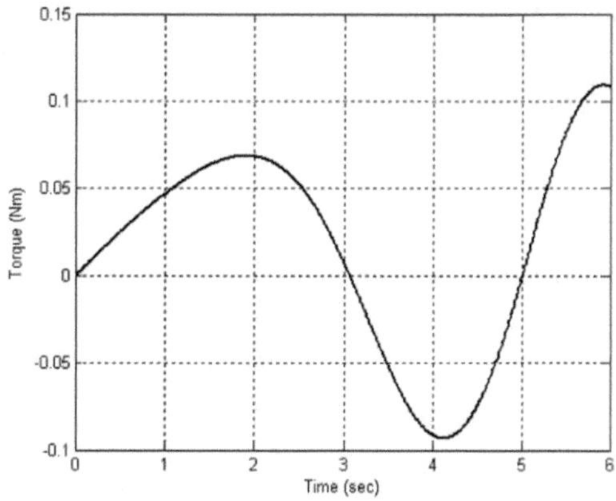

(d) Joint 4

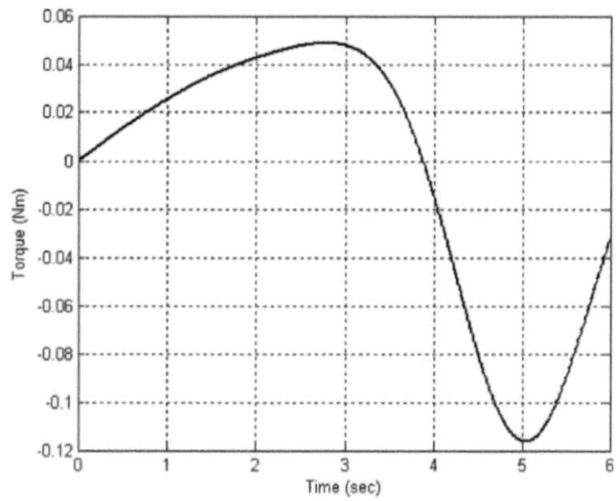

(e) Joint 5

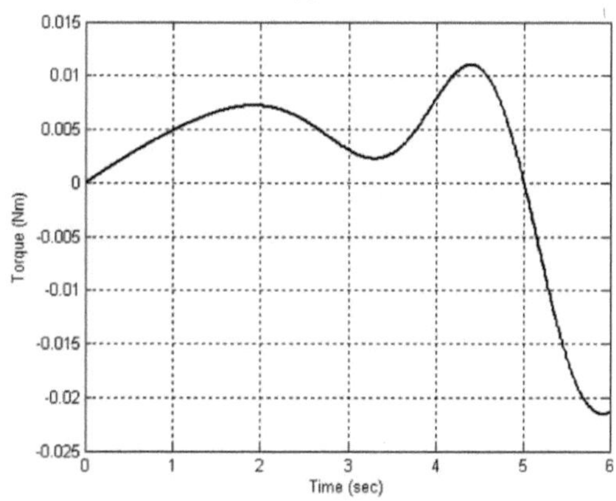

(f) Joint 6

Fig. 4.4 Comparison of joint responses of the PUMA robot

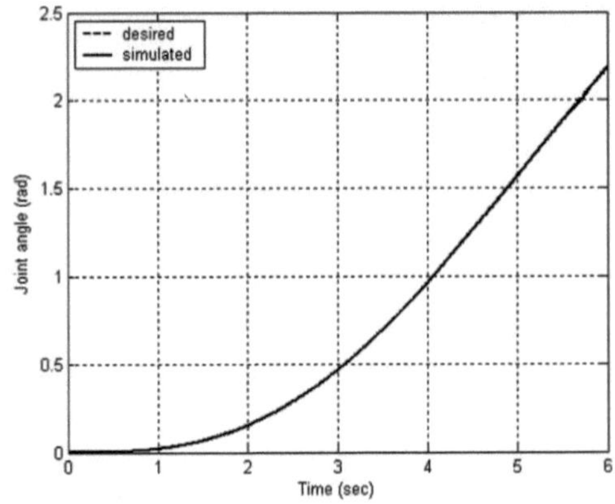

(a) Joint 1

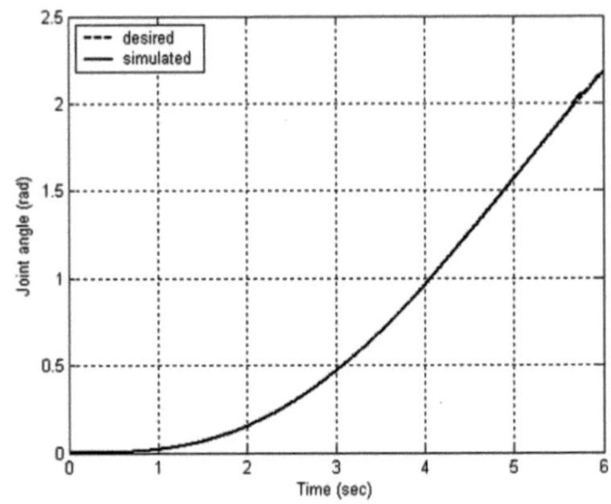

(b) Joint 2

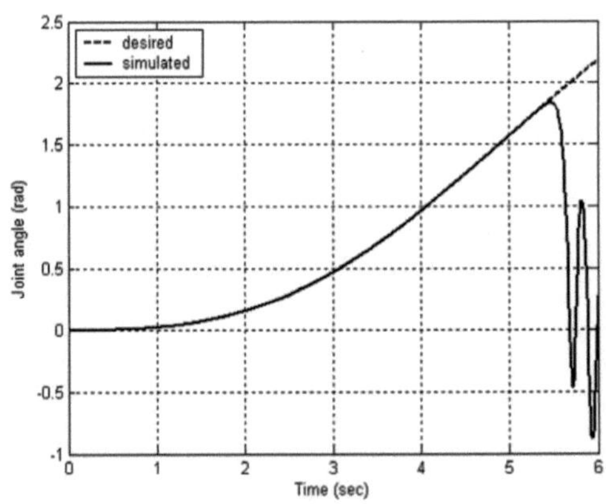

(c) Joint 3

Fig. 4.4 (continued)

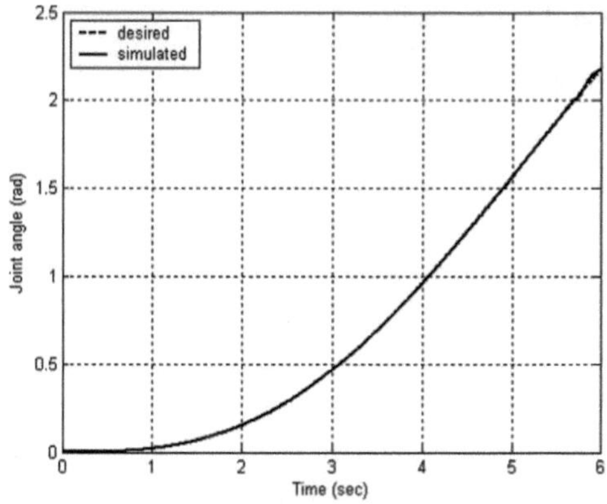

(d) Joint 4

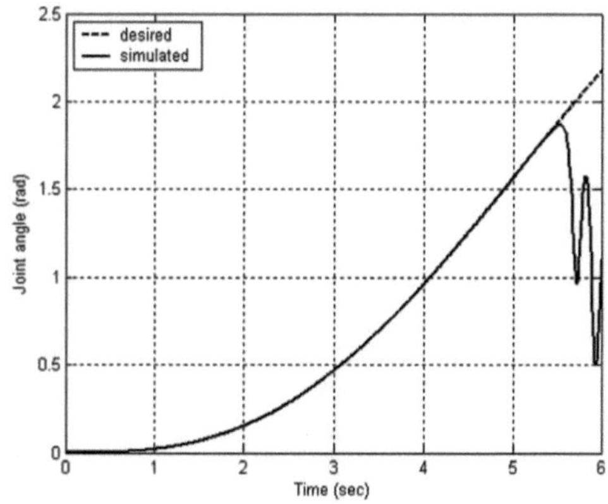

(e) Joint 5

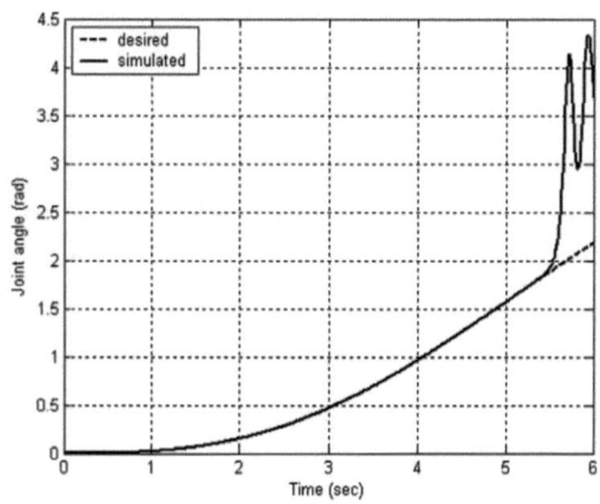

(f) Joint 6

Fig. 4.5 Comparison of joint rates of the PUMA robot

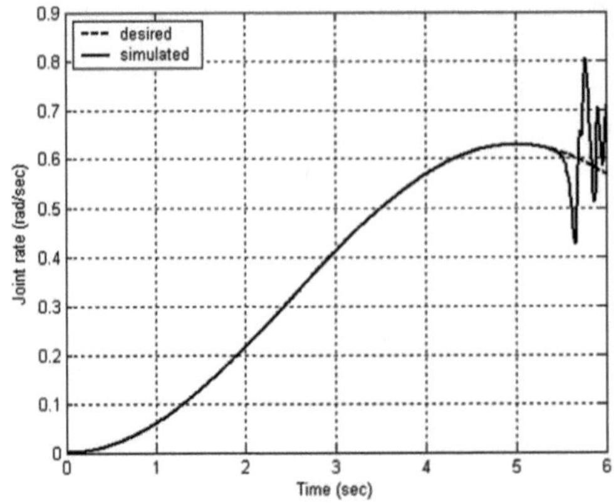

(a) Joint 1

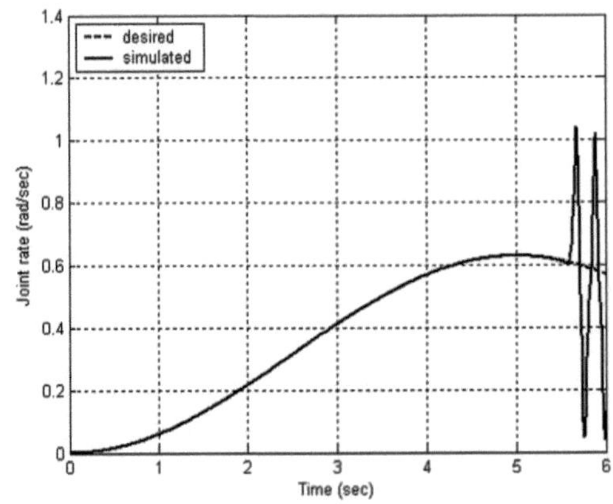

(b) Joint 2

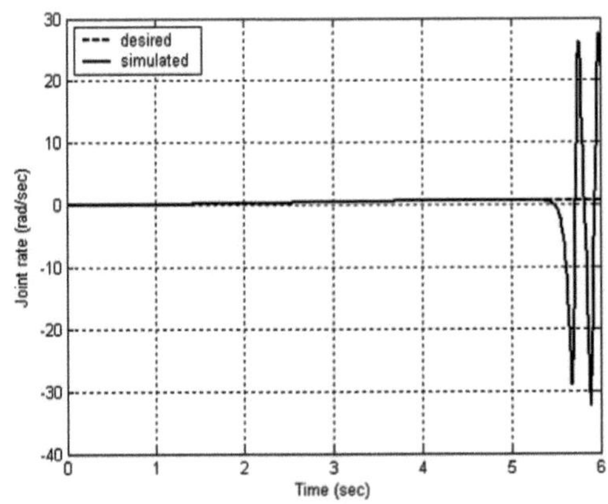

(c) Joint 3

Fig. 4.5 (continued)

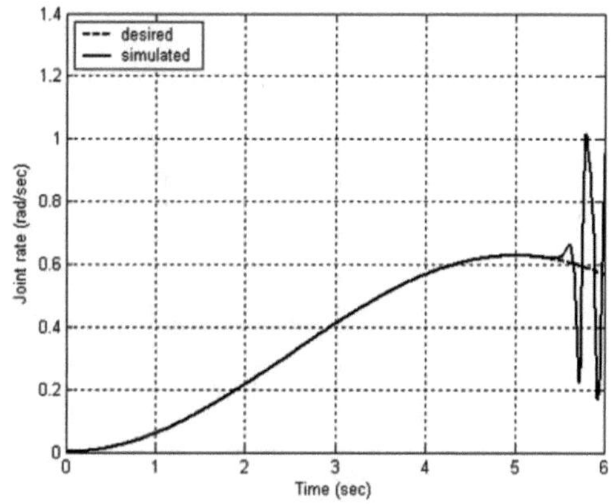

(d) Joint 4

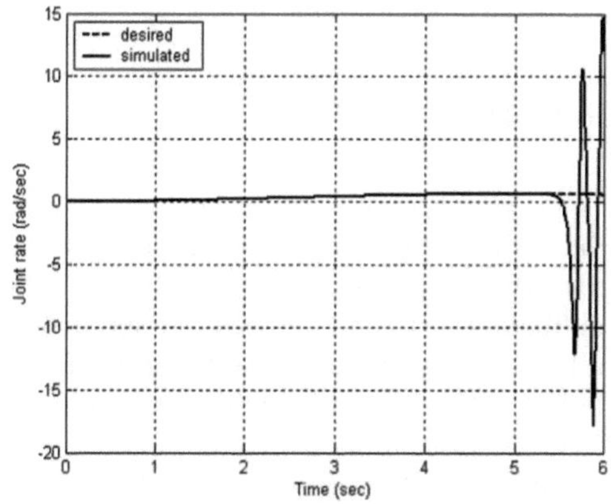

(e) Joint 5

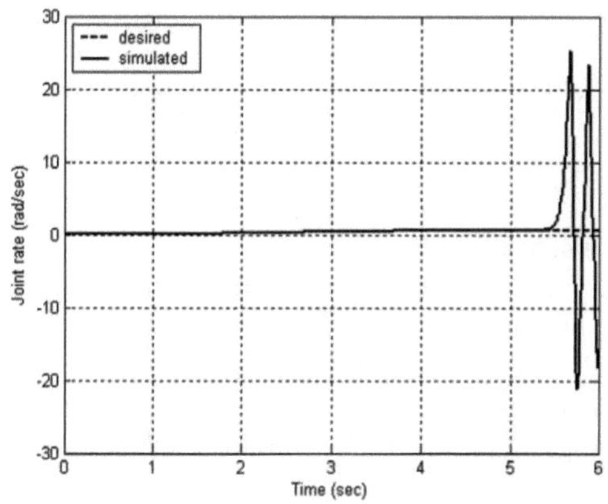

(f) Joint 6

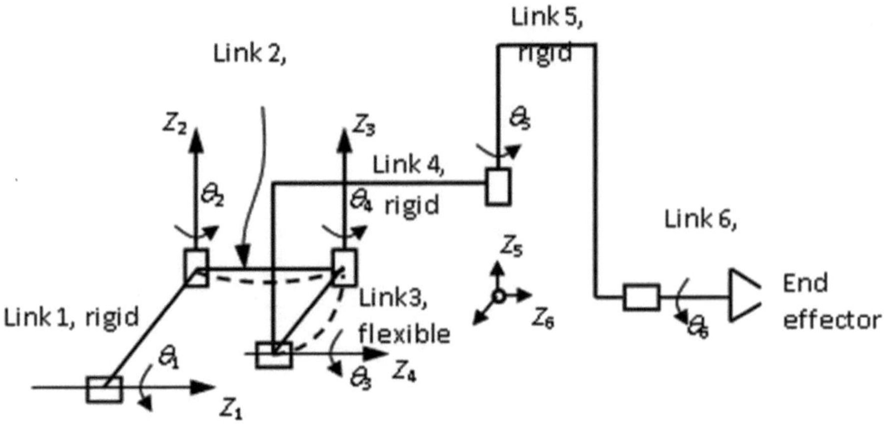

Fig. 4.6 Six-link SSRMS

4.3 Stanford Arm

So far, all the robotic arms considered consisted of only revolute pairs. In this section, the six-link robot, namely the Stanford arm, has one prismatic joint along with five revolute joints, as shown in Fig. 4.11. The DH and other parameters are given in Table 4.3. Initially, a free-fall simulation was performed similar to the PUMA robot to verify the results of this arm with one prismatic joint. The results are shown in Fig. 4.12 and verified with the algorithm of Bhangale [2]. Simulation results for 2.0 s duration are shown, as the results are unstable beyond this point. For the forced simulation study, the joint torques/force are evaluated for the trajectory shown below:

$$\theta_i = \left[\frac{\pi}{T}t - \frac{1}{2}\sin\left(\frac{2\pi}{T}t\right)\right];\ \theta_i(0) = 0 \text{ and } \theta_i(T) = \pi\ :\ \text{for } i = 1, 2, 4, 5, 6 \quad (4.3)$$

$$b_i = \left[\frac{\pi}{T}t - \frac{1}{2}\sin\left(\frac{2\pi}{T}t\right)\right];\ b_i(0) = 0 \text{ and } b_i(T) = \pi\ :\ \text{for } i = 3\ : \quad (4.4)$$

where $T = 10$.

The joint torques and force plots are shown in Fig. 4.13, whereas the corresponding simulation results are shown in Figs. 4.14 and 4.15. As an interest to see the effect of flexibility when a prismatic joint exists in a robot, the second link is assumed flexible from a practically feasible point of view. The simulation results are compared in Figs. 4.14 and 4.15, with those where all the links are rigid.

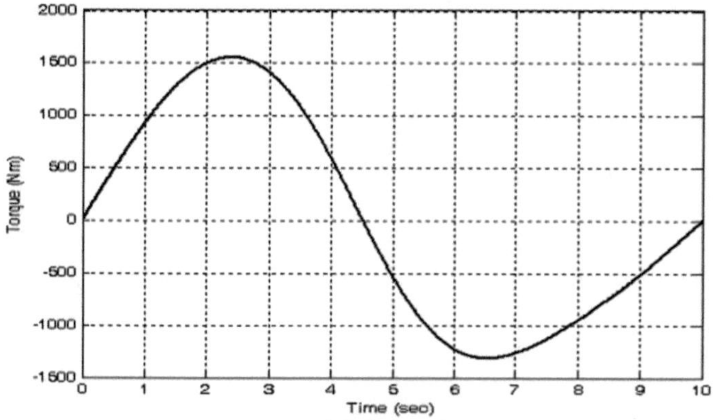

(a) Joint 1

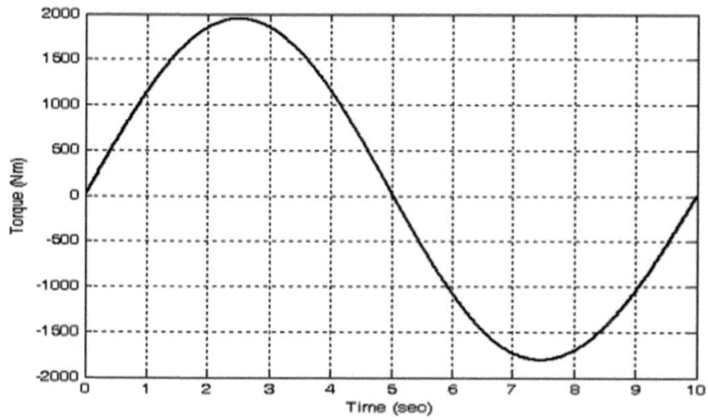

(b) Joint 2

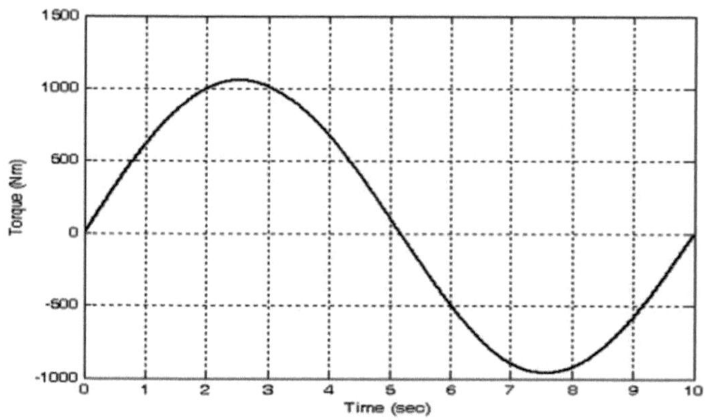

(c) Joint 3

Fig. 4.7 Joint torques for the SSRMS

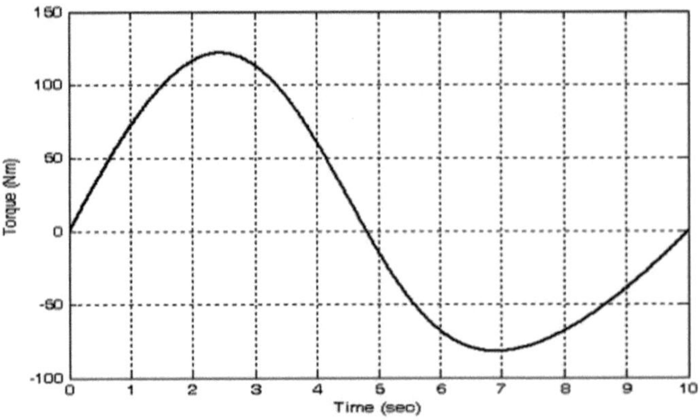

(d) Joint 4

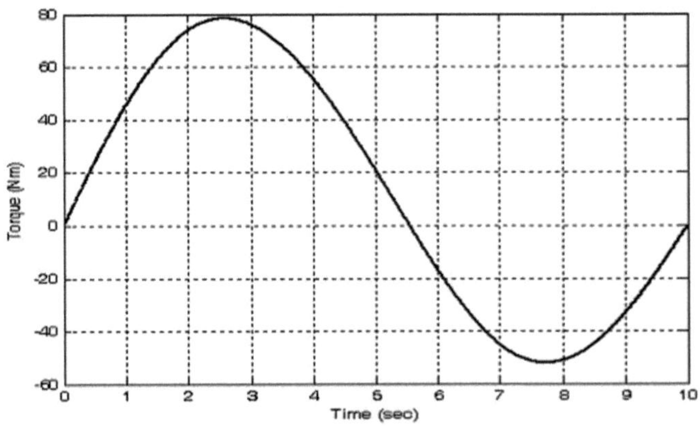

(e) Joint 5

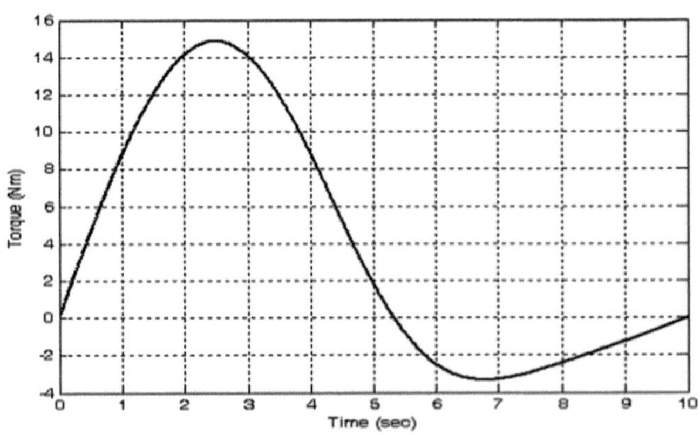

(f) Joint 6

Fig. 4.7 (continued)

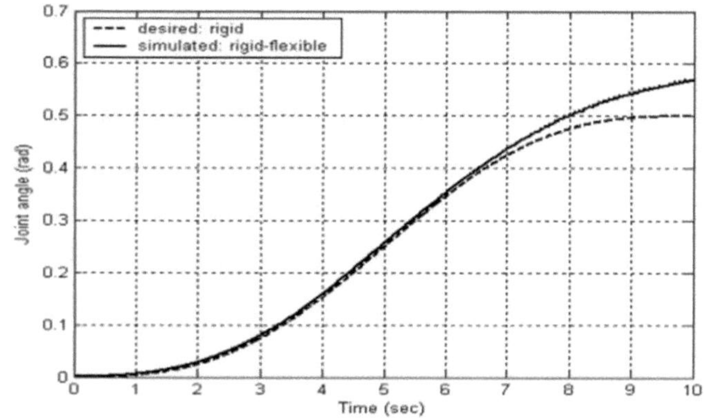

(a) Angular displacement of Joint-1

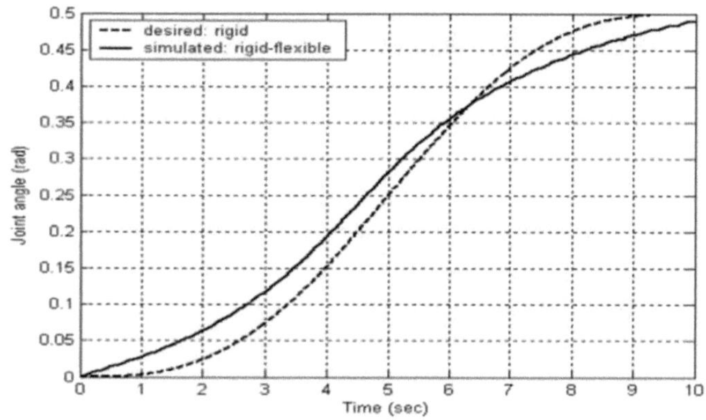

(b) Angular displacement of Joint 2

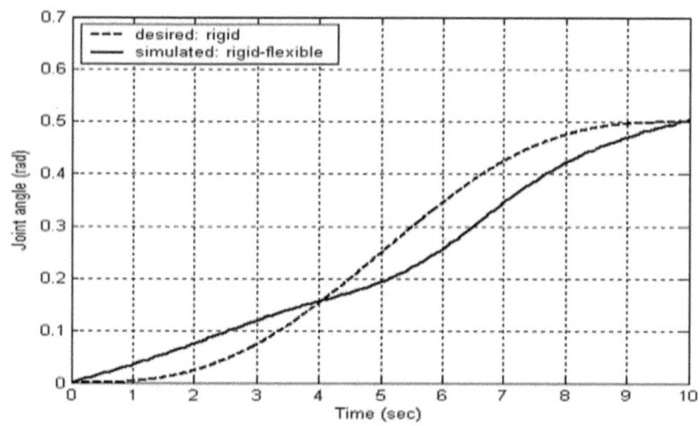

(c) Angular displacement of Joint-3

Fig. 4.8 Comparison of joint responses of the SSRMS

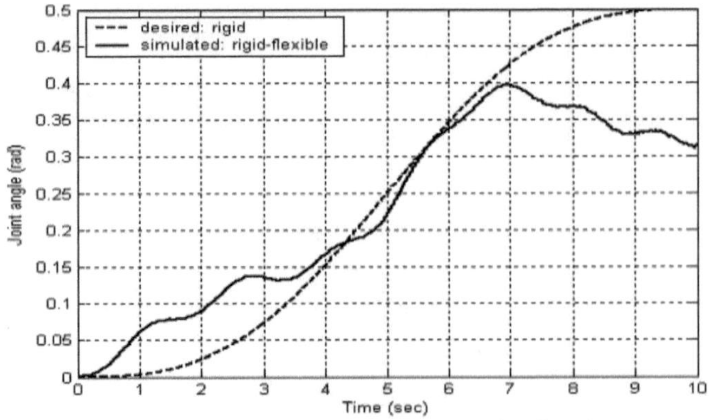

(d) Angular displacement of Joint 4

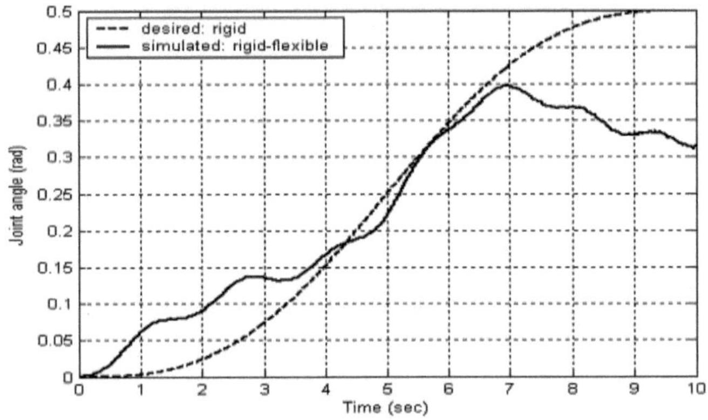

(e) Angular displacement of Joint-5

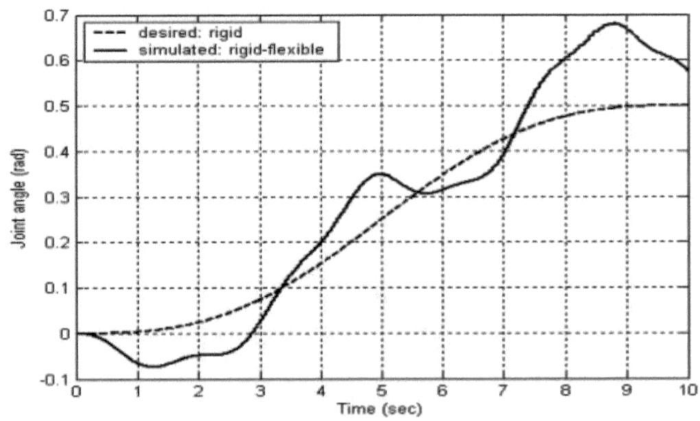

(f) Angular displacement of Joint 5

Fig. 4.8 (continued)

Fig. 4.9 Comparison of joint rates of the SSRMS

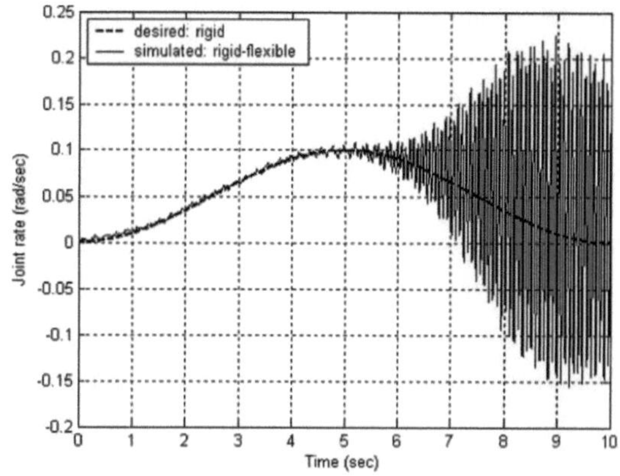

(a) Joint 1

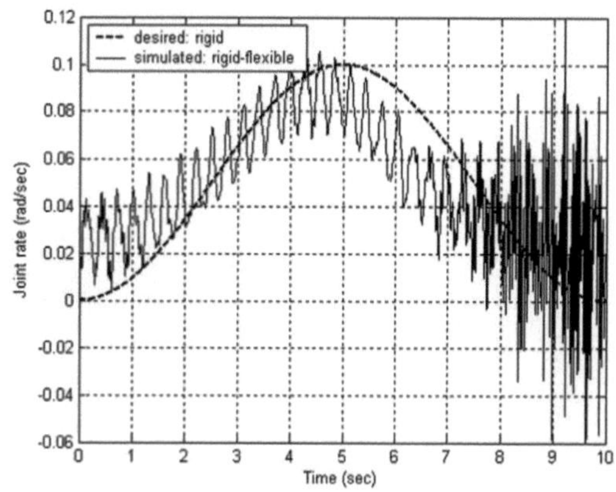

(b) Joint 2

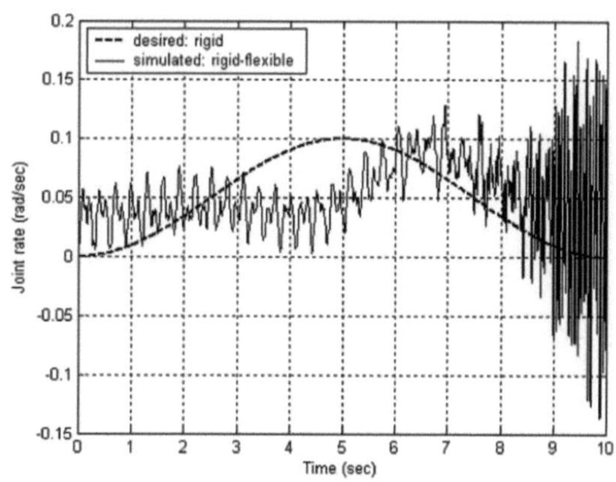

(c) Joint 3

Fig. 4.9 (continued)

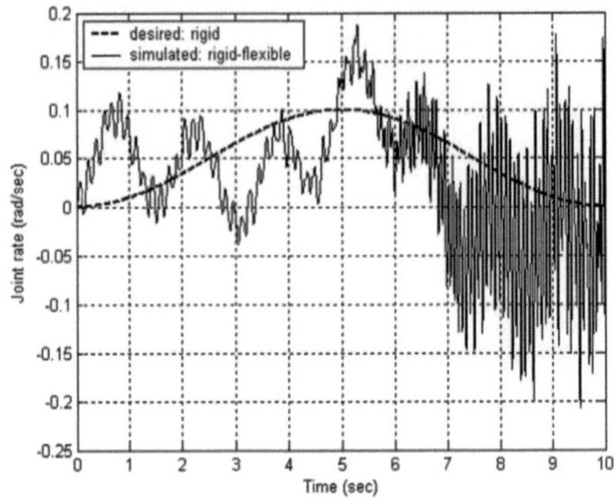

(d) Joint 4

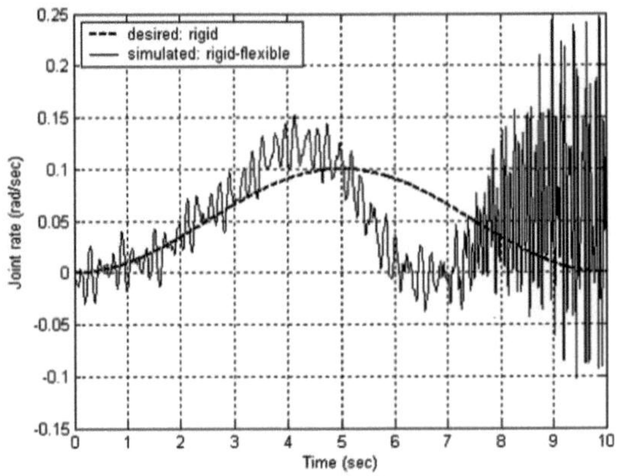

(e) Joint 5

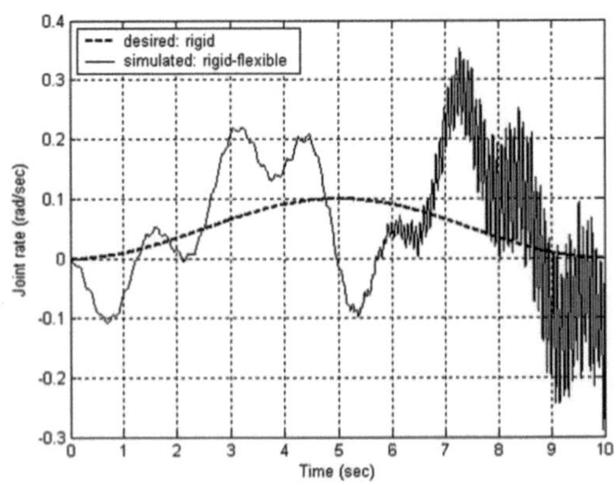

(f) Joint 6

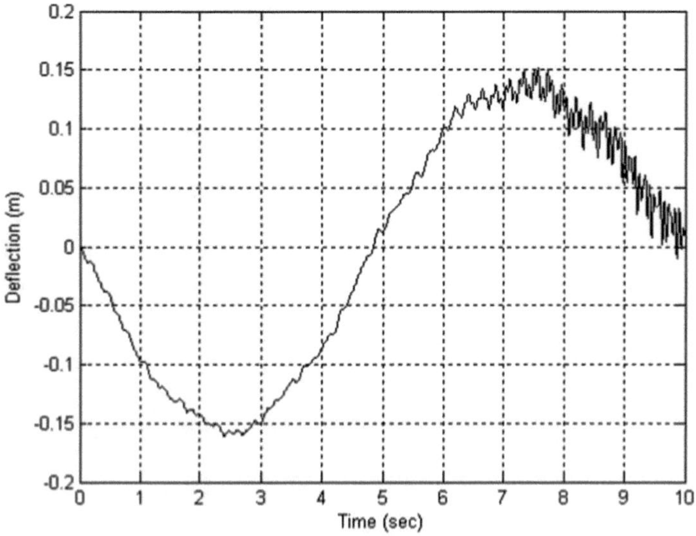

(a) Link 1

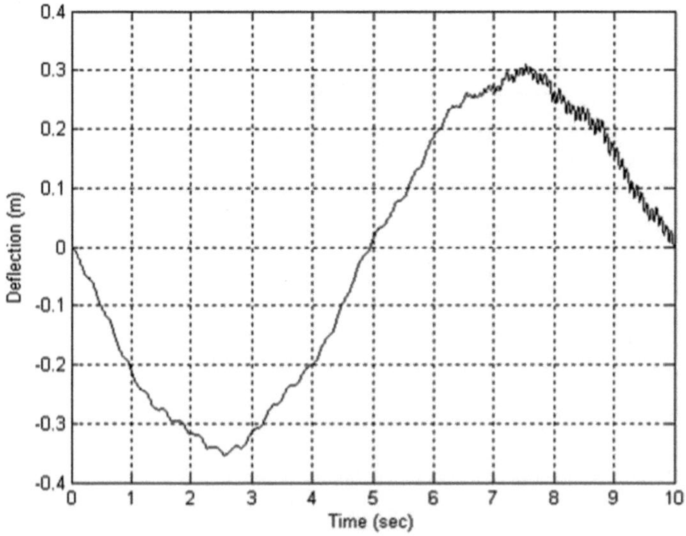

(b) Link 2

Fig. 4.10 Tip deflection of the flexible links of SSRMS

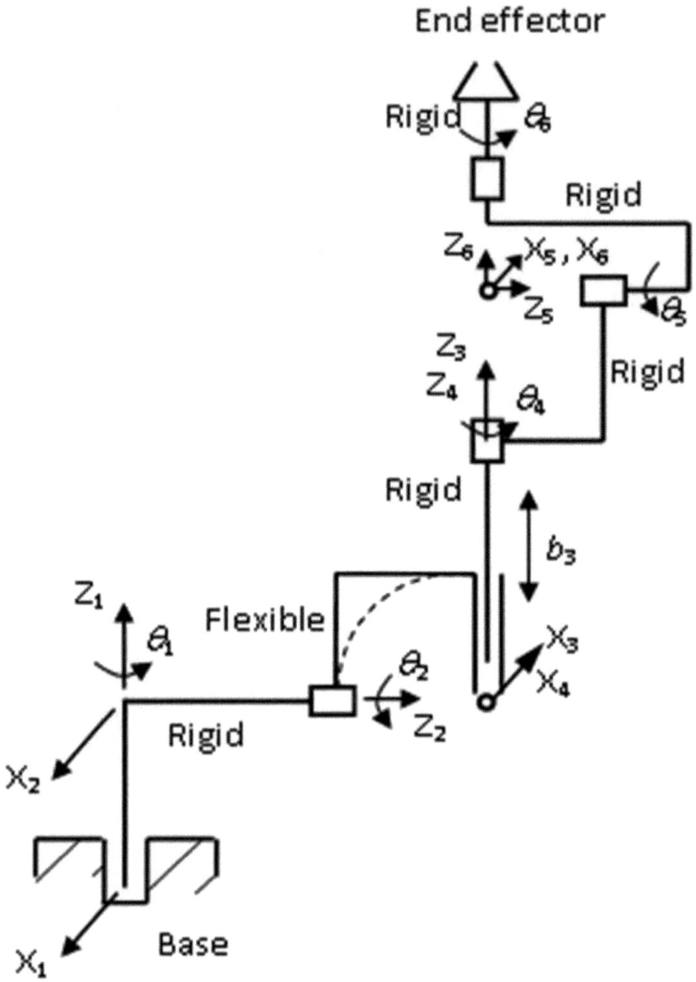

Fig. 4.11 Six-link Stanford arm

Table 4.3 DH and the other parameters of the Stanford arm

i	a_i (m)	b_i (m)	α_i (rad)	θ_i (rad)	E_iI_i (Nm^2)		m_i (Kg)	I_{zz}^{hub} (Kgm^2)
					Rigid	Rigid–flexible		
1	0	0.1	$-\pi/2$	$\theta_1[0]$[a]	–	–	9.0	0
2	0.432	0.1	$-\pi/2$	$\theta_2[0]$	–	1×10^3	6.0	0
3	1.5	$b_3[0]$	0	$\pi/2$	–	–	4.0	0
4	0	0.6	$\pi/2$	$\theta_4[0]$	–	–	1.0	0.181
5	0.2	0	$-\pi/2$	$\theta_5[0]$	–	–	0.6	0.0735
6	0	0	0	$\theta_6[0]$	–	–	0.5	0.0141

[a]The values in [and] show the initial configuration of robot

Fig. 4.12 Free-fall simulation of the Stanford arm

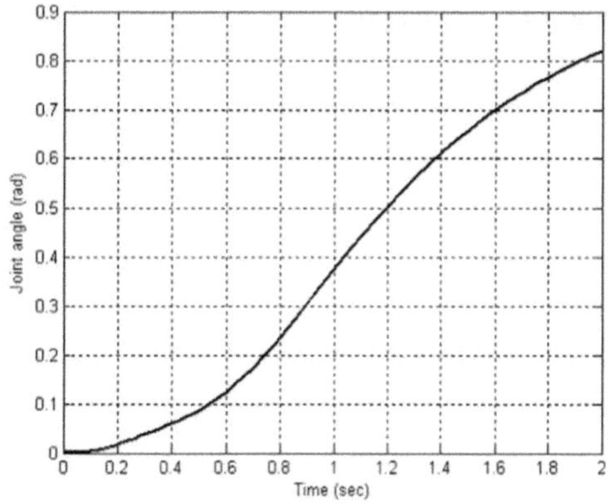

(a) Joint 1

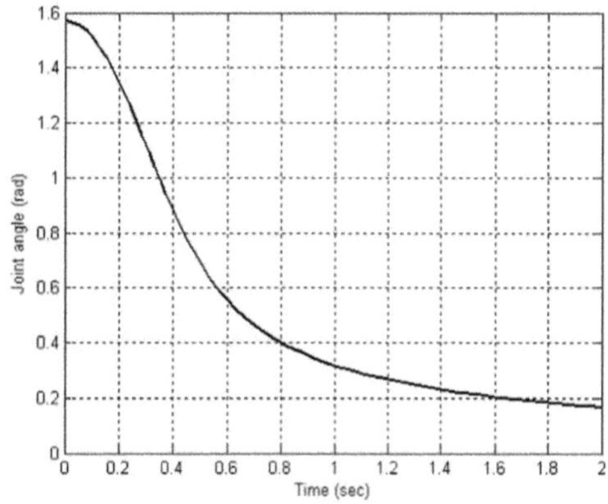

(b) Joint 2

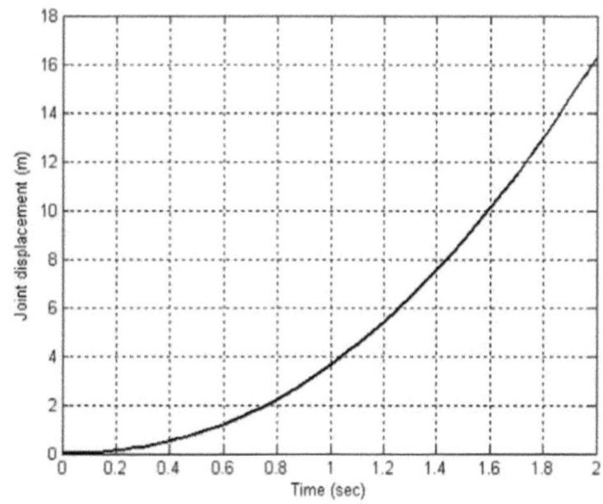

(c) Joint 3

Fig. 4.12 (continued)

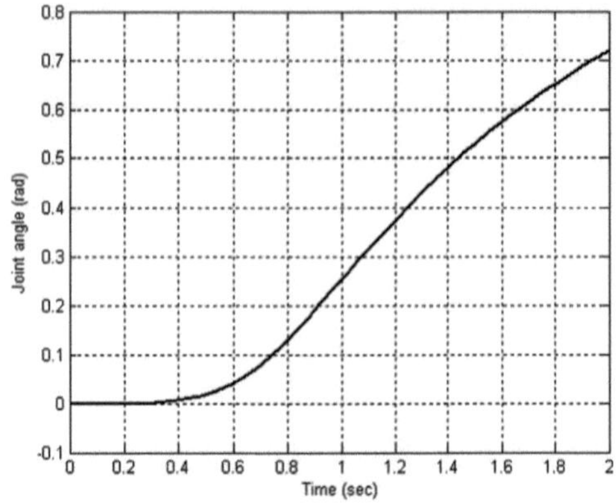

(d) Joint 4

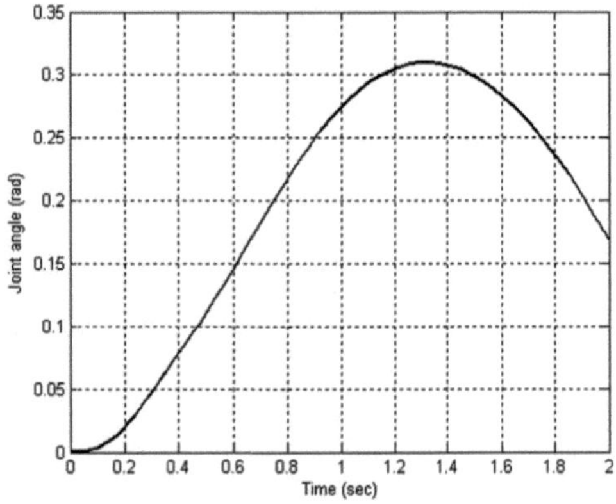

(e) Joint 5

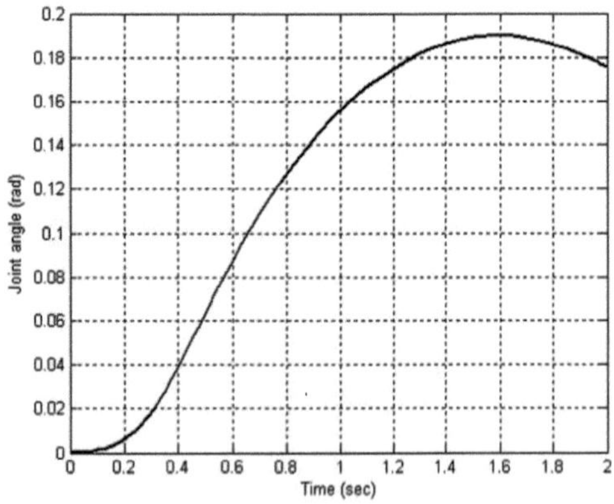

(f) Joint 6

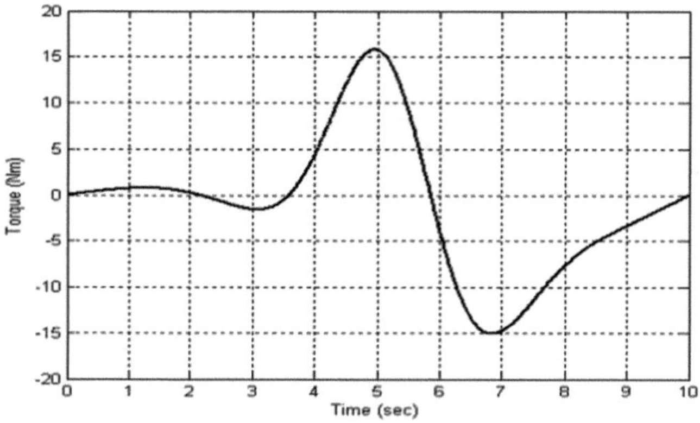

(a) Joint 1

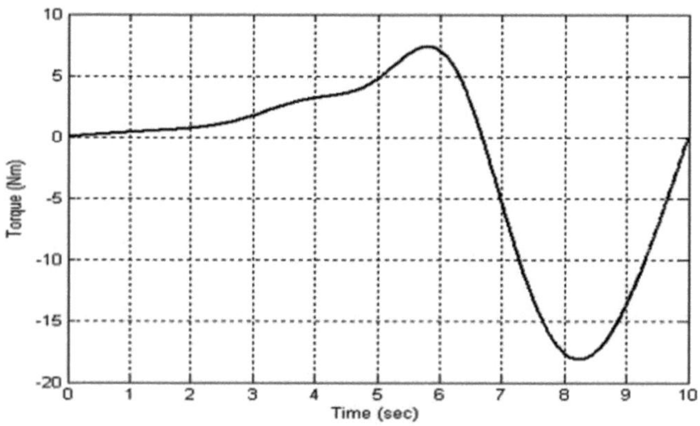

(b) Joint 2

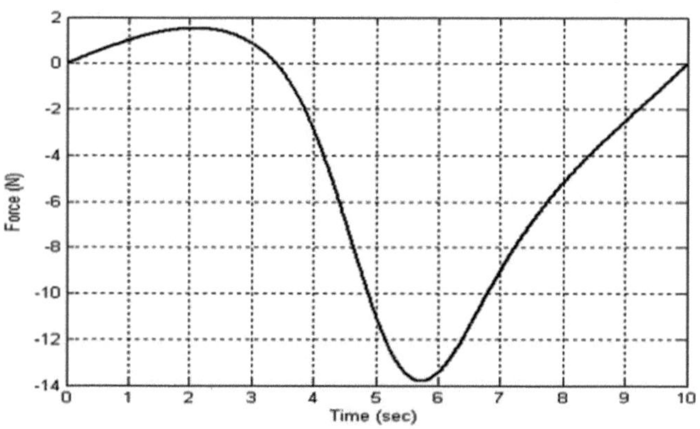

(c) Joint 3

Fig. 4.13 Joint torques and force for the Stanford arm

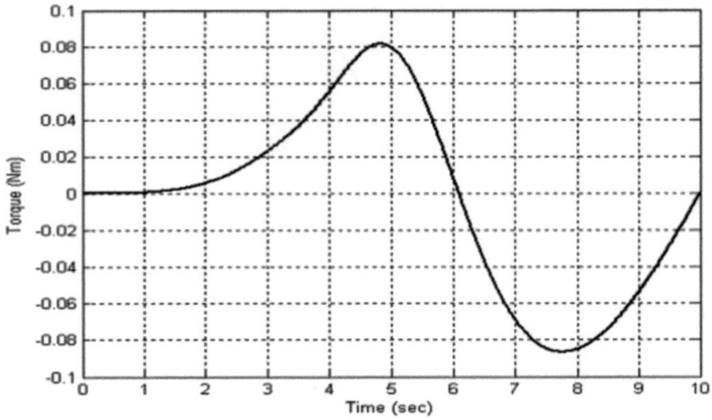

(d) Joint 4

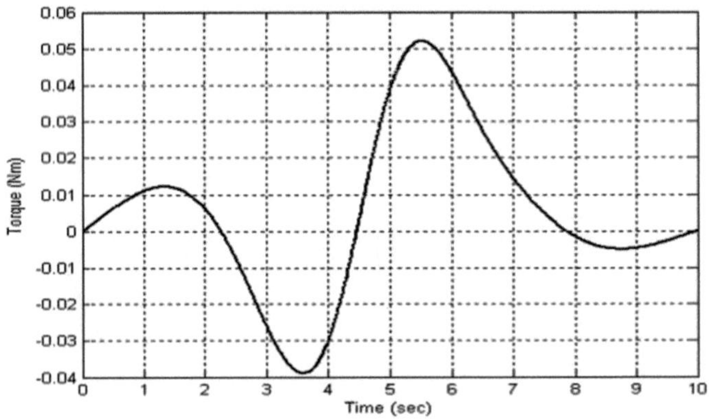

(e) Joint 5

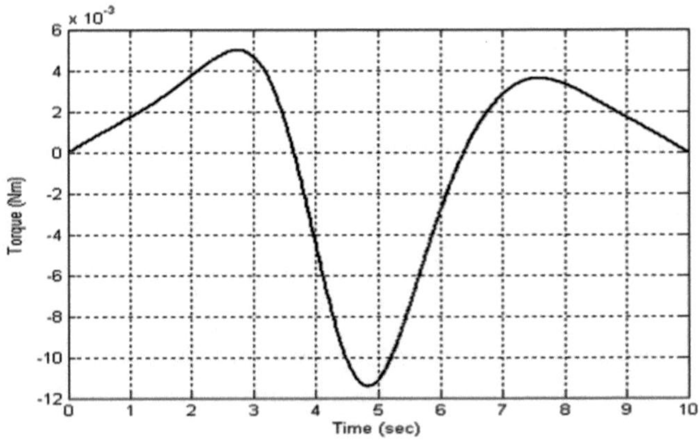

(f) Joint 6

Fig. 4.13 (continued)

Fig. 4.14 Comparison of joint responses of the Stanford arm

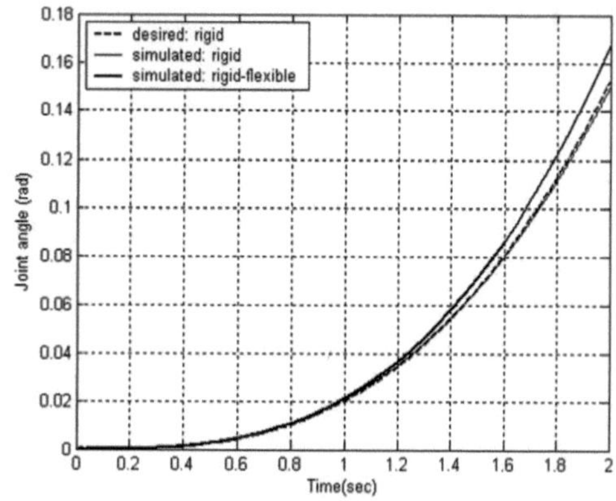

(a) Joint 1

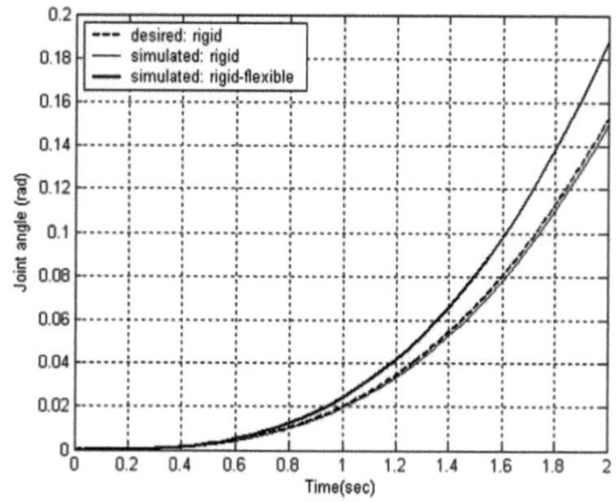

(b) Joint 2

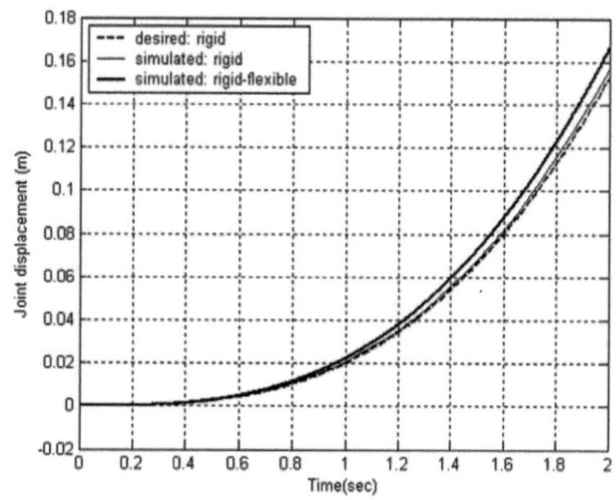

(c) Joint 3

Fig. 4.14 (continued)

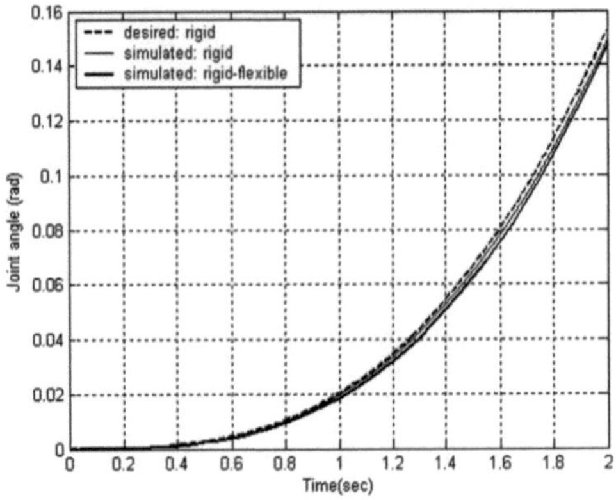

(d) Joint 4

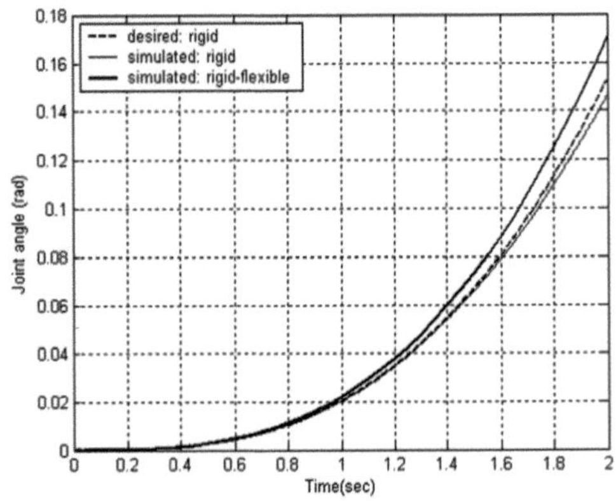

(e) Joint 5

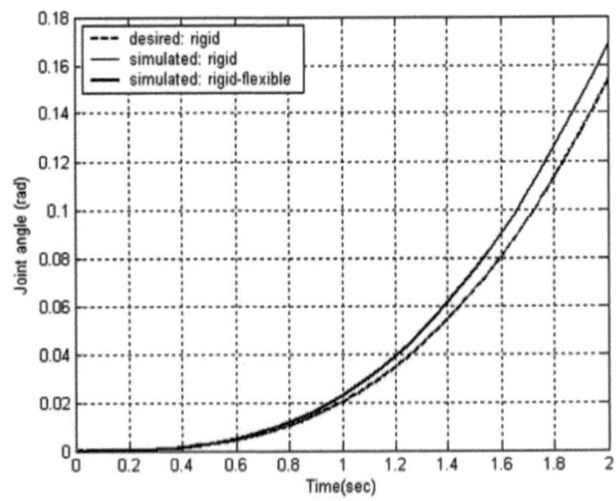

(f) Joint 6

Fig. 4.15 Comparison of joint rates of the Stanford arm

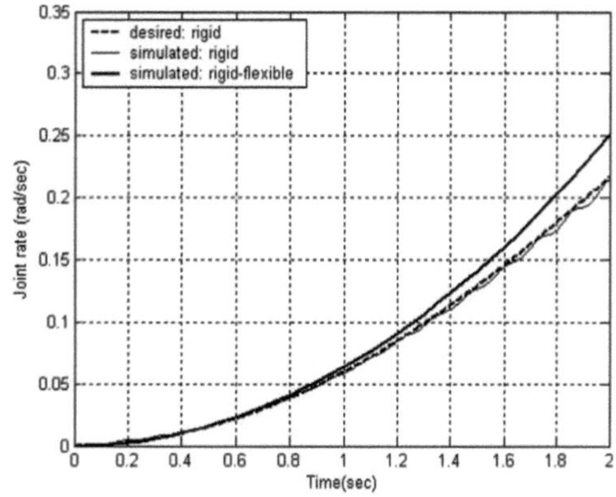

(a) Joint 1

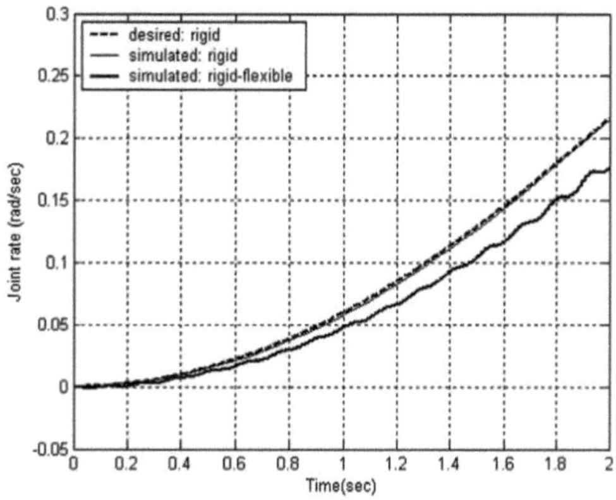

(b) Joint 2

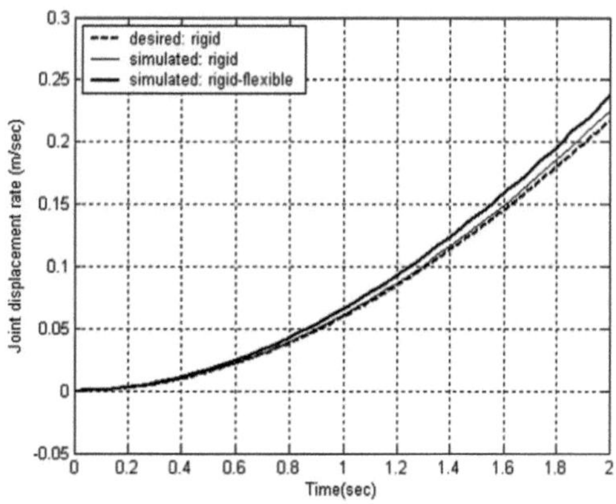

(c) Joint 3

Fig. 4.15 (continued)

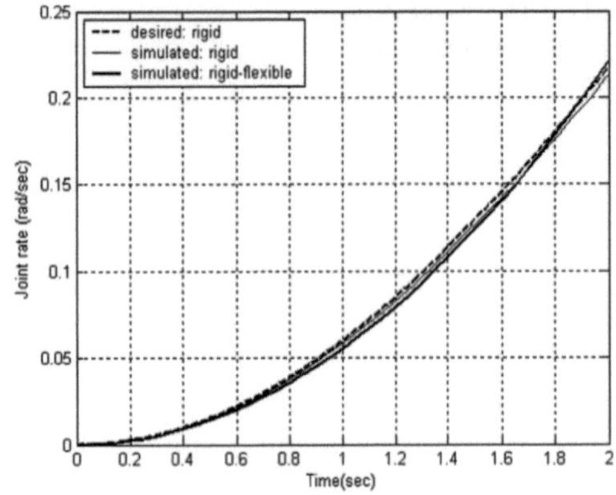

(d) Joint 4

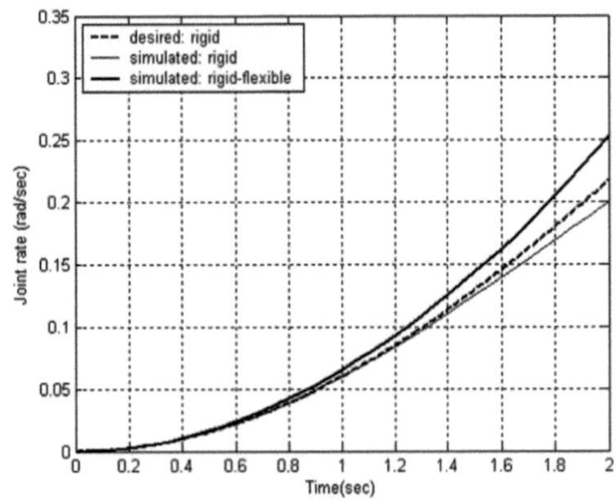

(e) Joint 5

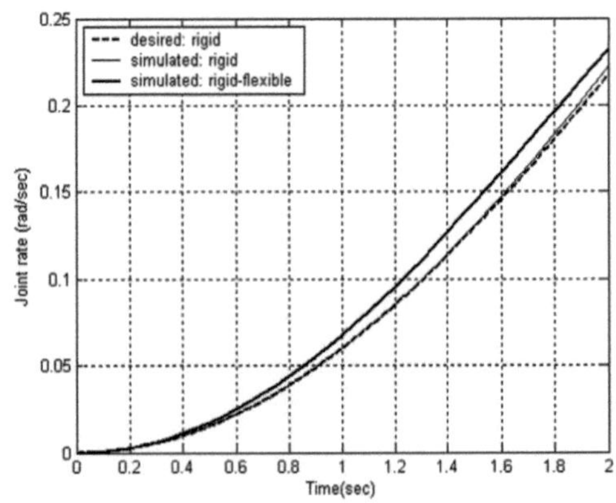

(f) Joint 6

4.4 Summary

This chapter presents the simulation results of various six-link spatial robots to validate the proposed dynamic modeling technique and the forward dynamics algorithms. Using the examples, the verification of the algorithm is carried out along with the study of the link flexibility. In some cases, e.g., Fig. 4.5, where simulation fails after 5.4 s, the reason has been attributed to numerical stability. This phenomenon of the algorithm is not robot or configuration specific. However, in certain configurations of robots, the effect of numerical instability could be more due to the ill-conditioning of associated matrices. These aspects are studied in Chap. 6. Hence, no further discussion is given here.

References

1. Angeles J (2003) Fundamentals of robotic mechanical system. Springer, New York
2. Bhangale P (2004) Dynamics-based modeling, computational complexity, and architecture selection of robot manipulators. Ph.D. Thesis, Indian Institute of Technology Delhi, India
3. Cyril X (1988) Dynamics of flexible-link manipulators. McGill University, Canada
4. Sharf I, Damaren C (1992) Simulation of flexible-link manipulators: basis functions and nonlinear terms in the motion equations, pp 1956–1962
5. Yoshikawa T (1990) Foundations of robotics: analysis and control. MIT Press

Summary

Part II
Dynamic Modeling of Closed-Loop Systems

Chapter 5
Dynamics of Closed-Loop Systems

In this chapter, dynamic modeling of closed-loop rigid–flexible systems comprising of both rigid and flexible links is presented. For this, the DeNOC-based formulation of motion for serial-chain, rigid–flexible multibody systems, which are introduced in the previous chapter, were used. The flexible links were modeled as Euler–Bernoulli beams, and their deformations were discretized using the assumed mode method (AMM) [4]. In order to be able to use the dynamic formulation for serial chains, the closed-loop system was cut at appropriate joints to form several open-loop, serial-chain subsystems. Then using the equations of motion of those subsystems, the dynamic equations of the closed-loop systems were obtained. The proposed approach allows one to solve closed-chain systems using the knowledge of serial-chain dynamics only. The concern is that one has to solve differential-algebraic equations (DAE), which may lead to constraint violations if the proper solver is not chosen for the forward dynamic simulation.

In this book, forward dynamics is solved using the reverse Gaussian elimination (RGE) of the generalized inertia matrix (GIM) [6, 7], appearing in the dynamic equations of motion of the serial-chain subsystems at hand. The RGE of the GIM was used to determine the joint accelerations, which led to numerically stable algorithms [1]. For the inverse dynamics, an alternate approach is proposed for the closed-loop system in which any joint in the system can be cut, unlike the methodology provided by [2]. The formulation is illustrated with several single and multiloop multibody systems. Each system considered several combinations of rigid and flexible links. In order to verify the results, RecurDyn software was used, which discretizes each flexible link using the finite element method (FEM).

P. V. Nandihal et al., *Dynamics of Rigid-Flexible Robots and Multibody Systems*, Intelligent Systems, Control and Automation: Science and Engineering 100,
https://doi.org/10.1007/978-981-16-2798-9_5

5.1 Dynamics of Closed-Loop Systems

The closed-loop system was analyzed by first cutting appropriate joints to form subsystems of simple open-loop serial-chain subsystems. Even though there is no restriction in deciding which joint(s) of the closed-loop system are to be cut, one may prefer to use the following guidelines to cut the joints:

1. cut higher degrees-of-freedom (DOF) joints in a system, e.g., spherical, universal, etc., to reduce the number of dynamic equations of motion (EOM);
2. cut joints such as to make the resulting open subsystems with minimum DOF.

In case the above guidelines are not followed, one may still derive the above advantages by segregating the EOMs corresponding to actuated and unactuated joints, as explained in Sect. 5.1.2. The cutting of a joint in a closed-loop system is illustrated by a four-bar mechanism as shown in Fig. 5.1. As per the guideline in (b), the system was divided into two open subsystems forming 2-DOF serial-chain manipulator and one-DOF manipulator by cutting joint 4 (the same could be achieved by cutting joint 2 also), instead of cutting joint 1 or 3 which would have resulted in only one 3-DOF manipulator. In the latter case, EOMs would be three instead of two and one EOMs in the former case, which is known to have led to computationally efficient algorithms [3].

The cut-opened joint was then substituted with suitable constraint forces denoted with λ's, which are known as Lagrange multipliers. These multipliers were treated as external forces to the resulting open-loop Subsystems-I and II. The two sets of equations of motion for serial-chain subsystems were then derived using the methodology presented in Sects. 2.2 and 3.2. These equations, along with the constraint forces, are rewritten below in compact form as

$$\mathbf{I}\ddot{\mathbf{q}} = \boldsymbol{\tau} + \boldsymbol{\tau}^{\lambda}$$
$$\text{Were, } \mathbf{I} \equiv \text{diag.}\begin{bmatrix} \mathbf{I}^{I} & \mathbf{O} \\ \mathbf{O} & \mathbf{I}^{II} \end{bmatrix}, \mathbf{q} \equiv \begin{bmatrix} \mathbf{q}^{I} \\ \mathbf{q}^{II} \end{bmatrix}, \text{ and } \boldsymbol{\tau} \equiv \begin{bmatrix} \boldsymbol{\tau}^{I} \\ \boldsymbol{\tau}^{II} \end{bmatrix} \tag{5.1}$$

Moreover, $\boldsymbol{\tau}^{\lambda} = \mathbf{J}^{T}\boldsymbol{\lambda}$ is the constraint forces due to the cut joints in which $\mathbf{J}$ is the constrained Jacobian matrix defined in the way such that $\mathbf{J}\dot{\mathbf{q}} = \mathbf{0}$. In Eq. (5.1), $\mathbf{I}$

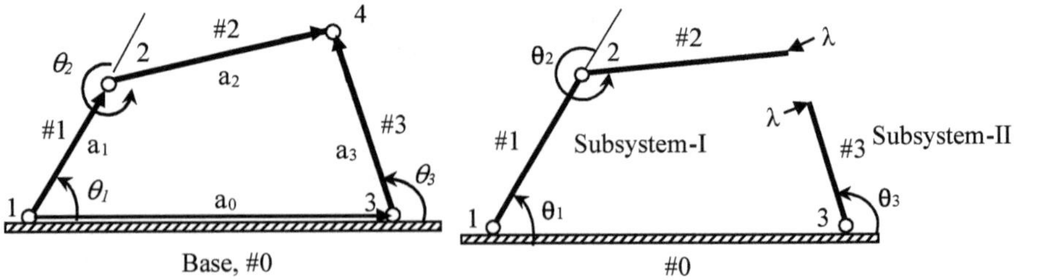

Fig. 5.1 Four-bar mechanism

is the generalized inertia matrix (GIM) of the closed-loop system at hand whose size corresponds to the generalized coordinates $\mathbf{q}$. Similarly, $\boldsymbol{\tau}$ is the vector of generalized forces of the system.

5.1.1 *Forward Dynamics*

Forward dynamics calculates the joint trajectories, i.e., the joint accelerations $\ddot{\mathbf{q}}$, the joint velocities $\dot{\mathbf{q}}$, and the joint angles $\mathbf{q}$, for a given set of torques/forces $\boldsymbol{\tau}$ at the joints. For the flexible multibody systems, the vectors $\mathbf{q}$, $\dot{\mathbf{q}}$, and $\ddot{\mathbf{q}}$ include the modal coordinates other than the joint motions. The associated constraint equations corresponding to a closed-loop system can be derived from the loop-closure constraints as

$$\mathbf{J}\dot{\mathbf{q}} = \mathbf{0} \text{ and } \mathbf{J}\ddot{\mathbf{q}} + \dot{\mathbf{J}}\dot{\mathbf{q}} = \mathbf{0} \tag{5.2}$$

where $\mathbf{J}$ is the Jacobian matrix of the system. Then, Eq. (5.1) along and the constraint equations, Eq. (5.2) is written in a matrix–vector form as

$$\begin{bmatrix} \mathbf{I} & \mathbf{J}^T \\ \mathbf{J} & \mathbf{O} \end{bmatrix} \begin{bmatrix} \ddot{\mathbf{q}} \\ \boldsymbol{\lambda} \end{bmatrix} = \begin{bmatrix} \boldsymbol{\tau} \\ -\dot{\mathbf{J}}\dot{\mathbf{q}} \end{bmatrix} \text{ or } \mathbf{I}_s\ddot{\mathbf{q}}_s = \mathbf{0} \tag{5.3}$$

where, $\mathbf{I}_s$ contains the GIM of the system, and the Jacobian $\mathbf{J}$, and $\boldsymbol{\varphi}$ is the right-hand side of Eq. (5.3). To solve for the accelerations $\ddot{\mathbf{q}}$ and the vector of Lagrange multipliers $\boldsymbol{\lambda}$, the GIM $\mathbf{I}_s$ needs to be inverted. Numerical integration is then generally carried out using the *ode45* function of MATLAB to obtain $\dot{\mathbf{q}}$ and $\mathbf{q}$ from calculated $\ddot{\mathbf{q}}$. This book will, however, introduce the reverse Gaussian elimination (RGE)-based forward dynamic algorithm in Chaps. 2 and 3, which will be compared against few classical approaches.

5.1.2 *Inverse Dynamics*

The determination of the joint torques/forces $\boldsymbol{\tau}$ for known trajectories, i.e., $\ddot{\mathbf{q}}$, $\dot{\mathbf{q}}$, $\mathbf{q}$ of the robot links, is known as inverse dynamics. For the inverse dynamics, the approach used here is similar to the one used in [2]. In the approach proposed by [2], the closed-loop system is cut-open in a manner such that at least one subsystem is *determinate.* A determinate subsystem is one whose number of equations of motion is equal to the number of unknowns associated with it. For example, in Fig. 5.1, if link #3 is driven by an actuator located at joint 3, then the Subsystem-I is determinate as its two dynamic equations of motion (EOM) corresponds to its two degrees of freedom (DOF) allows one to compute two components of the unknown Lagrange multiplier vector $\boldsymbol{\lambda}$. On the other hand, Subsystem-II is indeterminate as it has one-DOF, which can have one EOM but three unknowns, namely two components of $\boldsymbol{\lambda}$ and one driving torque at joint 3. Using the results calculated from the determinate subsystems, say, Subsystem-I of Fig. 5.1, the other subsystem unknowns, i.e., Subsystem-II of Fig. 5.1, can be calculated. If this is not the case, then more than one subsystem needs to be combined to form a larger determinate system. This is the case if the driving actuator is located at joint 1 of Fig. 5.1. Such a choice of joints to be cut to get at least one determinate subsystem is not straightforward. Therefore, an alternative approach is proposed in this book, in which no restriction on cutting a joint is put. One can cut any joint of a closed-loop system to obtained open-loop serial-chain subsystems. Here, the Lagrange multipliers are solved uniquely using the passive joint information of all the subsystems before their active joint torques/forces are computed. It provides the same computational benefit as determinate subsystems but without any restriction on cut joints.

The steps are demonstrated using the same planar four-bar mechanism as shown in Fig. 5.1. To perform the inverse dynamics, the system is cut at joint 4, which divides the system into a two-link subsystem and another one-link subsystem. If the joint 1 is assumed to be actuated, both the resulting subsystems are indeterminate, i.e., more unknowns associated with them compared to their number of equations. The Subsystems-I and II are 2-DOF and 1-DOF, respectively, and possess two and one equations of motion. The unknowns associated with Subsystem-I are two components of the Lagrange multiplier vector, i.e., λ_x, λ_y, and the driving torque τ, whereas Subsystem-II has only two components of Lagrange multipliers, i.e., λ_x, λ_y. Note here that λ_x and λ_y denote the reactions at joint 4. The equations of motion of both the subsystems can be rewritten similar to Eq. (5.1) as

$$\begin{aligned} \mathbf{I}^I\ddot{\mathbf{q}}^I + \mathbf{h}^I &= \boldsymbol{\tau}^I + \mathbf{J}^{I^T}\boldsymbol{\lambda} \\ I^{II}\ddot{q}^{II} + h^{II} &= \tau^{II} + \mathbf{J}^{II^T}\boldsymbol{\lambda} \end{aligned} \tag{5.4}$$

where $\boldsymbol{\tau}^I \equiv \begin{bmatrix}\tau \ 0\end{bmatrix}^T$, as only one joint is active and another one is passive in Subsystem-I, and $\tau^{II} = 0$, as the only joint in Subsystem-II is passive. Next, the actuated joint vector ($\boldsymbol{\eta}$) and unactuated joint vector ($\boldsymbol{\beta}$) for Subsystems-I and II are written as follows:

$$\begin{aligned} \boldsymbol{\eta}^I = \begin{bmatrix}1 \ 0\end{bmatrix}^T \text{ and } \boldsymbol{\eta}^{II} = [0] \\ \boldsymbol{\beta}^I = \begin{bmatrix}0 \ 1\end{bmatrix}^T \text{ and } \boldsymbol{\beta}^{II} = [1] \end{aligned} \tag{5.5}$$

The Lagrange multipliers associated with each subsystem are then be evaluated by pre-multiplying the EOM corresponding to each subsystem by $(\boldsymbol{\beta}^k)^T$, for $k = \text{I}$ and II. The resulting equations without the driving torque term $\boldsymbol{\tau}$ as $\boldsymbol{\beta}^{I^T}\boldsymbol{\tau}^I = 0$ are as follows:

$$(\boldsymbol{\beta}^I)^T(\mathbf{0}^I)^* = (\boldsymbol{\beta}^I)^T(\mathbf{J}^I)^T\boldsymbol{\lambda} \text{ and } (\boldsymbol{\beta}^{II})^T(\mathbf{0}^{II})^* = (\boldsymbol{\beta}^{II})^T(\mathbf{J}^{II})^T\boldsymbol{\lambda} \tag{5.6}$$

where $(\mathbf{0}^I)^* \equiv \mathbf{I}^I\ddot{\mathbf{q}}^I + \mathbf{h}^I$ and $(\mathbf{0}^{II})^* \equiv I^{II}\ddot{q}^{II} + h^{II}$. Note that, the two sets of the equation in Eq. (5.6) correspond to the two components of $\boldsymbol{\lambda}$, namely $\boldsymbol{\lambda}_x$ and $\boldsymbol{\lambda}_y$, respectively. Combining them one can compute $\boldsymbol{\lambda}$ as follows.

$$\boldsymbol{\lambda} = (\mathbf{J}_\beta^T)^{-1}\mathbf{0}_\beta^* \tag{5.7}$$

where $\mathbf{J}_\beta^T$ and $\mathbf{0}_\beta^*$ are defined as below:

$$\mathbf{J}_\beta \equiv \begin{bmatrix} \mathbf{J}^I\boldsymbol{\beta}^I \\ \mathbf{J}^{II}\boldsymbol{\beta}^{II} \end{bmatrix} \text{ and } \mathbf{0}_\beta^* \equiv \begin{bmatrix} (\boldsymbol{\beta}^I)^T(\mathbf{0}^I)^* \\ (\boldsymbol{\beta}^{II})^T(\mathbf{0}^{II})^* \end{bmatrix} \tag{5.8}$$

Finally, the actuated joint torques/forces, i.e., τ of $\boldsymbol{\tau}^I$ for the 4-bar mechanism, can be computed as

$$\begin{aligned} &\tau = \mathbf{0}_\eta^* - \mathbf{J}_\eta^T\boldsymbol{\lambda} \\ &\text{where, } \mathbf{0}_\eta^* \equiv \begin{bmatrix} (\boldsymbol{\eta}^I)^T(\mathbf{0}^I)^* \\ (\boldsymbol{\eta}^{II})^T(\mathbf{0}^{II})^* \end{bmatrix} \text{ and } \mathbf{J}_\eta \equiv \begin{bmatrix} \mathbf{J}^I\boldsymbol{\eta}^I \\ \mathbf{J}^{II}\boldsymbol{\eta}^{II} \end{bmatrix} \end{aligned} \tag{5.9}$$

The above formulation requires the inversion of the 2×2 matrix $\mathbf{J}_\beta$, which is similar to the inversion of GIM of the determinate Subsystem-I, i.e., $\mathbf{I}^I$. In the mechanism, while the joint 3 is unactuated but no restriction on which joint (2 or 4) to cut.

5.2 Rigid–Flexible Planar Four-Bar Mechanism

The planar four-bar mechanism shown in Fig. 5.2a has four links with one fixed link and three moving links. The four-bar mechanism was analyzed by first cutting it at a joint to form two open-loop, serial-chain subsystems. They are called Subsystems-I and II. Subsystem-I has two rigid or flexible links, whereas Subsystem-II has one

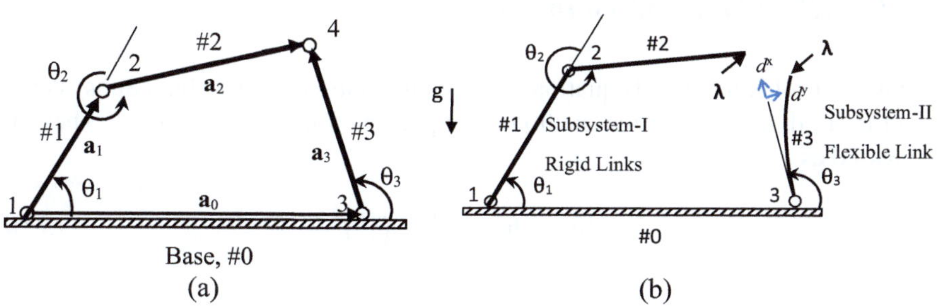

Fig. 5.2 Rigid–flexible four-bar mechanism

rigid or flexible link, as shown in Fig. 5.2b. The cut-opened joint was then substituted with the constraint forces. The dynamics of rigid–flexible four-bar mechanism was formulated for the following cases depending on whether a link in the mechanism is rigid or flexible:

Case 1: Rigid–Rigid–Rigid (*RRR*).

In this case, the dynamic analysis was done considering all three links, i.e., crank (#1), coupler (#2), and rocker (#3) as rigid (*R*). Figure 5.2a shows the configuration of *RRR*.

Case 2: Rigid–Rigid–Flexible (*RRF*).

In this case, the crank (#1) and the coupler (#2) which belong to Subsystem-I, were treated rigid (*R*), whereas the rocker (#3), the only link in Subsystem-II, was treated flexible (*F*). Figure 5.2b shows the configuration of *RRF* with the cut-opened joint.

Case 3: Rigid–Flexible–Flexible (*RFF*).

In this case, the coupler (#2) and rocker (#3) were considered flexible (*F*), whereas the crank (#1) was considered as rigid (*R*).

Case 4: Flexible–Flexible–Flexible (*FFF*).

Here all three links, i.e., crank (#1), coupler (#2), and rocker (#3), are considered flexible (*F*).

The steps involved in the dynamic formulation of the four-bar mechanism are summarized as follows:

- write the equations of motion for the two-link serial system for Subsystem-I;
- write the equation of motion for a single rigid link for *RRR* case and a single flexible link for *RRF, RFF,* and *FFF;*
- write the constraint equations; and
- finally, solve the system of equations of motion to get the required numerical results.

For illustration purposes, the dynamic formulation of the *RRR* is explained first, followed by the *RRF.* In addition, the results of the dynamic simulations of *RFF* and *FFF* are presented.

5.2.1 *RRR Planar Mechanism*

The two sets of constrained equations of motion of the two serial-chain Subsystems-I and II are derived using the methodology presented in Chaps. 2 and 3. They are expressed as

$$\mathbf{I}^{I}\ddot{\mathbf{q}}^{I} + \mathbf{h}^{I} = \boldsymbol{\tau}^{I} + \left(\boldsymbol{\tau}^{\lambda}\right)^{I} \tag{5.10}$$

$$\mathbf{I}^{II}\ddot{\mathbf{q}}^{II} + \mathbf{h}^{II} = \boldsymbol{\tau}^{II} + \left(\boldsymbol{\tau}^{\lambda}\right)^{II} \tag{5.11}$$

where $\mathbf{I}^I$ is the 2×2 GIM for Subsystem-I which is basically a two-link serial-chain rigid link manipulator which was obtained as

$$\mathbf{I}^I \equiv \begin{bmatrix} \frac{\rho_1 a_1^3}{3} + \frac{\rho_2 a_2^3}{3} + \rho_2 a_1^2 a_2 + \rho_2 a_1 a_2^2 c_2 & \frac{\rho_2 a_2^3}{3} + \frac{\rho_2 a_1 a_2^2}{2} c_2 \\ sym & \frac{\rho_2 a_2^3}{3} \end{bmatrix} \tag{5.12}$$

In Eq. (5.12), ρ_i and a_i are the mass per unit length and link length of the ith link, respectively. Moreover, $c_i \equiv \cos\theta_i$ and $s_i \equiv \sin\theta_i$, where θ_i is the independent joint angle for joints 1 and 2. Similarly in Eq. (5.10), the generalized coordinate vector $\mathbf{q}^I \equiv \begin{bmatrix} \theta_1 \; \theta_2 \end{bmatrix}^T$ and the vector of convective inertia terms $\mathbf{h}^I$ is given by

$$\mathbf{h}^I \equiv \begin{bmatrix} -\frac{\rho_2 a_1 a_2^2}{2} s_2 \dot{\theta}_2^2 - \rho_2 a_1 a_2^2 s_2 \dot{\theta}_1 \dot{\theta}_2 \\ \frac{\rho_2 a_1 a_2^2}{2} s_2 \dot{\theta}_1^2 \end{bmatrix} \tag{5.13}$$

Then, the generalized force vector $\boldsymbol{\tau}^I$ for Subsystem-I is given by, $\boldsymbol{\tau}^I = \boldsymbol{\tau}^{EI} + \boldsymbol{\tau}^{gI}$, where $\boldsymbol{\tau}^{EI}$ is the vector of external joint torques, and $\boldsymbol{\tau}^{gI}$ is the vector of generalized force due to gravity which is given by

$$\boldsymbol{\tau}^{gI} \equiv \begin{bmatrix} \frac{\rho_1 g a_1^2}{2} c_1 + \rho_2 g a_1 a_2 c_1 + \frac{\rho_2 g a_2^2}{2} c_{12} \\ \frac{\rho_2 g a_2^2}{2} c_{12} \end{bmatrix} \tag{5.14}$$

where g is the acceleration due to gravity, and $s_{12} \equiv \sin(\theta_1 + \theta_2)$ and $c_{12} \equiv \cos(\theta_1 + \theta_2)$. For Subsystem-II, the terms $\mathbf{I}^{II}$, $\mathbf{h}^{II}$, $\boldsymbol{\tau}^{gII}$ of Eq. (5.11) are scalars for the one-degree-of-freedom single-link rigid system which are given by

$$\mathbf{I}^{II} \equiv \frac{\rho_3 a_3^3}{3}, \quad \mathbf{h}^{II} \equiv 0, \quad \boldsymbol{\tau}^{gII} \equiv \frac{\rho_3 g a_3^2}{2} c_3 \tag{5.15}$$

Moreover, in Eqs. (5.10) and (5.11), $\boldsymbol{\tau}^{\lambda(k)}$ is the constraint forces due to the cut joint-4 of Fig. 5.2 (a) of the kth Subsystem. Note that $\boldsymbol{\tau}^{\lambda I} = (\mathbf{J}^I)^T \boldsymbol{\lambda}$ and $\boldsymbol{\tau}^{\lambda II} = (\mathbf{J}^{II})^T \boldsymbol{\lambda}$ in which $\mathbf{J}^I$ and $\mathbf{J}^{II}$ are the Jacobian's relating to the velocities of the tip of the manipulators, i.e., at joint-4 with the joint rates $\dot{\mathbf{q}}^I = [\dot{\theta}_1 \; \dot{\theta}_2]^T$ and $\dot{\mathbf{q}}^{II} = \dot{\theta}_3$, respectively. The Jacobian's $\mathbf{J}^I$ and $\mathbf{J}^{II}$ are given by

$$\mathbf{J}^I \equiv \begin{bmatrix} -a_1 s_1 - a_2 s_{12} & -a_2 s_{12} \\ a_1 c_1 + a_2 c_{12} & a_2 c_{12} \end{bmatrix}, \text{ and } \mathbf{J}^{II} \equiv \begin{bmatrix} -a_3 s_3 \\ a_3 c_3 \end{bmatrix} \tag{5.16}$$

The term $\boldsymbol{\lambda} = \begin{bmatrix} \lambda_1 \; \lambda_2 \end{bmatrix}^T$ is the two-dimensional vector of Lagrange multipliers that needs to be evaluated from the dynamic equations of motion. Next, the appropriate constraints are introduced. In this case, the velocity of the tips of the Subsystems-I and II at joint-4 must be the same. In other words,

$$\mathbf{J}^I\dot{\mathbf{q}}^I = \mathbf{J}^{II}\dot{\mathbf{q}}^{II} \text{ or } \mathbf{J}^I\dot{\mathbf{q}}^I - \mathbf{J}^{II}\dot{\mathbf{q}}^{II} = 0 \tag{5.17}$$

Using Eqs. (5.10), (5.11), and (5.17), the equations of motion of the *RRR* four-bar mechanism are written in compact form as

$$\mathbf{I}\ddot{\mathbf{q}} = \mathbf{0} + \mathbf{J}^T\boldsymbol{\lambda} \tag{5.18}$$

where $\mathbf{I} \equiv diag.\left[\mathbf{I}^I \ \mathbf{I}^{II}\right]$ and $\mathbf{J} \equiv \left[\mathbf{J}^{I^T} \ -\mathbf{J}^{II^T}\right]^T$, whereas vectors $\ddot{\mathbf{q}} = \left[\ddot{\mathbf{q}}^{I^T} \ \ddot{\mathbf{q}}^{II^T}\right]^T$, and $\mathbf{0} = \boldsymbol{\tau} - \mathbf{h}$, in which $\boldsymbol{\tau} \equiv \left[\boldsymbol{\tau}^{I^T} \ \boldsymbol{\tau}^{II^T}\right]^T$ and $\mathbf{h} \equiv \left[\mathbf{h}^{I^T} \ \mathbf{h}^{II^T}\right]^T$. Equations (5.17) and (5.18) are combined written in a matrix–vector form as

$$\begin{bmatrix} \mathbf{I} & \mathbf{J}^T \\ \mathbf{J} & \mathbf{O} \end{bmatrix} \begin{bmatrix} \ddot{\mathbf{q}} \\ \boldsymbol{\lambda} \end{bmatrix} = \begin{bmatrix} \boldsymbol{\tau} \\ -\dot{\mathbf{J}}\dot{\mathbf{q}} \end{bmatrix} \tag{5.19}$$

The forward dynamics simulation results for the *RRR* four-bar mechanism were generated and are presented in the following subsection.

5.2.1.1 Numerical Simulation

For the numerical simulation, lengths of crank (a_1), coupler (a_2), rocker (a_3), and fixed base (a_0) were taken as 0.5 m, 1 m, 1.3 m, and 1.5 m, respectively. The masses of the crank (m_1), coupler (m_2), and rocker (m_3) were taken as 1.5 kg, 3 kg, and 5 kg, respectively. The links have a circular cross section with 20 mm diameter. The initial configuration of the *RRR* four-bar mechanism is defined in Table 5.1. For simulation, the mechanism was allowed to fall freely due to gravity, with no external torque. The joint accelerations $\ddot{\mathbf{q}}$ were obtained using RGE-based forward dynamics algorithm. Numerical integrations were carried out using the *ode45* function of MATLAB. Figure 5.3 shows the plot of the variations of the joint angles of the crank (θ_1), coupler (θ_2), and rocker (θ_3) with respect to time during the free fall, and Fig. 5.4 shows the corresponding joint rates. The *RRR* four-bar mechanism was also modeled in the commercial software RecurDyn as shown in Fig. 5.5, to verify the results. The mass and inertia properties were kept the same while modeling. The forward dynamics results are plotted in Figs. 5.3 and 5.4. The results are matching, thus verifies the results obtained using the proposed modeling technique.

Table 5.1 Initial conditions of the *RRR* four-bar mechanism

	Joint 1	Joint 2	Joint 3
Joint angles	$\theta_1 = 0°$	$\theta_2 = 81.08°$	$\theta_3 = 130.54°$
Joint velocities	$\dot{\theta}_1 = 0$ rad/s	$\dot{\theta}_2 = 0$ rad/s	$\dot{\theta}_3 = 0$ rad/s

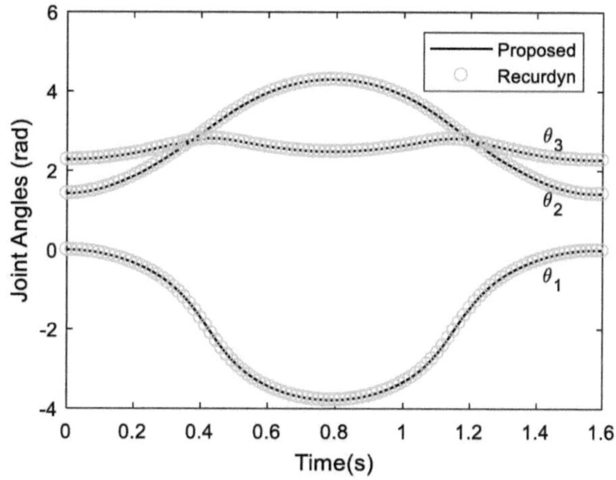

Fig. 5.3 Joint angles of *RRR* four-bar mechanism

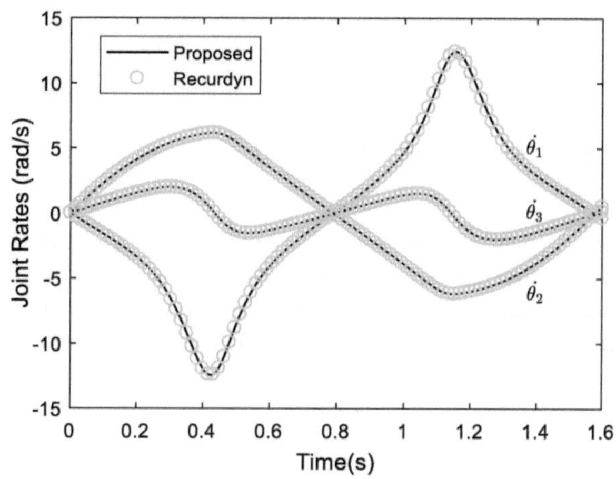

Fig. 5.4 Joint rates of *RRR* four-bar mechanism

5.2.2 *RRF Planar Mechanism*

The *RRF* four-bar mechanism is shown in Fig. 5.2b. The crank (#1) and coupler (#2) were treated as rigid, whereas the output link (#3) was considered as flexible. Here also, the mechanism is cut at joint-4 and is substituted with the constraint force $\boldsymbol{\lambda}$, as shown in Fig. 5.2b. For the dynamic simulation of the *RRF* case, the only difference is in the dynamic equations of motion of Subsystem-II, i.e., Eq. (5.11), which has a single flexible link, and the associated constraint Jacobian. In this illustration, the deflection due to bending, i.e., transverse displacement (d_y), with 2-modes are considered for the flexible link while the axial deflection and the torsional effect are neglected. The expressions for the generalized inertia matrix $\mathbf{I}^{II}$ and the vector of convective inertia terms $\mathbf{h}^{II}$ for Subsystem-II were obtained in Eqs. (3.49), and (3.51), respectively. Then, the generalized force vector $\boldsymbol{\tau}^{II}$ for Subsystem-II is given

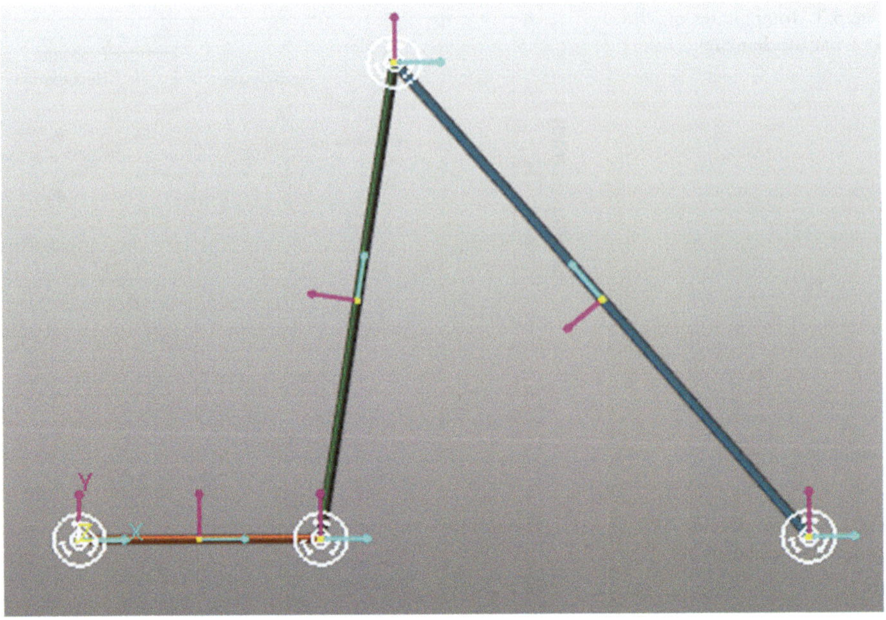

Fig. 5.5 RecurDyn model of the *RRR* four-bar mechanism

by $\boldsymbol{\tau}^{II} = \boldsymbol{\tau}^{EII} + \boldsymbol{\tau}^{gII} + \boldsymbol{\tau}^{sII}$, where $\boldsymbol{\tau}^{EII}$ is a vector of external joint torques, $\boldsymbol{\tau}^{gII}$ is the vector of generalized force due to gravity, and $\boldsymbol{\tau}^{sII}$ is the vector of internal forces due to strain. The vectors $\boldsymbol{\tau}^{gII}$ and $\boldsymbol{\tau}^{sII}$ were obtained in Eqs. (3.52) and (3.53), respectively. Then, the Jacobian $\mathbf{J}^{II}$ and the vector $\dot{\mathbf{q}}^{II}$ for a single flexible link were derived as

$$\mathbf{J}^{II} \equiv \begin{bmatrix} -\left(s_1^{y3} dy_{31} + s_2^{y3} dy_{32}\right) c_3 & -s_1^{y3} s_3 & s_2^{y3} s_3 \\ -\left(s_1^{y3} dy_{31} + s_2^{y3} dy_{32}\right) s_3 & s_1^{y3} c_3 & s_2^{y3} c_3 \end{bmatrix}, \text{ and } \dot{\mathbf{q}}^{II} \equiv \begin{bmatrix} \dot{\theta}_3 \\ \dot{d}y_{31} \\ \dot{d}y_{32} \end{bmatrix} \tag{5.20}$$

Substituting the above equations in Eq. (5.19), the forward dynamics results of the *RRF* four-bar mechanism were generated and presented in the following subsection.

5.2.2.1 Numerical Simulation

The lengths of crank (a_1), coupler (a_2), output link (a_3), and fixed base (a_0) were taken as 0.5 m, 1 m, 1.3 m, and 1.5 m, respectively. The masses of the crank (m_1), coupler (m_2), and output link (m_3) were taken as 1.5 kg, 3 kg, and 5 kg, respectively. The links were assumed to have a circular cross section of 20 mm diameter. The modulus of elasticity of the flexible link was taken as 2×10^{11} N/m^2. For simulation purposes, the mechanism was allowed to fall freely under gravity, with no external torque. For this illustration, the fixed-free boundary condition (i.e., cantilever beam assumption) was considered for the selection of the modal function. In order to perform the simulation, the $\ddot{\mathbf{q}}_{(k)}$s for the two serial subsystems were solved using the scheme presented in Sect. 5.1.1.

Numerical integrations were done using the *ode45* function of MATLAB with the same initial conditions as shown in Table 5.1, along with the initial values of the variables dy_{31}, dy_{32}, and its first derivative equals to zero. Figures 5.6 and 5.7 show the joint angles and the joint rates of the *RRF* mechanism for all the joints. Figure 5.8 shows the transverse displacement u_y of the flexible output link (rocker).

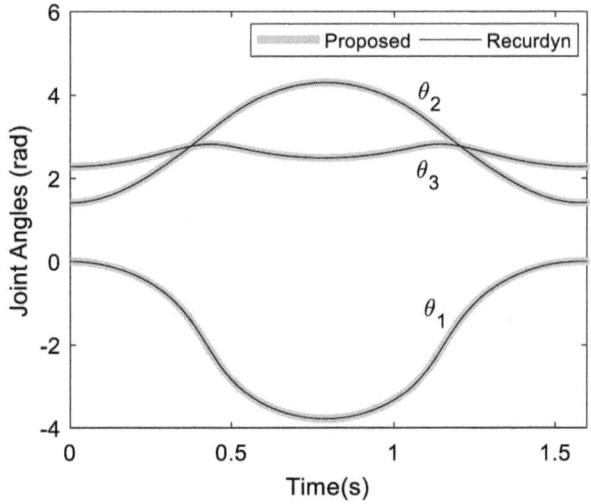

Fig. 5.6 Joint angles of the *RRF* four-bar mechanism

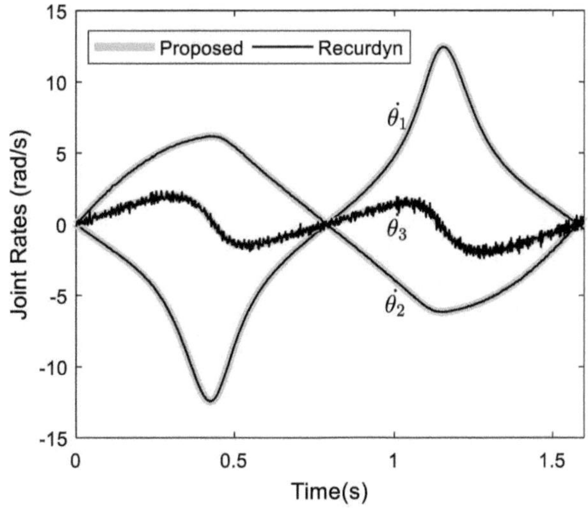

Fig. 5.7 Joint rates of the *RRF* four-bar mechanism

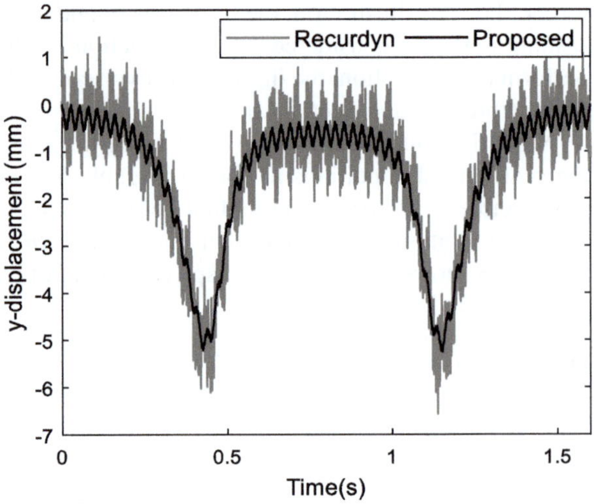

Fig. 5.8 Transverse displacement of the tip of flexible rocker (location of joint 4)

Though the joint angles and joint rates in Figs. 5.6 and 5.7 show the same variation as in Figs. 5.3 and 5.4, respectively, of the *RRR* case, the joint rate corresponding to flexible link ($\dot{\theta}_3$) indicates the small fluctuations. It is apparently due flexibility of the rocker (#3). FE modeling of the rigid–flexible four-bar mechanism was done using RecurDyn software, as shown in Fig. 5.9. The detailed steps of modeling and simulation are explained in Appendix E. The results of RecurDyn software were

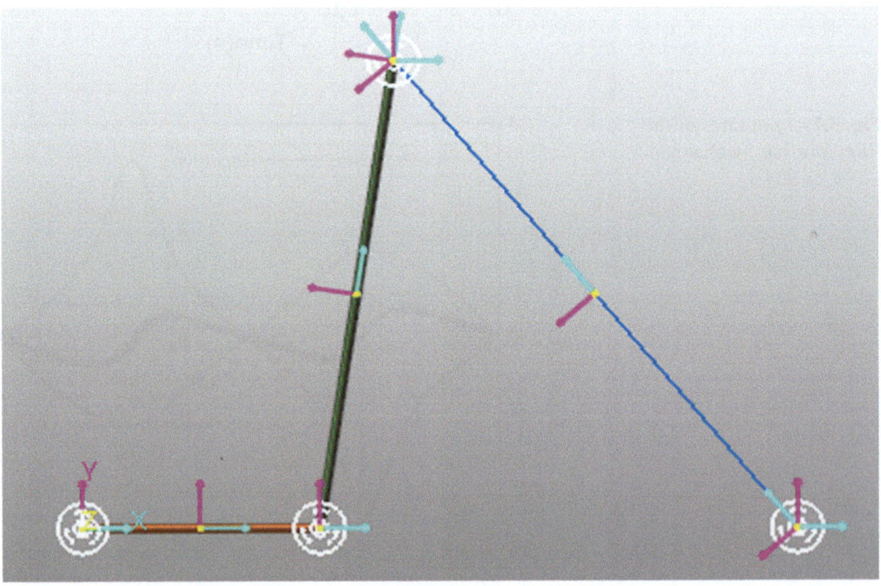

Fig. 5.9 RecurDyn model of *RRF* four-bar mechanism

overlapped with the results from the proposed formulation. The results show a close match.

Whatever variation observed in Fig. 5.8 could be due to the different methodology adopted by RecurDyn for dynamic modeling and the successive algorithm used to solve them. Irrespective of the variations, the results using the proposed method show low-frequency variations compared to those obtained using RecurDyn. Hence, the better numerical stability of the proposed algorithm is established.

5.2.3 *RFF* Planar Mechanism

The dynamic equations of motion for the *RFF* four-bar mechanism were formulated similar to the *RRF* case. The difference exists only in Subsystem-I, where the equations of motion are considered for a two-link *RF* manipulator instead of an *RR* manipulator. The dynamic formulation of the *RFF* mechanism is not explained here in detail to avoid monotonicity. The dynamic formulation of the *RF* manipulator is given in [5] and re-derived for the simulation purposes of this book. The numerical values are taken the same as those for the *RRF* mechanism for rigid and flexible links. The plots in Figs. 5.10 and 5.11, respectively, show the joint angles and joint rates of the *RFF* mechanism. Also, the transverse displacements of the flexible coupler and the flexible rocker are shown in Figs. 5.12 and 5.13, respectively. The *RFF* results generated using RecurDyn were also overlapped with the results obtained from the proposed formulation. This shows the close match, thus verifying the results. Note the smaller variation of the results for the proposed algorithm in Fig. 5.13, which is due to more numerical stability.

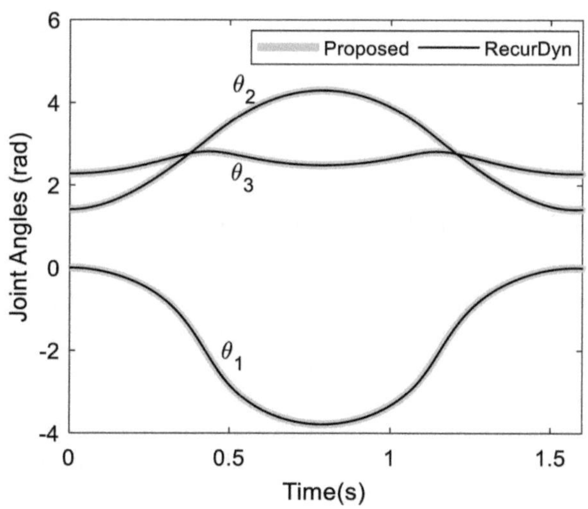

Fig. 5.10 Joint angles of *RFF* four-bar mechanism

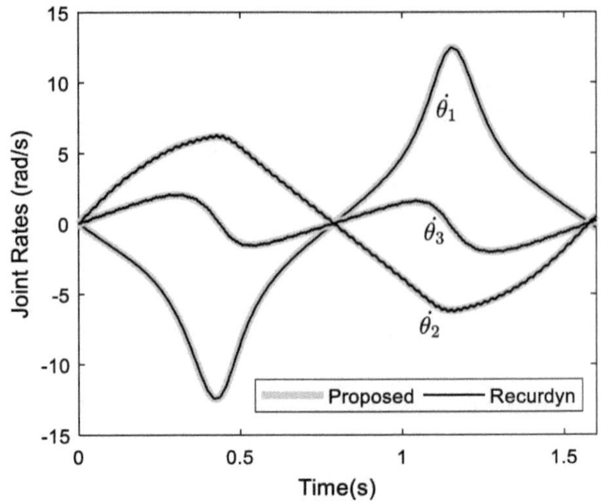

Fig. 5.11 Joint rates of *RFF* four-bar mechanism

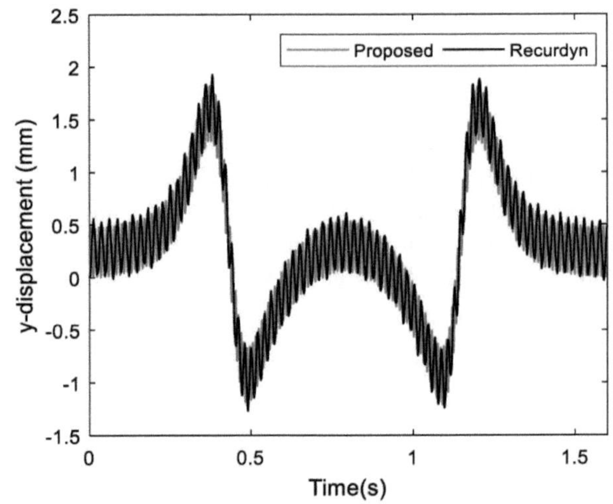

Fig. 5.12 Transverse displacement of flexible coupler (location of joint 4)

5.2.4 *FFF* Planar Mechanism

Here, all three links, i.e., crank (#1), coupler (#2), and rocker (#3) of the four-bar mechanism, were considered flexible (F). The parameters of flexible links were considered same as explained in previous subsections. The plots in Figs. 5.14 and 5.15, respectively, show the joint angles and joint rates of *FFF* mechanism. Figures 5.16, 5.17, and 5.18 show the transverse displacements of the tips of the flexible coupler, rocker, and crank, i.e., locations of joint 4, 4, and 2, respectively.

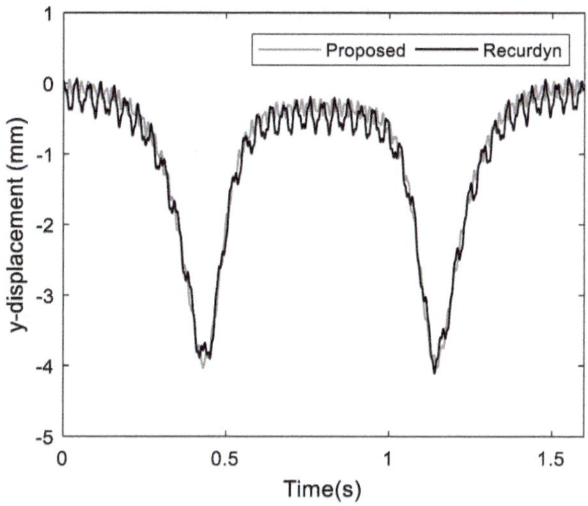

Fig. 5.13 Transverse displacement of flexible rocker (location of joint 4)

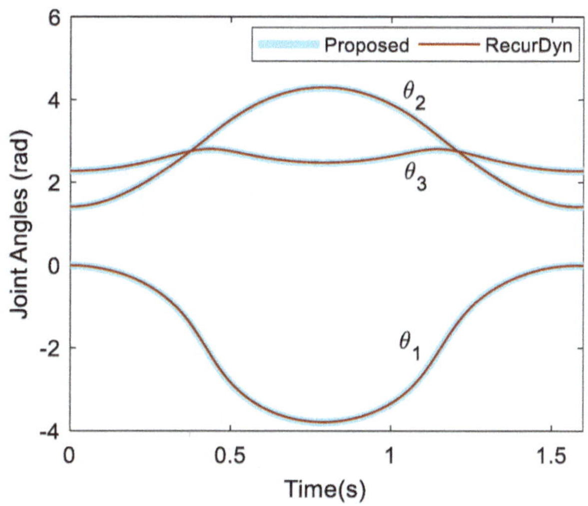

Fig. 5.14 Joint angles of *FFF* four-bar mechanism

5.3 Rigid–Flexible Five-Bar Mechanism

The closed-loop five-bar mechanism shown in Fig. 5.19a consists of five links with one rigid base and four moving links which are connected by revolute joints. It has five joints numbered 1–5, in which joints 1 and 3 being the actuated joints and the remaining are passive joints. The mechanism was analyzed by cutting it at joint 3 to form two open-loop, serial-chain, two-link subsystems, as shown in Fig. 5.19b. The cut-opened joint was then substituted with unknown constraint forces $\boldsymbol{\lambda}$'s, i.e., Lagrange multipliers, which need to be evaluated using dynamic equations

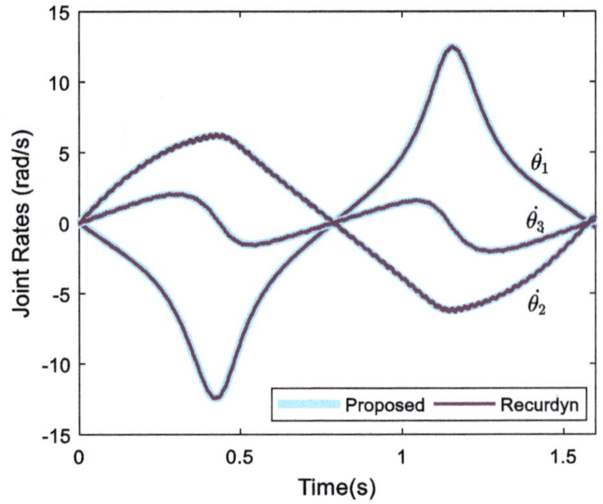

Fig. 5.15 Joint rates of *FFF* four-bar mechanism

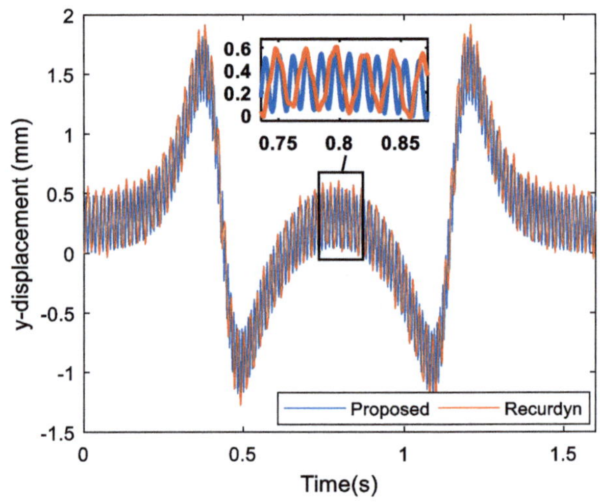

Fig. 5.16 Transverse displacement of flexible coupler (location of joint 4)

of motion. Once calculated, these multipliers were treated as external forces to the resulting open-loop Subsystems-I and II. The dynamics of the rigid–flexible five-bar mechanism was formulated for the following cases:

Case 1: Rigid–Rigid–Rigid–Rigid (*RRRR*).

In this case, the dynamic analysis was done considering all four links as rigid.

Case 2: Rigid–Flexible–Flexible–Rigid (*RFFR*).

In this case, the links 1 and 3 were considered rigid (*R*), whereas the links 2 and 4 were treated flexible (*F*).

Case 3: Flexible–Rigid–Rigid–Flexible (*FRRF*).

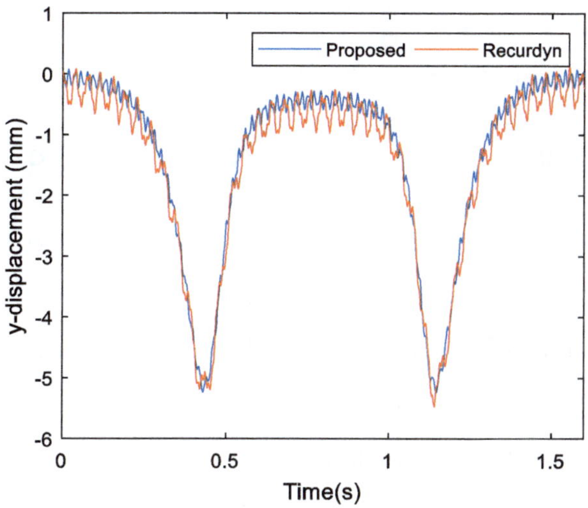

Fig. 5.17 Transverse displacement of flexible rocker (location of joint 4)

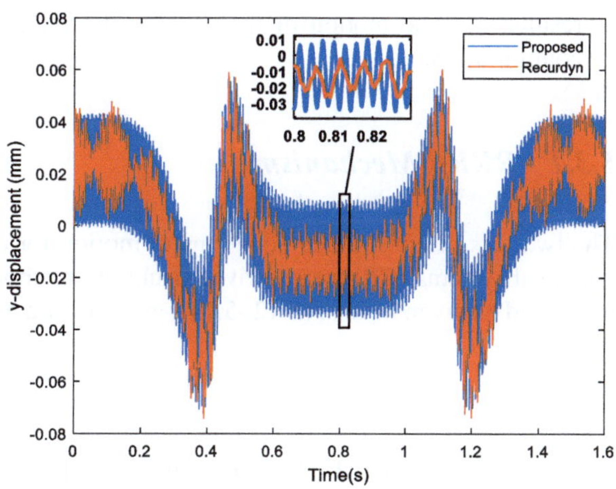

Fig. 5.18 Transverse displacement of flexible crank (location of joint 2)

In this case, the links 1 and 3 were considered flexible (F), whereas the links 2 and 4 were rigid (R).

Case 4: Flexible–Flexible–Flexible–Flexible ($FFFF$).

In this case, all four links were considered flexible (F).

The steps involved in the dynamic formulation of the five-bar mechanism are summarized as follows:

- write the equations of motion for a two-link serial manipulator for both subsystems;
- write the constraint equations; and

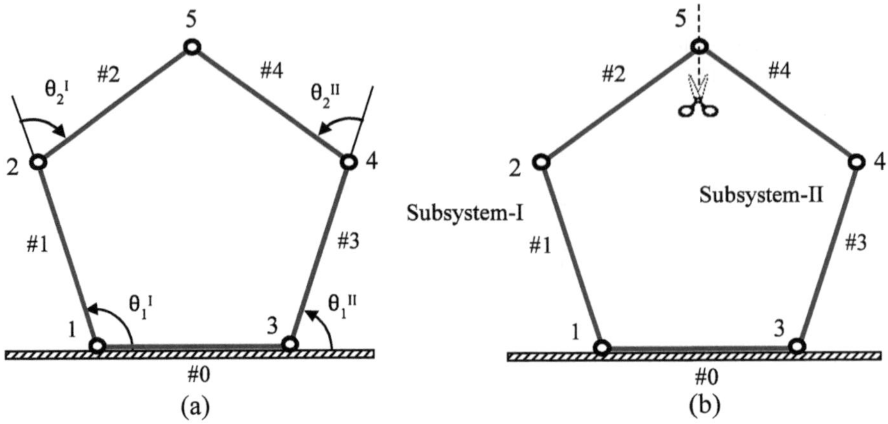

Fig. 5.19 Five-bar mechanism

- finally, solve the system of equations of motion to get the required numerical results, i.e., the motion of the five-bar mechanism for a given initial state or configuration.

5.3.1 *RRRR Mechanism*

The two sets of constrained equations of motion for an open-loop, two-link serial-chain subsystems (*RR*) were derived using the methodology presented in Chaps. 2 and 3 and as given in Eqs. (5.12–5.14), and they are expressed as

$$\mathbf{I}\ddot{\mathbf{q}} = \mathbf{0} + \boldsymbol{\tau}^{\lambda} \tag{5.21}$$

were, $\mathbf{I} \equiv \text{diag.}\left[\mathbf{I}^{I}, \mathbf{I}^{II}\right]$, $\mathbf{q} \equiv \left[\mathbf{q}^{I^T}, \mathbf{q}^{II^T}\right]^T$, and $\mathbf{0} = \boldsymbol{\tau} - \mathbf{h}$, in which
$\boldsymbol{\tau} \equiv \left[\boldsymbol{\tau}^{I^T}\ \boldsymbol{\tau}^{II^T}\right]^T$ and $\mathbf{h} \equiv \left[\mathbf{h}^{I^T}\ \mathbf{h}^{II^T}\right]^T$

Moreover, $\boldsymbol{\tau}^{\lambda} = \mathbf{J}^T\boldsymbol{\lambda}$ is the constraint forces due to the cut joint in which $\mathbf{J}$ is the constrained Jacobian matrix defined in the way such that that $\mathbf{J}\dot{\mathbf{q}} = \mathbf{0}$. In Eq. (5.21), $\mathbf{I}^{(k)}$ for k = I, II is the GIM of the kth Subsystem whose size corresponds to the generalized coordinates $\mathbf{q}^{(k)}$. Similarly, $\mathbf{0}^{(k)}$ and $\boldsymbol{\tau}^{(k)}$ are defined. The equations of motion and the constraint equations for the *RRRR* five-bar mechanism can be given in the form of Eq. (5.3) as

$$\begin{bmatrix} \mathbf{I} & \mathbf{J}^T \\ \mathbf{J} & \mathbf{O} \end{bmatrix}\begin{bmatrix} \ddot{\mathbf{q}} \\ \boldsymbol{\lambda} \end{bmatrix} = \begin{bmatrix} \boldsymbol{\tau} \\ -\dot{\mathbf{J}}\dot{\mathbf{q}} \end{bmatrix} \tag{5.22}$$

The above equation was used to solve the joint accelerations $\ddot{\mathbf{q}}$. Then, these $\ddot{\mathbf{q}}$ were integrated numerically to obtain the joint velocities $\dot{\mathbf{q}}$ and the joint angles $\mathbf{q}$.

5.3.1.1 Numerical Simulation

For the numerical simulation, lengths and masses of all the links, i.e., #1 (a_1), #2 (a_2), #3 (a_3), #4 (a_4), and the fixed link, #5 (a_5), were taken the same as 1 m and 0.6 kg. The links have a circular cross section with a 20 mm diameter. The initial configuration of the *RRRR* five-bar mechanism is defined in Table 5.2. For simulation, the mechanism was allowed to fall freely under gravity, with no external torque. Numerical integrations were done using the *ode45* function of MATLAB. Then, the forward dynamic simulation results were generated using the RGE-based algorithm for the calculation of joint accelerations $\ddot{\mathbf{q}}$. Figure 5.20 shows the joint angles $\theta_1(\equiv \theta_1^I)$ and $\theta_3(\equiv \theta_1^{II})$, whereas Fig. 5.21 shows the joint angles $\theta_2(\equiv \theta_2^I)$ and $\theta_4(\equiv \theta_2^{II})$ during the free fall. Similarly, Fig. 5.22 shows the joint rates $\dot{\theta}_1(\equiv \dot{\theta}_1^I)$ and $\dot{\theta}_3(\equiv \dot{\theta}_1^{II})$, whereas Fig. 5.23 shows the joint rates $\dot{\theta}_2(\equiv \dot{\theta}_2^I)$ and $\dot{\theta}_4(\equiv \dot{\theta}_2^{II})$. The *RRRR* five-bar mechanism was then modeled in the commercial software RecurDyn to verify the results, as shown in Fig. 5.5. The results are a close match with the proposed formulation (Fig. 5.24).

Table 5.2 Initial conditions of the *RRRR* mechanism

	Joint 1	Joint 2	Joint 3	Joint 4
Joint angles	$\theta_1^I = 108°$	$\theta_2^I = -72°$	$\theta_1^{II} = 72°$	$\theta_2^{II} = 72°$
Joint velocities	$\dot{\theta}_1^I = 0\,\text{rad/s}$	$\dot{\theta}_2^I = 0$ rad/s	$\dot{\theta}_1^{II} = 0$ rad/s	$\dot{\theta}_2^{II} = 0$ rad/s

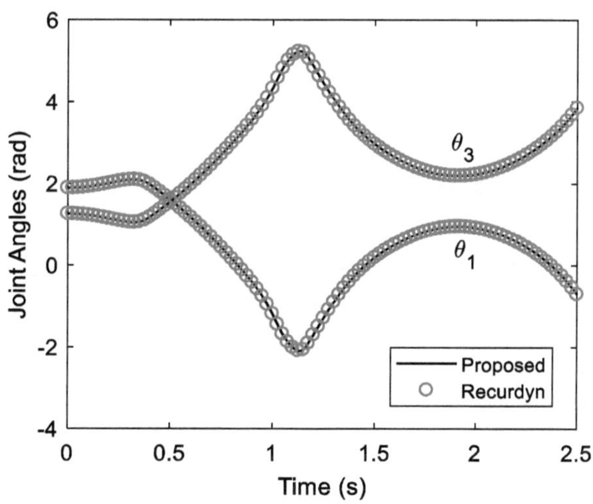

Fig. 5.20 Joint angles of *RRRR* mechanism

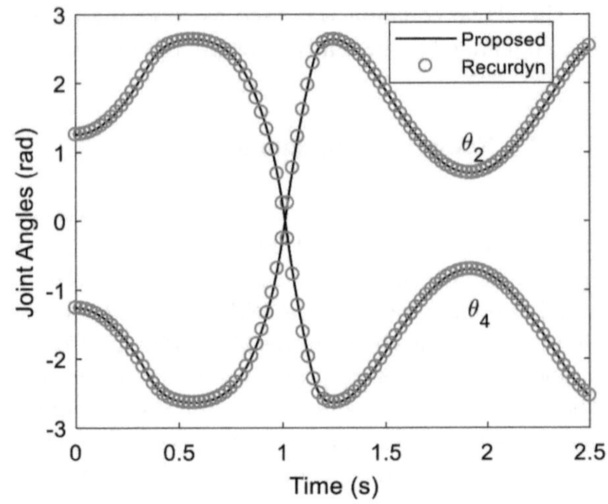

Fig. 5.21 Joint angles of *RRRR* mechanism

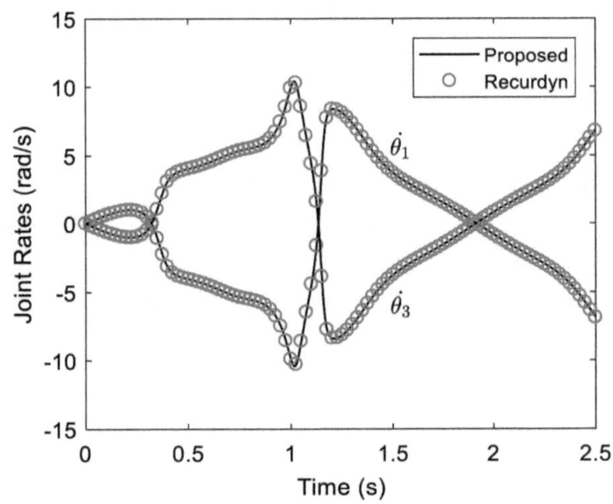

Fig. 5.22 Joint rates of *RRRR* mechanism

5.3.2 *RFFR* Mechanism

In *RFFR* five-bar mechanism, the links connected to the base were treated rigid (*R*), and the other two links were considered flexible (*F*). The dynamic equations of motion of the *RFFR* five-bar mechanism were formulated similar to the *RRRR*. Here, the equations of motion were derived for the two-link *RF* manipulator instead of the *RR* manipulator for Subsystems-I and II. The constraint was the same tip velocities of the subsystems; i.e., the velocity components at the location of joint 5 are the same. The dynamic formulation of the *RFFR* mechanism is not explained here in detail to avoid monotonicity.

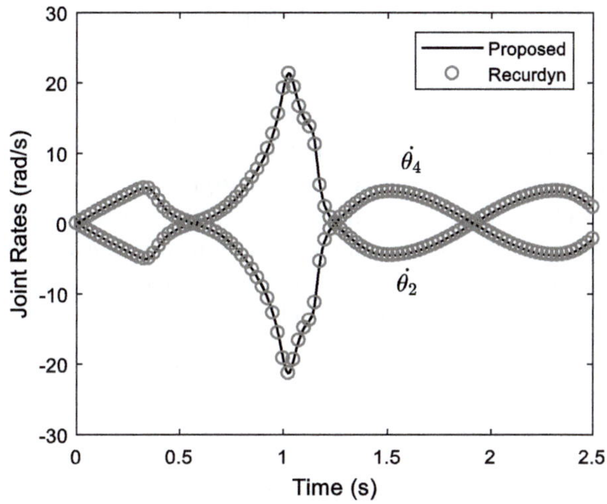

Fig. 5.23 Joint rates of *RRRR* mechanism

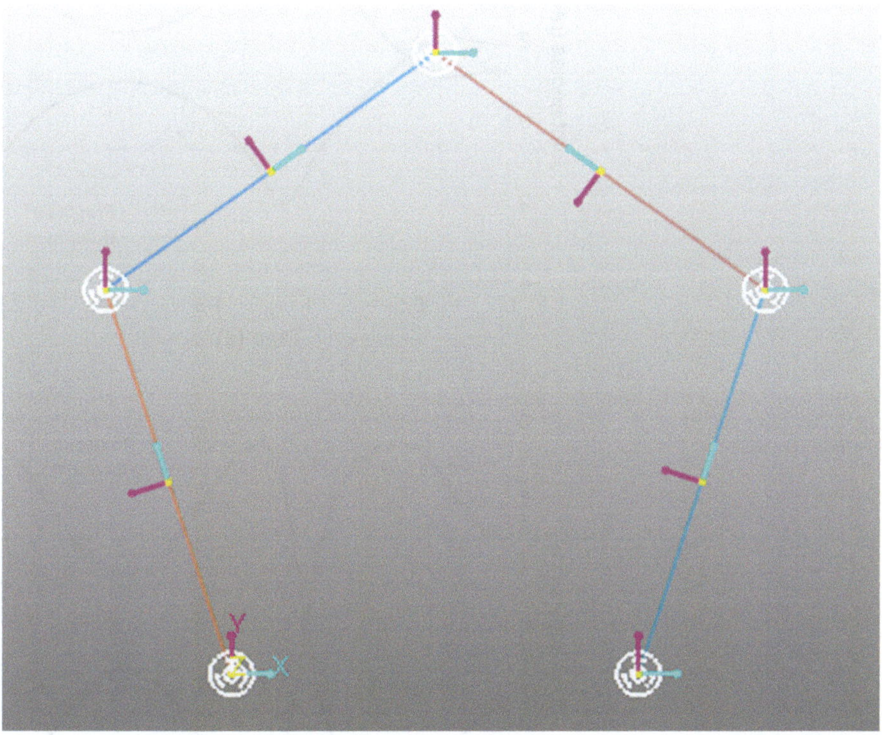

Fig. 5.24 RecurDyn model of the five-bar mechanism

5.3.2.1 Numerical Simulation

The link lengths and the masses for all the links were taken as 1 m and 0.6 kg, respectively. The modulus of elasticity of the flexible links was taken as 2×10^{11} N/m^2. For simulation purposes, the mechanism was allowed to fall freely under gravity, with no external torque. The initial configuration was taken the same as that of the *RRRR* mechanism, as given in Table 5.2. For illustration, the fixed-free boundary condition (i.e., cantilever beam assumption) was considered for the selection of the modal functions. The forward dynamics simulation was performed for a duration of 2.5 s, and the results are plotted in Figs. 5.25, 5.26, 5.27, 5.28, 5.29, and 5.30. Figure 5.25

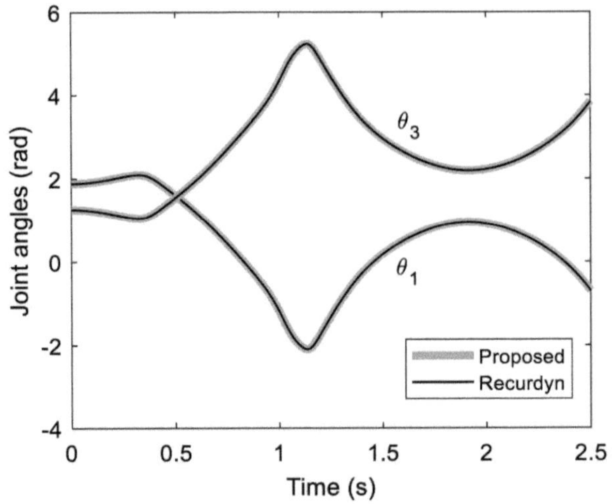

Fig. 5.25 Joint angles of *RFF* mechanism

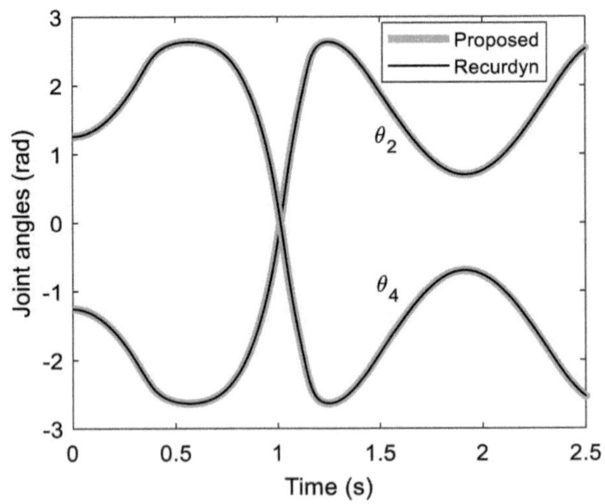

Fig. 5.26 Joint angles of *RFFR* mechanism

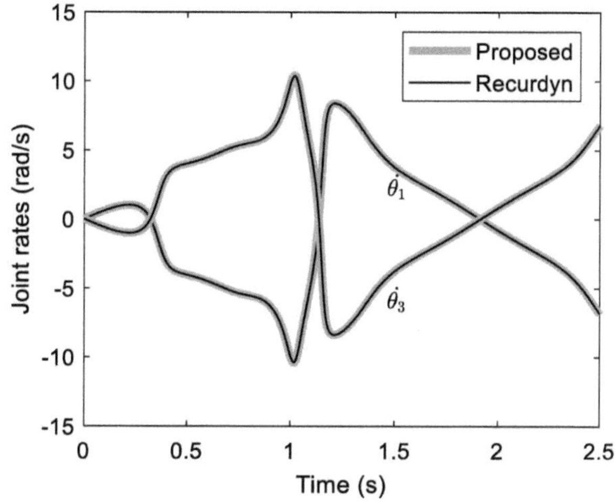

Fig. 5.27 Joint rates of *RFFR* mechanism

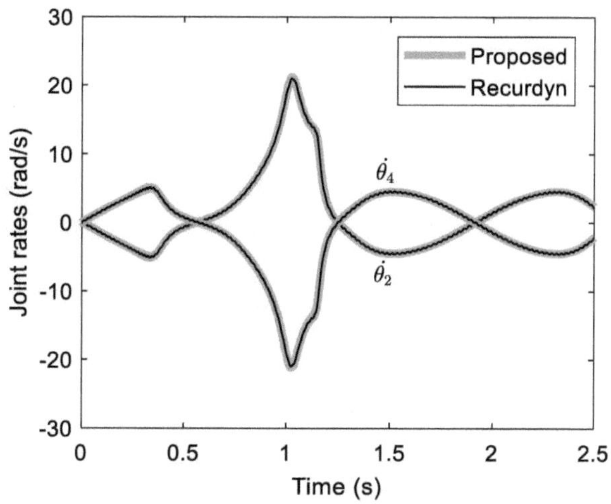

Fig. 5.28 Joint rates of *RFFR* mechanism

shows the joint angles of rigid links, whereas Fig. 5.26 shows the joint angles of flexible links. Similarly, Fig. 5.27 shows the joint rates of rigid links, whereas Fig. 5.28 shows the joint rates of flexible links. Also, the transverse displacements of flexible links, #2 and #4 are shown in Figs. 5.29 and 5.30, respectively. The *RFFR* results were generated using RecurDyn, which shows a close match with the results of the proposed formulation. Note that, the plots in Figs. 5.29 and 5.30 are the mirror reflection of each other about a horizontal line, i.e., $y = 0$. This is true as the two subsystems at joint 5 should have the same displacements but measured from two oppositely placed subsystems.

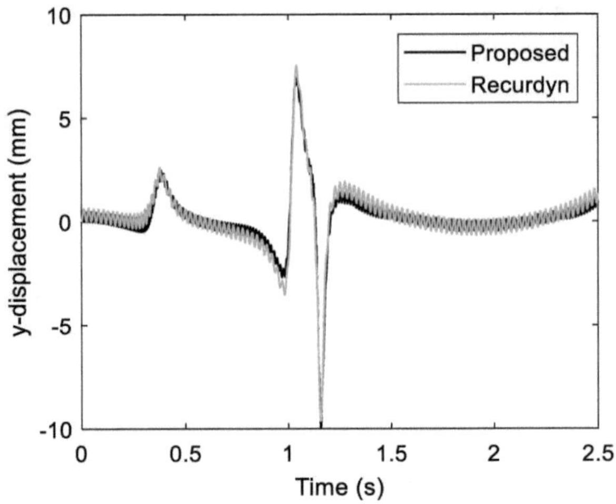

Fig. 5.29 Transverse displacement of #2 (location of joint 5) for *RFFR* mechanism

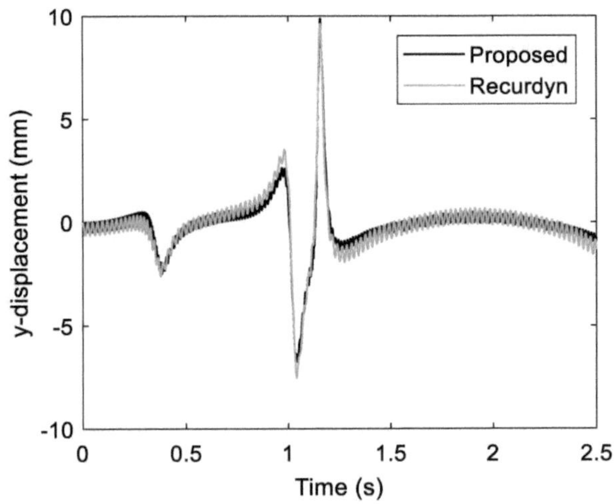

Fig. 5.30 Transverse displacement of #4 (location of joint 5) for *RFFR* mechanism

5.3.3 *FFFF* Mechanism

In the *FFFF* five-bar mechanism, all the links except the fixed link were treated as flexible (*F*). The dynamic equations of motion for the *FFFF* five-bar mechanism were formulated similarly to the *RRRR* case. Here, the equations of motion were derived for the full flexible two-link manipulator (*FF*) instead of the *RR* manipulator for Subsystems-I and II. The constraint provided was the same tip velocities of both the subsystems.

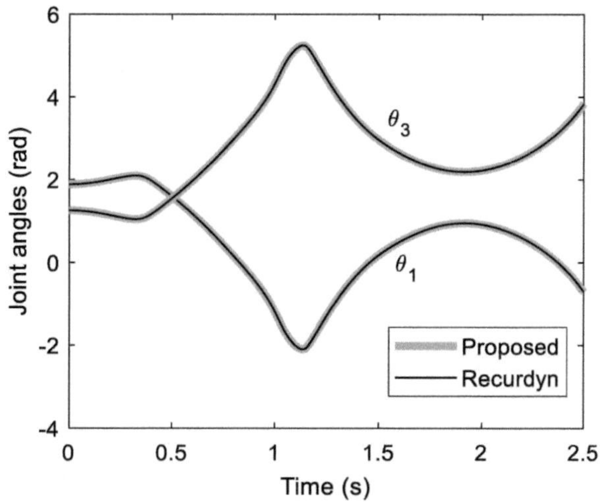

Fig. 5.31 Joint angles of *FFFF* five-bar mechanism

5.3.3.1 Numerical Simulation

The link lengths and the masses for all links were taken as 1 m and 0.6 kg, respectively. The modulus of elasticity of the flexible links was taken as 2×10^{11} N/m^2. The mechanism was allowed to fall freely under gravity without any external torque. The initial configuration was taken the same as that of the *RRRR* mechanism as given in Table 5.2. The initial conditions corresponding to the modal coordinates were taken as zeros. For this illustration, the modal function was selected based on the fixed-free boundary condition (i.e., cantilever beam assumption). The forward dynamics simulation was performed for a duration of 2.5 s. The results are plotted in Figs. 5.31, 5.32, 5.33, 5.34, 5.35, 5.36, 5.37, and 5.38. The results obtained from the proposed approach and RecurDyn software are close to each other, which verifies the results. Comparing the plots in Figs. 5.35, 5.36, 5.37, and 5.38, one can easily interpret the better numerical stability of the proposed algorithm. Besides, the plots in Figs. 5.37 and 5.38 actually represent the same displacement but are seen from two oppositely placed systems.

5.4 Rigid–Flexible 3-DOF Planar Parallel Manipulator

The previous sections dealt with the single-loop systems, in this section, a multiclosed-loop system, namely the rigid–flexible, three-degrees-of-freedom (3-DOF) planar parallel manipulator shown in Fig. 5.39 is analyzed. It has a fixed base and a moving platform connected by three legs. Each leg has two links connected by revolute joints. The tip of each leg is connected to the moving platform with another revolute joint. Thus, each leg has three revolute joints with 3-DOF. Similar to the previous sections, the 3-DOF parallel manipulator was cut at the vertices

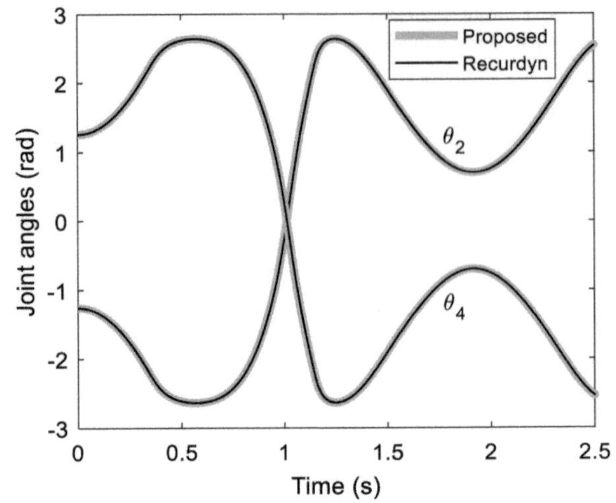

Fig. 5.32 Joint angles of *FFFF* five-bar mechanism

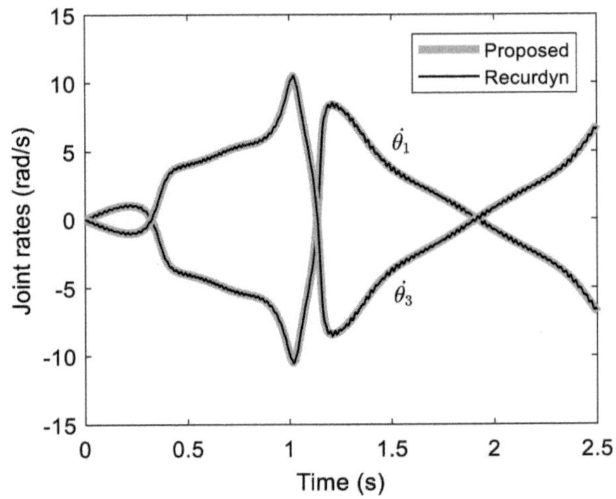

Fig. 5.33 Joint rates of *FFFF* five-bar mechanism

of the moving platform, as shown in Fig. 5.40. It has resulted in four subsystems, namely three two-link rigid–flexible serial manipulators and one moving platform. The cut joints were replaced with suitable constraint forces denoted with $\boldsymbol{\lambda}$'s which are known as Lagrange multipliers. These multipliers were treated as external forces to the resulting open-loop subsystems. The dynamics of the rigid–flexible, 3-DOF parallel manipulator was derived and simulated for the following cases:

Case 1: *3RR* (Rigid–Rigid) manipulator.

In this case, both the links of the two-link serial subsystems were considered rigid (*R*).

Case 2: *3RF* (Rigid–Flexible) manipulator.

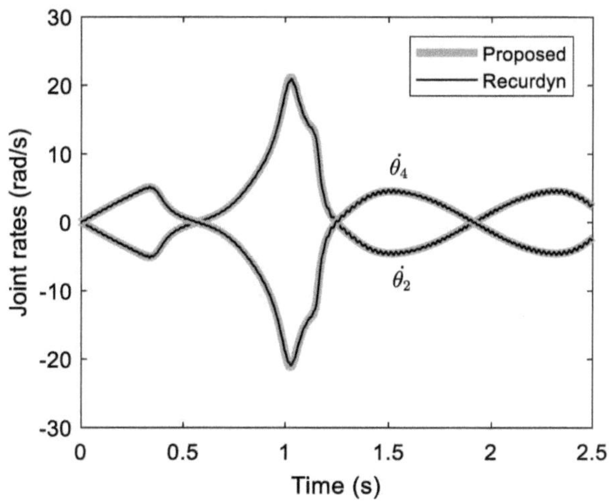

Fig. 5.34 Joint rates of *FFFF* five-bar mechanism

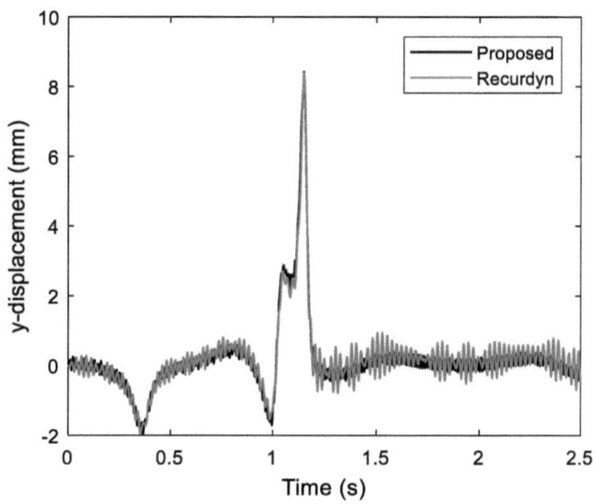

Fig. 5.35 Transverse displacement of #1 (location of joint 2) for *FFFF* five-bar mechanism

In this case, the two-link rigid–flexible serial subsystems were considered to have a rigid (R) link connected to the base, whereas the link connected to the platform was assumed flexible (F).

Case 3: *3FF* (Flexible–Flexible) manipulator.

In this case, both the links of two-link serial subsystems are flexible (F) links.

The dynamics of the rigid–flexible, 3-DOF planar parallel manipulator was then resolved in the following steps:

- write equations of motion for three two-link serial rigid–flexible subsystems;
- write equations of motion for the moving platform;
- write the constraint equations; and

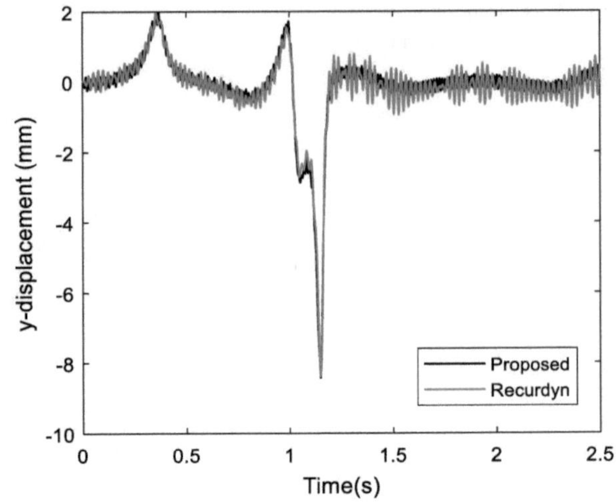

Fig. 5.36 Transverse displacement of #3 (location of joint 4) for *FFFF* five-bar mechanism

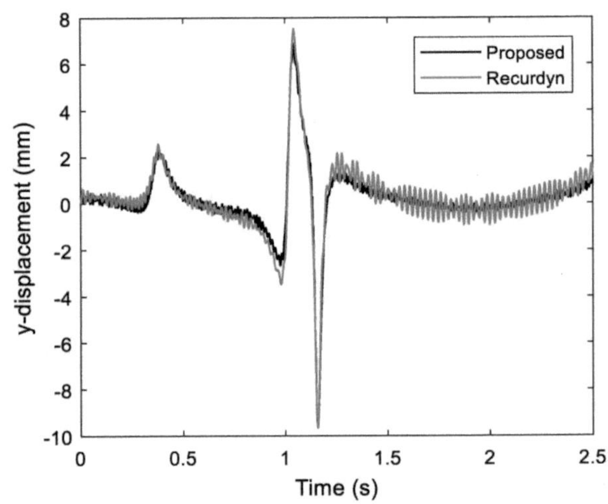

Fig. 5.37 Transverse displacement of #2 (location of joint 5) for *FFFF* five-bar mechanism

- finally, solve the system of equations of motion to get the required simulation results.

The derivations of the equations of motion of the two-link serial rigid–flexible manipulator were done using the methodology presented in Sects. 2.2 and 3.2 which were also used in both the four-bar and five-bar mechanisms. The derivation of the equations of motion of the moving platform is explained next.

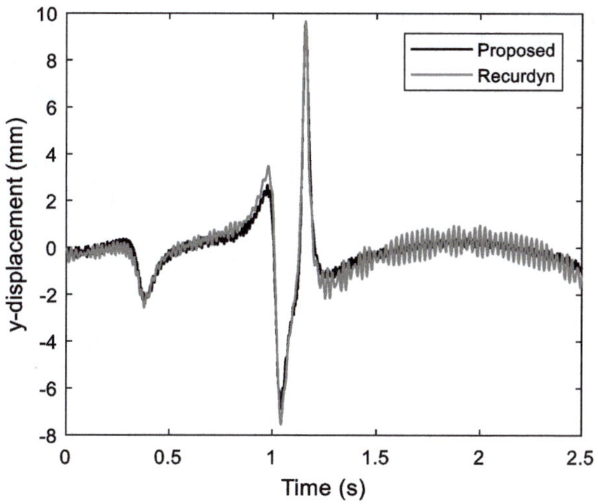

Fig. 5.38 Transverse displacement of #2 (location of joint 5) for *FFFF* five-bar mechanism

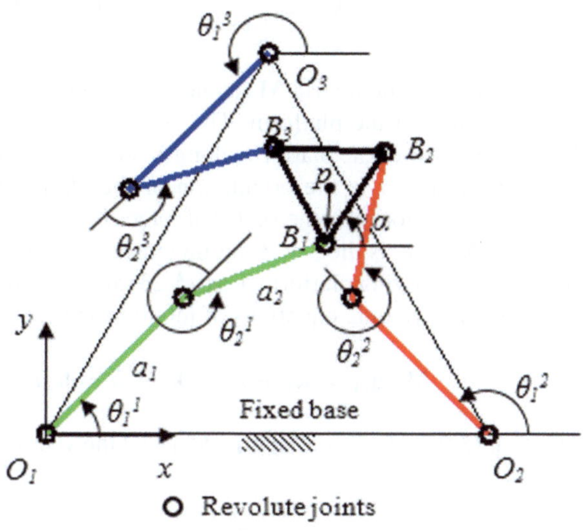

Fig. 5.39 Rigid–flexible 3-DOF parallel manipulator

5.4.1 Equations of Motion of the Moving Platform

Note that the freely moving platform has 6-DOF in space and 3-DOF in a plane, which can be expressed as

$$\mathbf{M}_e\dot{\mathbf{t}}_e + \mathbf{W}_e\mathbf{M}_e\mathbf{t}_e = \mathbf{w}_e \tag{5.23}$$

where the 6×6 matrices $\mathbf{M}_e$, $\mathbf{W}_e$, and the 6-dimensional vectors $\mathbf{t}_e$ and $\mathbf{w}_e$ are given as

Fig. 5.40 Subsystems of 3-DOF parallel manipulator

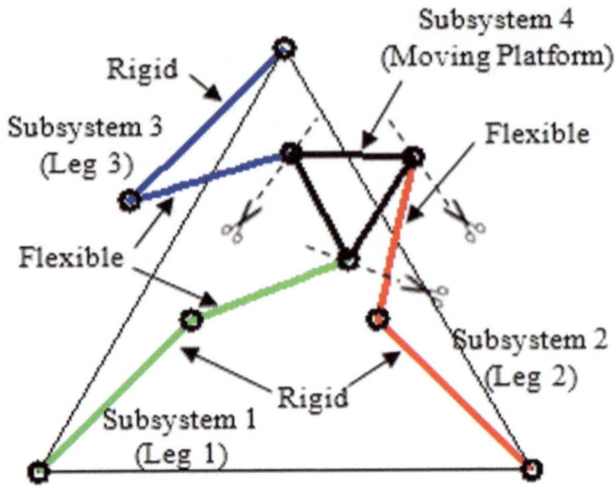

$$\mathbf{M}_e = \begin{bmatrix} m_e\mathbf{1} & \mathbf{O} \\ \mathbf{O} & \mathbf{I}_e \end{bmatrix}, \ \mathbf{W}_e = \begin{bmatrix} \mathbf{O} & \mathbf{O} \\ \mathbf{O} & \boldsymbol{\omega}_e \times 1 \end{bmatrix}, \ \mathbf{t}_e = \begin{bmatrix} \dot{\mathbf{q}}_e \\ \boldsymbol{\omega}_e \end{bmatrix}, \text{and } \mathbf{w}_e = \begin{bmatrix} \mathbf{f}_e \\ \mathbf{n}_e \end{bmatrix} \tag{5.24}$$

In Eq. (5.24), the matrix $\mathbf{M}_e$ is the mass matrix of the moving platform in which m_e is the mass of the platform. The vector $\dot{\mathbf{q}}_e \equiv [\ddot{x}_e \ \ddot{y}_e \ 0]^T$, where $\ddot{x}_e$ and $\ddot{y}_e$ are the accelerations of the platform along X- and Y-directions, respectively. Also, $\boldsymbol{\omega}_e \equiv [0 \ 0 \ \dot{\alpha}]^T$, where α is the orientation of the platform shown in Fig. 5.39. The 6-dimensional vector $\mathbf{w}_e$ is the vector of forces $\mathbf{f}_e$ and moments $\mathbf{n}_e$ acting on the moving platform. Also, $\mathbf{I}_e$ is the 3 × 3 inertia tensor of the moving platform. For the planar motion of the moving platform, Eqs. (5.23) and (5.24) can be significantly simplified. The resultant dynamic equations of motion can then be represented as

$$\mathbf{M}_{e1}\ddot{\mathbf{q}}_{e1} = \boldsymbol{\lambda}_1 + \boldsymbol{\lambda}_2 + \boldsymbol{\lambda}_3; \mathbf{I}_e\ddot{\alpha} = \tilde{\mathbf{b}}_1\boldsymbol{\lambda}_1 + \tilde{\mathbf{b}}_2\boldsymbol{\lambda}_2 + \tilde{\mathbf{b}}_3\boldsymbol{\lambda}_3 \tag{5.25}$$

In Eq. (5.25), the 2 × 2 matrix $\mathbf{M}_{e1}$ and the vector of accelerations $\ddot{\mathbf{q}}_{e1}$ are given by

$$\mathbf{M}_{e1} \equiv \begin{bmatrix} m_e & 0 \\ 0 & m_e \end{bmatrix}; \ and \ \ddot{\mathbf{q}}_{e1} \equiv \begin{bmatrix} \ddot{x}_e \\ \ddot{y}_e \end{bmatrix} \tag{5.26}$$

Also, $\boldsymbol{\lambda}_1$, $\boldsymbol{\lambda}_2$, and $\boldsymbol{\lambda}_3$ are the 2-dimensional vectors of constraint reaction forces arising due to the cut joints at points B_1, B_2, and B_3, as shown in Fig. 5.39. They are also known as Lagrange multipliers. Moreover, $\tilde{\mathbf{b}}_1$, $\tilde{\mathbf{b}}_2$, and $\tilde{\mathbf{b}}_3$ are the skew-symmetric matrices associated with the vectors $\mathbf{b}_1$, $\mathbf{b}_2$, and $\mathbf{b}_3$ which denote the platform center P from the vertices B_1, B_2, and B_3, respectively.

5.4.2 Kinematic Constraints

The kinematic constraints to ensure the connectivity of the cut joints at points B_1, B_2, and B_3 are used for the dynamic modeling of the rigid–flexible 3-DOF planar parallel manipulator. They are written in the acceleration level as

$$\mathbf{J}^k\ddot{\mathbf{q}}^k + \tilde{\mathbf{b}}^k\ddot{\alpha} - \ddot{\mathbf{q}}_e = -\dot{\mathbf{J}}^k\dot{\mathbf{q}}^k - \dot{\tilde{\mathbf{b}}}^k\dot{\alpha}^2, \text{ for } k = 1, 2, 3 \quad (5.27)$$

where $\mathbf{J}^k$ is the Jacobian matrix of the two-link serial rigid–flexible subsystem, and $\ddot{\mathbf{q}}^k \equiv \left[\ddot{\theta}_1\ \ddot{\theta}_2\ \ddot{\mathbf{d}}\right]^k$ for the kth subsystem. The final equations of motion including the equations of motion of the moving platform and the kinematic constraints are assembled below in a matrix–vector form as

$$\begin{bmatrix} \mathbf{I}_1 & \mathbf{O} & \mathbf{O} & \mathbf{O} & \mathbf{O} & \mathbf{J}_1^T & \mathbf{O} & \mathbf{O} \\ & \mathbf{I}_2 & \mathbf{O} & \mathbf{O} & \mathbf{O} & \mathbf{O} & \mathbf{J}_2^T & \mathbf{O} \\ & & \mathbf{I}_3 & \mathbf{O} & \mathbf{O} & \mathbf{O} & \mathbf{O} & \mathbf{J}_3^T \\ & & & \mathbf{M}_e & \mathbf{O} & -\mathbf{1} & -\mathbf{1} & -\mathbf{1} \\ & & & & \mathbf{I}_e & \tilde{\mathbf{b}}_1^T & \tilde{\mathbf{b}}_2^T & \tilde{\mathbf{b}}_3^T \\ & & & & & \mathbf{O} & \mathbf{O} & \mathbf{O} \\ & sym & & & & & \mathbf{O} & \mathbf{O} \\ & & & & & & & \mathbf{O} \end{bmatrix} \begin{bmatrix} \ddot{\mathbf{q}}_1 \\ \ddot{\mathbf{q}}_2 \\ \ddot{\mathbf{q}}_3 \\ \ddot{\mathbf{q}}_e \\ \ddot{\alpha}_e \\ \boldsymbol{\lambda}_1 \\ \boldsymbol{\lambda}_2 \\ \boldsymbol{\lambda}_3 \end{bmatrix} = \begin{bmatrix} \phi_1 \\ \phi_2 \\ \phi_3 \\ \mathbf{0} \\ \mathbf{0} \\ \phi_4 \\ \phi_5 \\ \phi_6 \end{bmatrix} \quad (5.28)$$

Equation (5.28) is the desired equations of motion of the system at hand that was used for the dynamic simulation of a rigid–flexible planar 3-DOF parallel manipulator.

5.4.3 3RR Manipulator

Freefall simulation of the 3RR 3-DOF parallel manipulator at hand was conducted; i.e., the manipulator in vertical configuration was allowed to fall freely under gravity. Tables 5.3 and 5.4 show the input parameters and the initial conditions of the 3-DOF rigid–flexible planar manipulator. The flexible link was assumed to have transverse displacement; i.e., the Euler–Bernoulli beam was used.

The simulation was run for a period of 0.5 s. Figures 5.41 and 5.42 show the joint angles for Subsystems-I–III, whereas Figs. 5.43 and 5.44 show the joint rates for the same subsystems. The *3RR* 3-DOF parallel manipulator was also modeled

Table 5.3 Input parameters of 3-DOF planar parallel manipulator

O_{1x}	O_{1y}	O_{2x}	O_{2y}	O_{3x}	O_{3y}	a_1	a_2	m_1	m_2	l	m_e
0	0	0.50 m	0	0.25 m	0.433 m	0.22 m	0.17 m	2.81 kgs	1.41 kgs	0.0722 m	8 kgs

Table 5.4 Initial values of the rigid–flexible 3RRR manipulator

	Subsystem-I	Subsystem-II	Subsystem-III	Subsystem-IV
θ_1	45°	135°	225°	–
θ_2	335.87°	302.28°	152.02°	–
d	0	0	0	–
x_e	–	–	–	0.3178 m
y_e	–	–	–	0.2882 m
α	–	–	–	57.33°

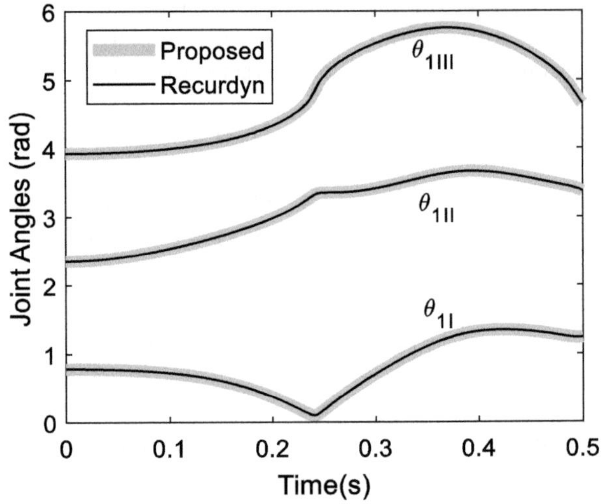

Fig. 5.41 Joint angles of 3*RR* manipulator

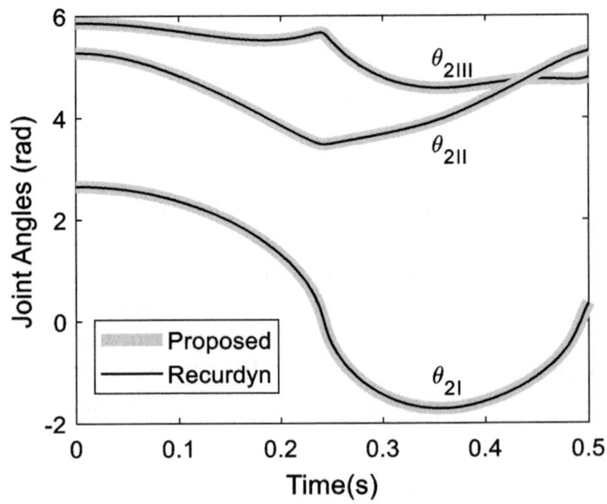

Fig. 5.42 Joint angles of 3*RR* manipulator

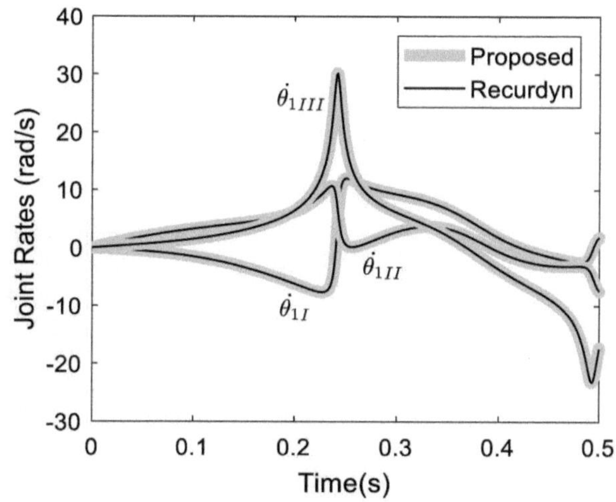

Fig. 5.43 Joint rates of 3*RR* manipulator

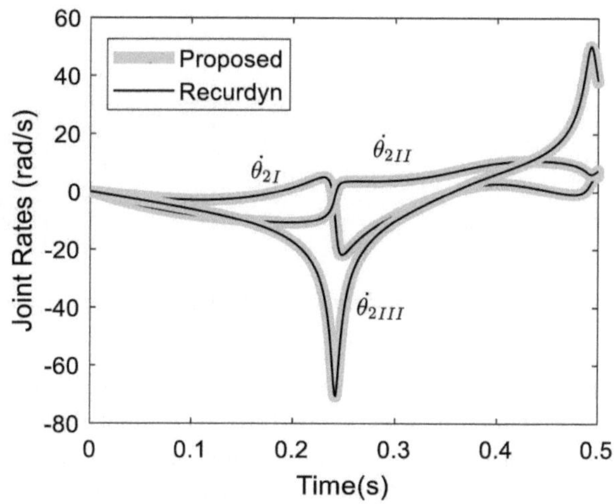

Fig. 5.44 Joint rates of 3*RR* manipulator

in RecurDyn software. The simulated results show a close match with the proposed DeNOC-based approach (Fig. 5.45).

5.4.4 3RF Manipulator

In the 3*RF* 3-DOF parallel manipulator, the links connected to the base were considered rigid (*R*), whereas those connected to the moving platform were considered flexible (*F*). The dynamic equations of motion for the 3*RF* 3-DOF parallel manipulator were formulated similar to the 3*RR* 3-DOF parallel manipulator. Here, the equations

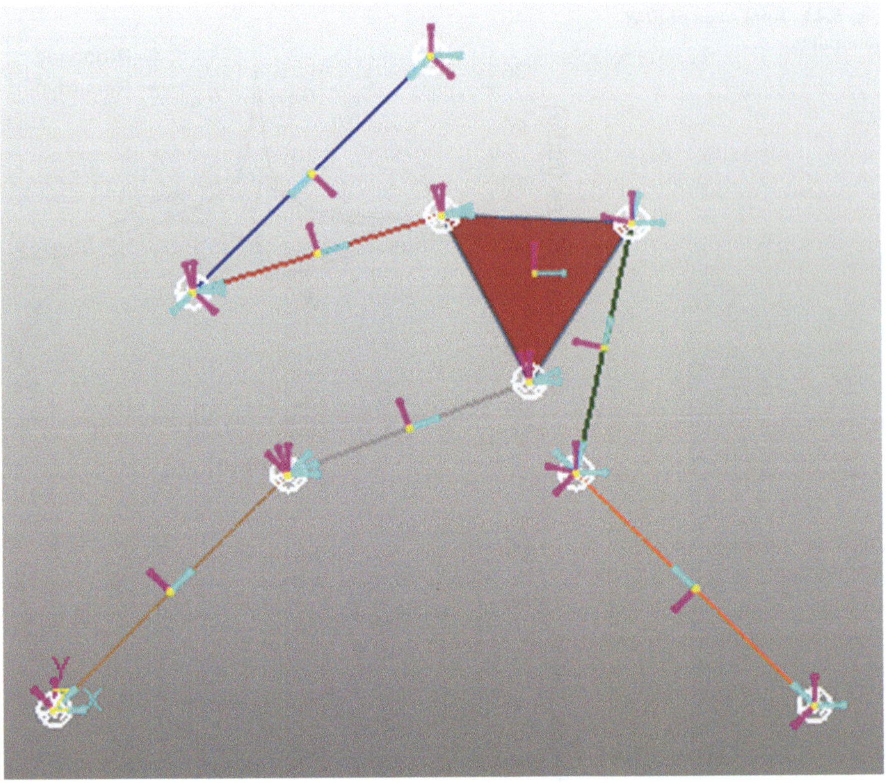

Fig. 5.45 RecurDyn model for the 3-DOF parallel manipulator

of motion were derived for rigid–flexible (*RF*) two-link manipulator instead of *RR* manipulator for Subsystems-I–III. The input parameters and initial configuration for forward dynamic simulation are considered the same as given in Tables 5.3 and 5.4, respectively. For simulation, the manipulator was allowed to fall freely under gravity, with no external torque. The simulation was run for a time period of 0.5 s. Figures 5.46 and 5.47 show the joint angles, and Figs. 5.48 and 5.49 show the joint rates. Also, the transverse displacements for the flexible links of Subsystems-I–III are shown in Figs. 5.50, 5.51, and 5.52. The results show very good agreement between the results of the DeNOC formulation and RecurDyn.

5.4.5 3*FF* Manipulator

In the 3*FF* 3-DOF parallel manipulator, all the links connecting the fixed and moving platform were treated as flexible (*F*). The dynamic equations of motion for the 3*FF* 3-DOF parallel manipulator were formulated similar to the 3*RR* or 3*RF* 3-DOF parallel

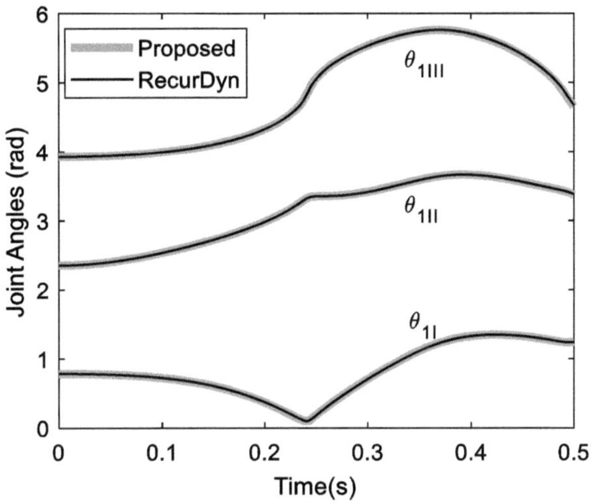

Fig. 5.46 Joint angles of 3*RF* manipulator

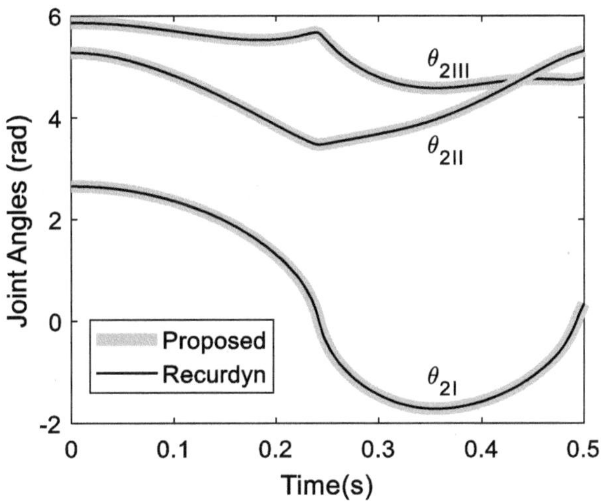

Fig. 5.47 Joint angles of 3*RF* manipulator

manipulator. Here, the equations of motion were derived for the full flexible two-link manipulator (*FF*) instead of the *RR* manipulator for Subsystems-I–III. The platform equations of motion remain the same. The velocity constraints were provided at the vertices of the platform, i.e., at points B_1, B_2, and B_3. The forward dynamics simulation was done for free fall. The results are plotted in Figs. 5.53, 5.54, 5.55, 5.56, 5.57, 5.58, 5.59, 5.60, 5.61, and 5.62, where $\theta_{1,1}$, $\theta_{1,2}$, and $\theta_{1,3}$ represent the first joint angles of Subsystems-I, II, and III, respectively. The same nomenclature was used for the joint rates and accelerations. The results were also obtained using the finite element model in RecurDyn, which shows a good match with those obtained from the DeNOC-based formulation.

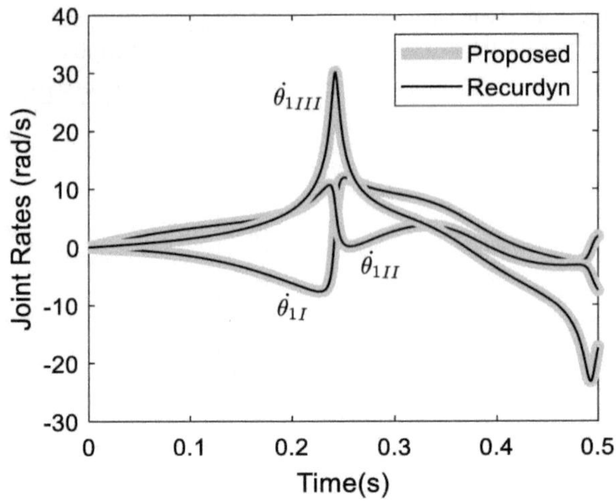

Fig. 5.48 Joint rates of $3RF$ manipulator

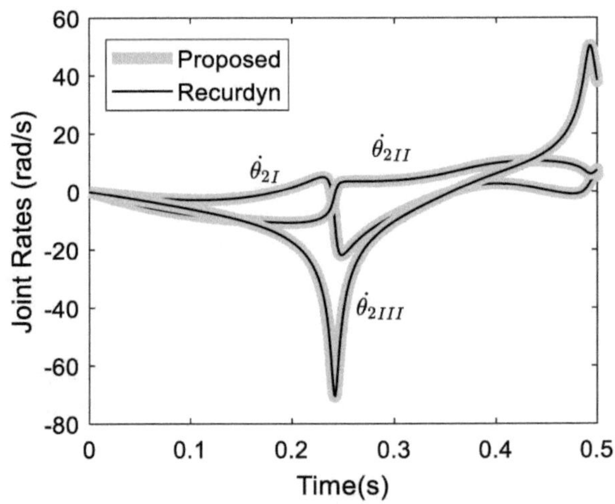

Fig. 5.49 Joint rates of $3RF$ manipulator

5.5 Forced Simulation

The forced simulation is explained in this session to illustrate the path tracking performance of a closed-loop system. It is assumed that the end-effector of a manipulator needs to follow the desired trajectory. For that, joint torques/forces can be obtained from the proposed inverse dynamics algorithm. Simulations of both rigid and rigid–flexible links in a manipulator are presented here to compare the effect of flexible links on the behavior of the manipulator, which would help designers to select the links or choose a suitable controller. Figure 5.63 shows the scheme for studying the flexibility effect, where a set of input joint torques/forces to follow a specified joint trajectory was calculated using the inverse dynamics algorithm of a

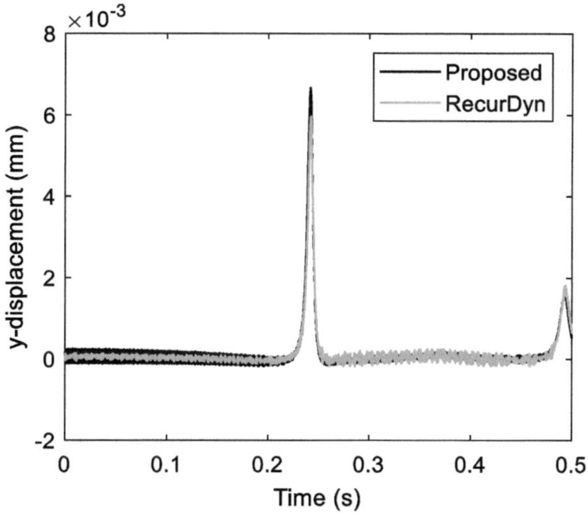

Fig. 5.50 Transverse displacement of the #2 of Subsystem-I

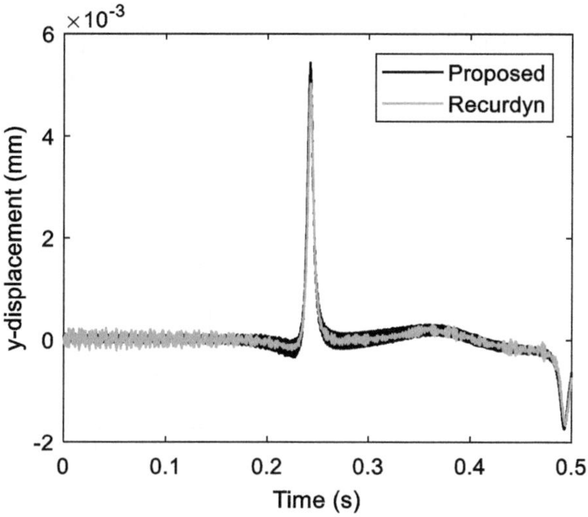

Fig. 5.51 Transverse displacement of the #2 of Subsystem-III

rigid system. These torques and forces were then given as inputs to two forward dynamics algorithms, one for rigid and another one for the rigid–flexible manipulator. The simulated joint trajectories were next compared with the specified input joint trajectories, which would show the effect of flexibility in the links, provided the numerical integration errors are negligible.

Fig. 5.52 Transverse displacement of the #2 of Subsystem-III

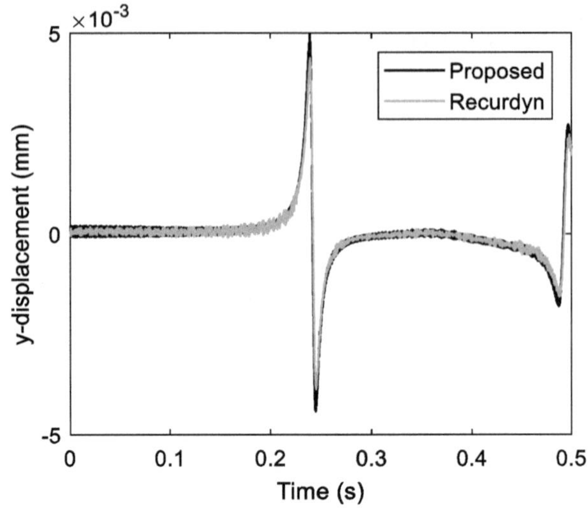

Fig. 5.53 Joint angles of 3*FF* manipulator

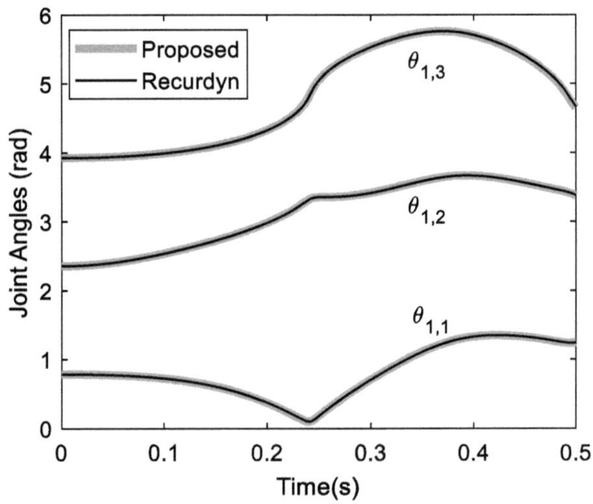

5.5.1 *Forced Simulation of 3-DOF Parallel Manipulator*

Here, the illustration is done with respect to the 3-DOF parallel manipulator. The tracking of a circular trajectory of 120 mm diameter with the center located at 0.25 m and 0.1443 m along X- and Y-directions, respectively, is desired by the 3RR and 3FF 3-DOF parallel manipulators as shown in Fig. 5.64. To find the required input torques, the inverse dynamics algorithm proposed in session 5.1.2 was used for the rigid 3RR 3-DOF parallel manipulator. The system was cut at the vertices of the moving platform as shown in Fig. 5.40. Subsystems-I–III have two-links each, connected by revolute

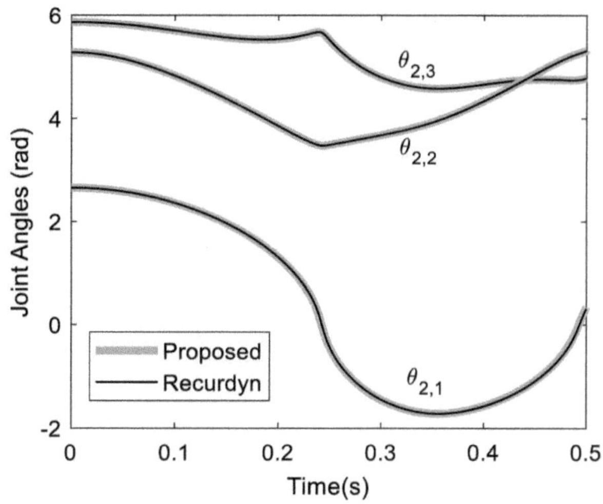

Fig. 5.54 Joint angles of $3FF$ manipulator

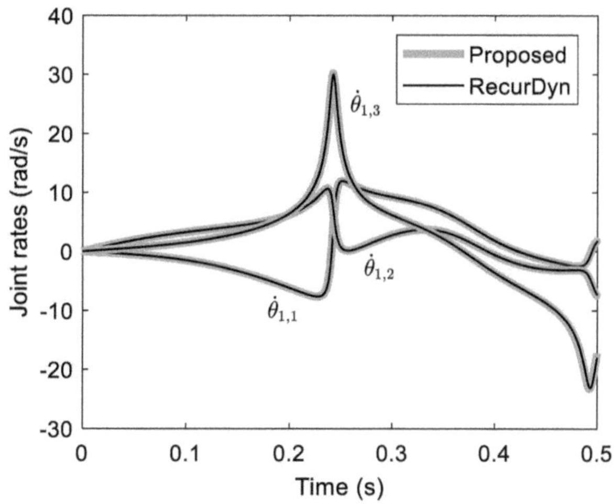

Fig. 5.55 Joint rates of $3FF$ manipulator

joints. The unknowns associated with the Subsystems-I–III have two components of Lagrange multipliers, λ_x, λ_y and the driving torque τ. They have two equations of motion for each subsystem. On the other hand, Subsystem-IV has three equations of motion, and the unknowns associated with it are the two components of the Lagrange multipliers, λ_x and λ_y at its vertices. Thus, the total number of unknowns are six for Subsystem-IV. This makes the system indeterminate. All four subsystems together, however, are determinate. To take advantage of the smaller subsystems, the alternative approach of Sect. 5.1.2 was adopted here, where the passive joint information was used to solve for the Lagrange multipliers, and the joint torques were computed for the specified trajectory of the end-effector. The computed joint torques/forces as shown in Fig. 5.64 were then fed to the forward dynamics algorithms of both $3RR$

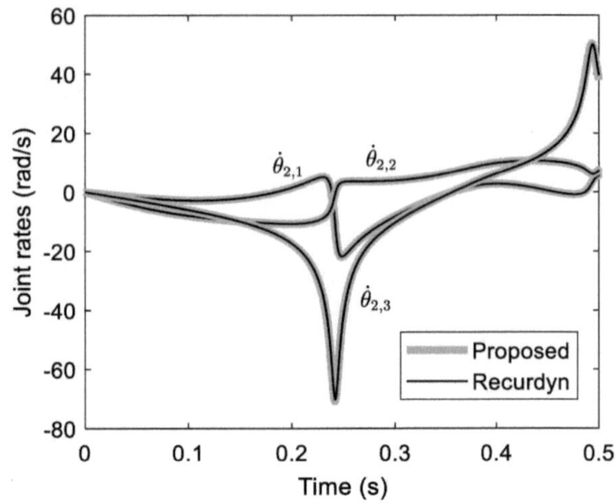

Fig. 5.56 Joint rates of *3FF* manipulator

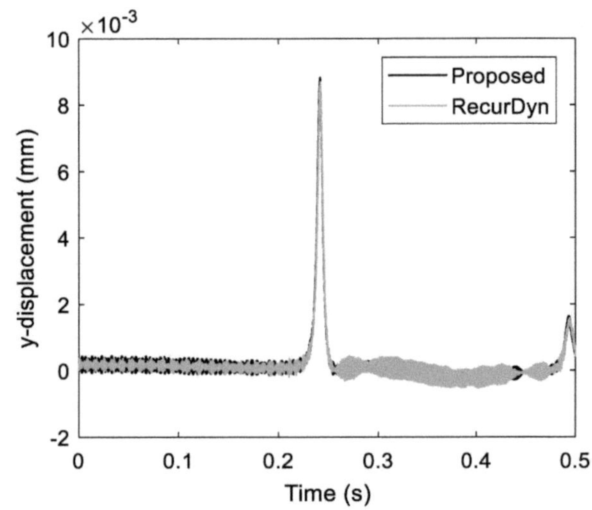

Fig. 5.57 Transverse displacement of Subsystem-I

and *3FF* 3-DOF parallel manipulators. For the computation purpose, the mass and inertia parameters were taken the same, as presented in the preceding sections. The joint angles of link 1 and link 2 for Subsystems-I–III are plotted in Figs. 5.65 and 5.66, respectively, for the *3RR* 3-DOF parallel manipulators along with and *3FF* manipulator to study the effect of flexibility.

The joint angles of the *3RR* system for tracking the circle show the smooth motion, whereas *3FF* system shows the fluctuations in the joint motions. Figures 5.67 and 5.68 show the errors in *x*- and *y*-coordinates of the end-effector while tracking the desired circle. The higher errors are certainly due to the flexibility of the links, which

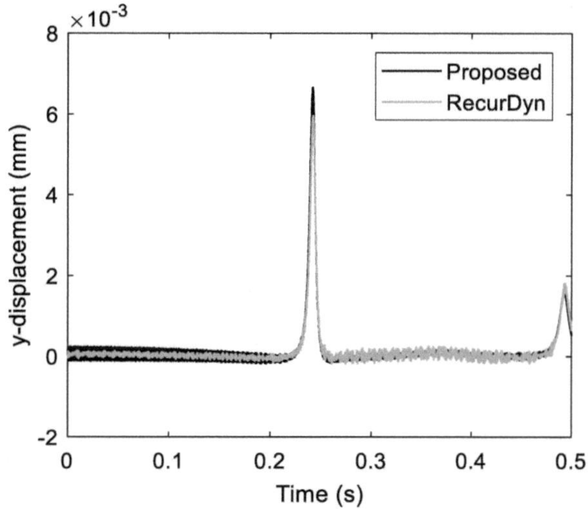

Fig. 5.58 Transverse displacement of Subsystem-I

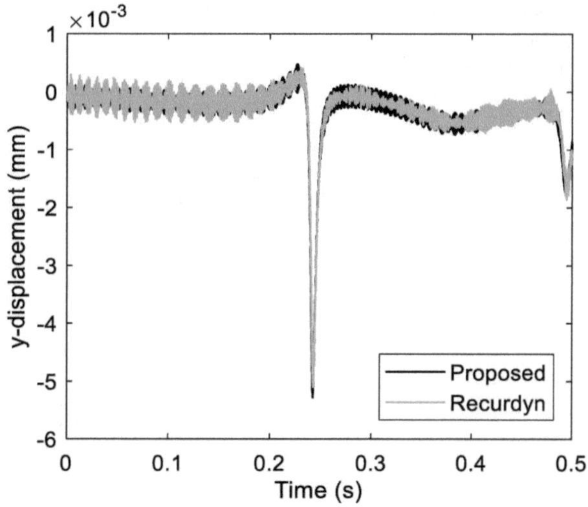

Fig. 5.59 Transverse displacement of Subsystem-II

then needed to be accounted for either during the mechanical design of the links or to choose to appropriate control algorithm to stabilize the motion of the manipulator.

5.6 Summary

Dynamic modeling and simulation of a closed-loop, rigid–flexible multibody system using the DeNOC matrices were proposed in this chapter. The methodology allows one to exploit the existing dynamic algorithms for the open-loop, serial-chain systems

Fig. 5.60 Transverse displacement Subsystem-II

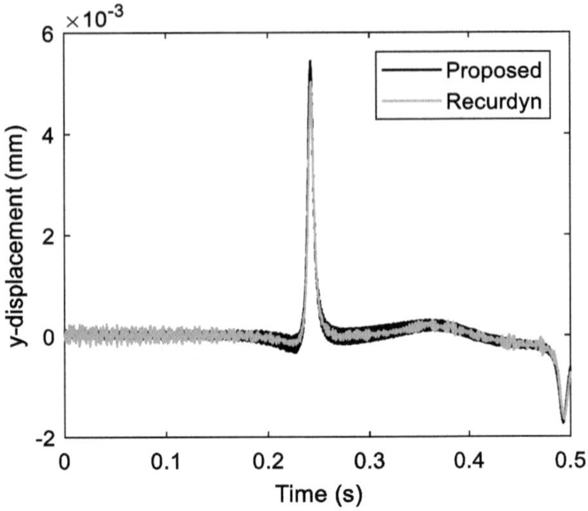

Fig. 5.61 Transverse displacement of Subsystem-III

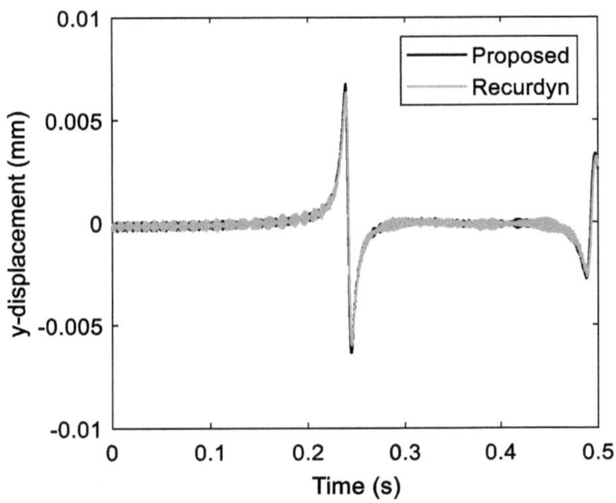

to solve the closed-loop systems. To illustrate the concept and methodology, several planar closed-loop systems, with a single loop, namely four-bar and five-bar mechanisms and multiclosed loops, such as 3-DOF parallel manipulator were undertaken. The results were verified with the results from a commercial software RecurDyn. Further, the forced simulation was carried out to study the effect of structural flexibility in the link. The illustration of the 3-DOF parallel manipulator was shown for the forced simulation. The study gives the direction to design and control engineers to develop a suitable design and/or control strategy to minimize tracking errors.

Fig. 5.62 Transverse displacement Subsystem-III

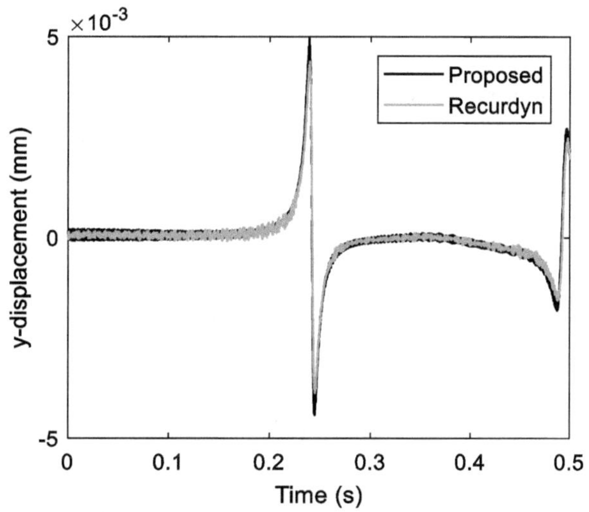

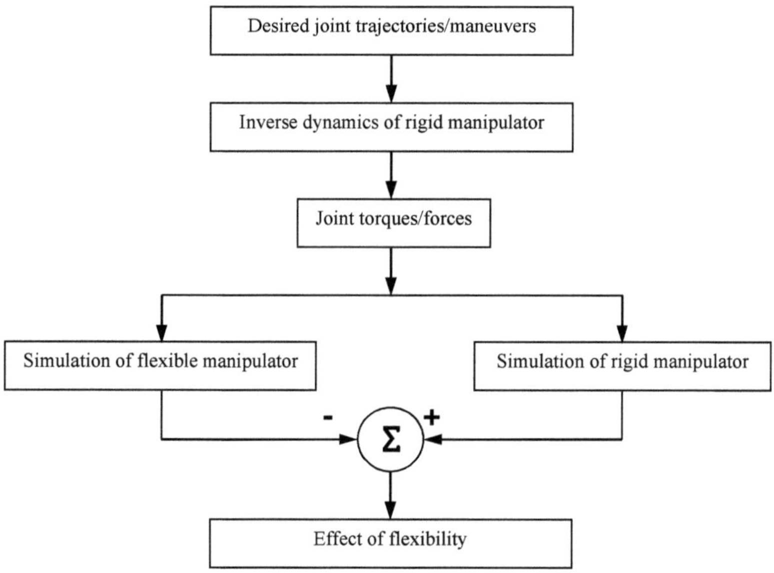

Fig. 5.63 Forced simulation for studying flexibility effect

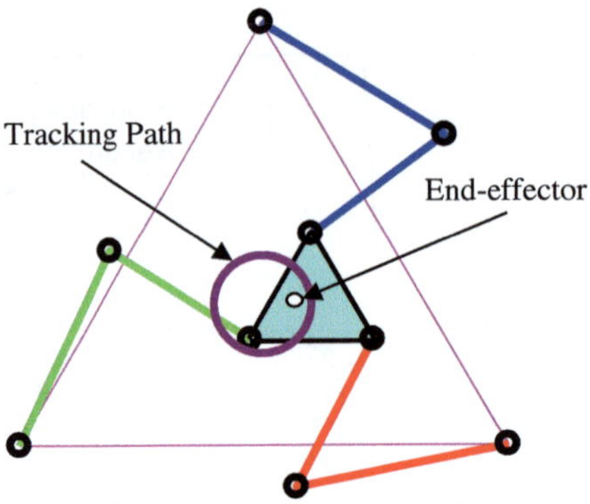

(a) The 3-DOF parallel manipulator

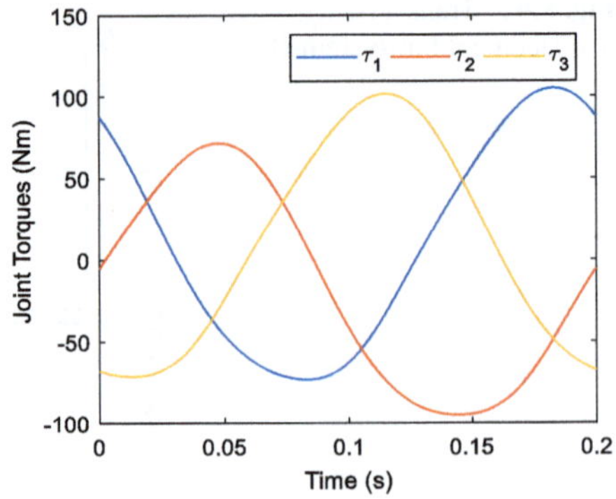

(b) Inverse dynamic results

Fig. 5.64 Path tracking by 3-DOF parallel manipulator and its associated joint torques

Fig. 5.65 Joint 1 angles of 3-DOF Parallel manipulator

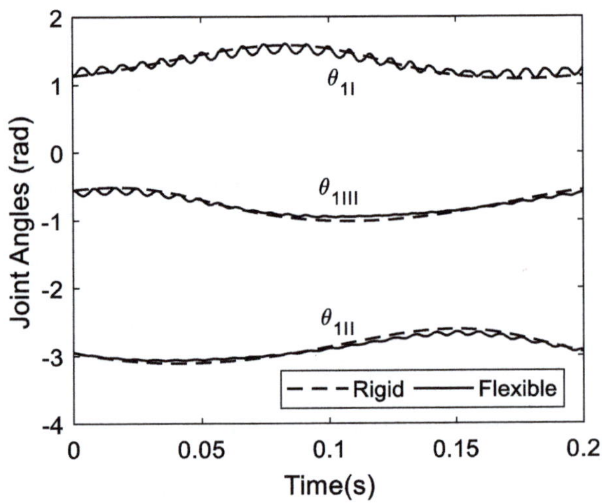

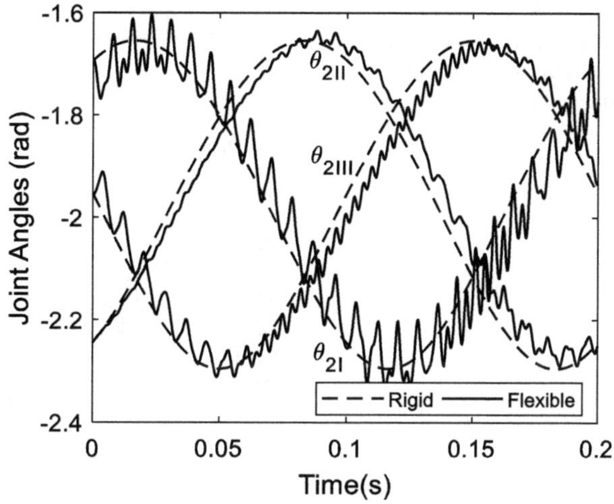

Fig. 5.66 Joint 2 angles of 3-DOF Parallel manipulator

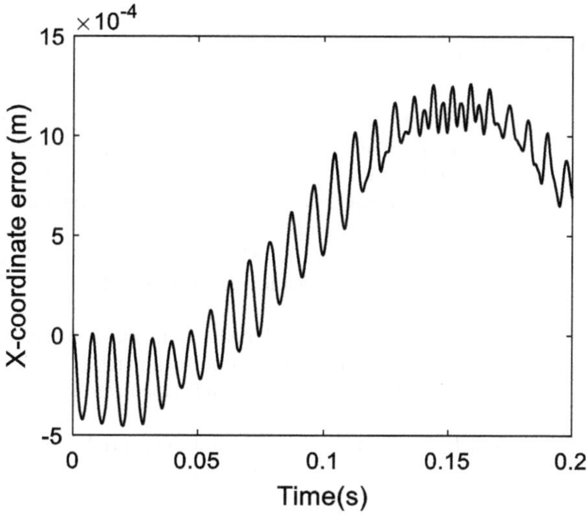

Fig. 5.67 Error along *x*-coordinate error of the end-effector of flexible manipulator

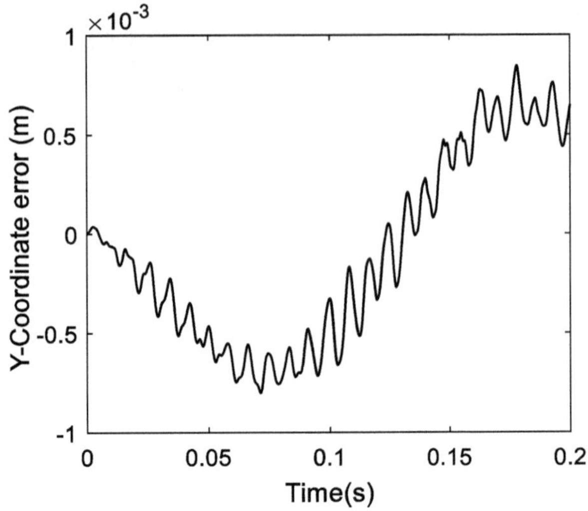

Fig. 5.68 Error along *y*-coordinate error of the end-effector of flexible manipulator

References

1. Agarwal A, Shah SV, Bandyopadhyay S, Saha SK (2014) Dynamics of serial kinematic chains with large number of degrees-of-freedom. Multibody SysDyn 32(3):273–298. https://doi.org/10.1007/s11044-013-9386-3
2. Chaudhary H, Saha SK (2007) Constraint force formulation for closed-loop multibody systems. Trans ASME J Mech Des (129,):1234–1242
3. Koul M, Shah SV, Saha SK, Manivannan M (2014) Reduced-order forward dynamics of multiclosed-loop systems. Multibody SysDyn 31(4):451–476. https://doi.org/10.1007/s11044-013-9379-2
4. Meirovitch L (1967) Analytical methods in vibration. The Mcmillan Company, New York, NY
5. Mohan A (2007) Dynamic analysis of flexible multibody robotic systems. IIT, Delhi
6. Saha SK (1999) Analytical expression for the inverted inertia matrix of serial robots. Int J Robot Res 18(1):116–124
7. Shah S, Saha S, Dutt J (2011) Modular framework for dynamic modeling and analyses of legged robots. Mechanism and Machine Theory

Chapter 6
Dynamics of Spatial Four-Bar Mechanism

This chapter presents the formulation of dynamic equations of motion of the closed-loop spatial four-bar mechanism with rigid and flexible links using the decoupled natural orthogonal complement (DeNOC) matrices. The flexible link was discretized using the assumed mode method to get the link deflection. The closed-loop rigid–flexible spatial four-bar mechanism was analyzed by first cutting it at appropriate joints to form several open-loop serial-chain subsystems. The opened joints were substituted with suitable constraint forces denoted with λ's, which are known as Lagrange multipliers. These multipliers were treated as external forces to the resulting open-loop subsystems. The Lagrange multipliers were then calculated at the subsystem level to reduce the complexity of the overall formulation.

6.1 Spatial Four-Bar Mechanism

The special four-bar mechanism is used to transmit motion between non-parallel axes in 3-dimensional space. Many configurations of the spatial four-bar were found in the literature, such as skew four-bar mechanism, Bennett mechanism, Hooke joint, spherical four-bar, spatial slider-crank mechanism [3, 5], etc. The RSSR spatial four-bar mechanism is most popular among all for which vast literature are available for its motion study, synthesis, and dynamic analysis [1, 2, 4, 6]. Most of the paper cited above considers the link of the mechanism as rigid. Very few paper deals with the flexibility of the spatial four-bar mechanism such as Liou and Erdman [4]. In this chapter, we consider the full rigid link mechanism as well as different combination of rigid and flexible links of a special four-bar mechanism for analysis.

P. V. Nandihal et al., *Dynamics of Rigid-Flexible Robots and Multibody Systems*, Intelligent Systems, Control and Automation: Science and Engineering 100,
https://doi.org/10.1007/978-981-16-2798-9_6

6.2 Rigid–Flexible Spatial Four-Bar Mechanism

A spatial four-bar mechanism comprises of four links with one fixed link and three moving links having revolute (R), spherical (S), spherical (S), and revolute (R) joints at joints 1, 2, 4, and 5, respectively, as shown in Fig. 6.1a.

Hence, it is also called RSSR mechanism. In this mechanism, the revolute joints, joints 1 and 4, are in different planes, and their axes are non-parallel. The RSSR mechanism has kinematically two degrees of freedom (DOF). One of them is redundant. It is the rotation of the coupler about its own axis. The RSSR mechanism was analyzed by cutting it at joint 5, i.e., the spherical joint, to form two open-loop serial-chain subsystems, as shown in Fig. 6.1b. The resulting open-loop subsystems are the two-link serial-chain system and a single-link system. The cutting of the spherical joint compared to any other joint reduced the extra effort to write more equations of motion. Hence, it is advantageous to cut at multi-DOF joints such as spherical and universal joint instead of one-DOF joints, i.e., revolute or prismatic.

One of the subsystems, namely, Subsystem-I of Fig. 6.1b, consists of two links with one revolute joint and one spherical joint. In contrast, the other subsystem, i.e., Subsystem-II, consists of only one link with one revolute joint. Note that the axis of the revolute joint of the single-link subsystem, i.e., Subsystem-II, is inclined with respect to the Z-axis of the fixed frame, say z_0. This inclination is defined as the rotation about fixed X or x_0-axis by an angle α_4. As per the notation of the Denavit–Hartenberg (DH) parameter, the angle α_4 is the twist angle. The cut-opened joint was then substituted with suitable constraint forces denoted by $\boldsymbol{\lambda}$'s, which are known as Lagrange multipliers. The Lagrange multipliers for spatial mechanism have three components, namely, λ_x, λ_y, and λ_z. For the dynamic computations using the recursive algorithms for spatial-chain systems, the Lagrange multipliers were treated as external forces to the subsystems-I and II. The following analyses were conducted for the comparison of results:

Case 1: Rigid–Rigid–Rigid (*RRR*): In this case, the dynamic analyses were done considering all three links, i.e., crank (#4) for Subsystem-II, and coupler (#3),

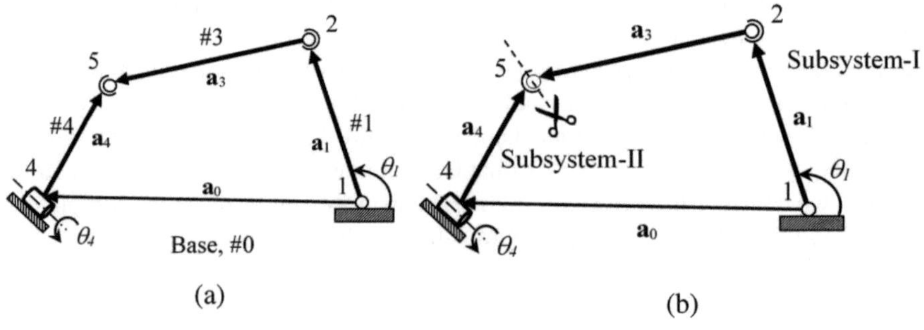

Fig. 6.1 RSSR mechanism

and rocker (#1) for Subsystem-I, as rigid (*R*). Figure 6.1 shows the *RRR*-RSSR mechanism.

Case 2: Rigid–Rigid–Flexible (*RRF*): Here, the crank (#4) of Subsystem-II and the coupler (#3) of Subsystem-I were considered rigid (*R*), while the rocker (#1) of Subsystem-I was considered flexible (*F*).

Case 3: Flexible–Rigid–Rigid (*FRR*): The coupler (#3) and rocker (#1) of Subsystem-I were treated as rigid (*R*). Only the crank (#4) of Subsystem-II, was considered flexible (*F*).

Case 4: Flexible–Flexible–Flexible (*FFF*): Here, all the three links of both subsystems were treated as flexible (*F*).

The dynamic formulations of all the above cases of the spatial four-bar mechanism were explained in succeeding subsections. The important steps involved in the dynamic formulation of a typical rigid–flexible spatial four-bar mechanism are as follows:

- write the equations of motion of a two-link serial rigid or flexible manipulator with one revolute and one spherical joint;
- write the equations of motion for a single-link rigid or flexible manipulator, as carried out in Chaps. 2 and 3;
- write the constraint equations;
- finally, solve the system of equations of motion to get the required solution, i.e., the motion of the four-bar mechanism for a given initial state or configuration.

6.2.1 RRR Spatial Mechanism

A spatial four-bar RRR-RSSR mechanism is shown in Fig. 6.1. All the links, i.e., crank (#4), coupler (#3), and rocker (#1), were treated as rigid. The spherical joints with three degrees of freedom (DOF) were modeled equivalent to three mutually orthogonal and interesting one-DOF revolute joints. Note that, the RSSR mechanism has one redundant degree of freedom which is the rotation of the coupler (#3) about its own axis. The same can be eliminated in the formulation of the equations of motion by replacing the spherical joint with a universal one. It will not change the input–output rotation of the spatial four-bar mechanism. For analysis purposes, the mechanism was partitioned into two separate subsystems, as shown in Fig. 6.1b.

The dynamic equations of motion of the two-link manipulator with a universal joint, i.e., Subsystem-I, were modeled as an equivalent three-link manipulator system with all joints as revolute. The length of the vanishing second link has an axis whose twist angle is $\alpha_2 = 90°$. The two sets of constrained equations of motion for the two serial-chain subsystems were then derived using the methodology presented in Chap. 2. They are expressed as

$$\begin{aligned} \mathbf{I}^I\ddot{\mathbf{q}}^I + \mathbf{h}^I &= \boldsymbol{\tau}^I + \left(\boldsymbol{\tau}^\lambda\right)^I \\ \mathbf{I}^{II}\ddot{\mathbf{q}}^{II} + \mathbf{h}^{II} &= \boldsymbol{\tau}^{II} + \left(\boldsymbol{\tau}^\lambda\right)^{II} \end{aligned} \quad (6.1)$$

The 3 × 3 GIM $\mathbf{I}^I$ for Subsystem-I with link length $a_2 = 0$, and twist angle $\alpha_2 = -\pi/2$ is written as

$$\mathbf{I}^I \equiv \begin{bmatrix} i_{11} & i_{12} & i_{13} \\ & i_{22} & i_{23} \\ sym & & i_{33} \end{bmatrix} \tag{6.2}$$

where

$$i_{11} \equiv \frac{\rho_1 a_1^3}{3} + \rho_3 a_1^2 a_3 + \left(\frac{\rho_3 a_3^3 c_3}{3} + \rho_3 a_1 a_3^2 c_2\right) c_3;$$

$$i_{12} \equiv \left(\frac{\rho_3 a_3^3 c_3}{3} + \frac{\rho_3 a_1 a_3^2 c_2}{2}\right) c_3;\ i_{13} \equiv -\frac{\rho_3 a_1 a_3^2 s_2}{2} s_3;\ i_{22} \equiv \frac{\rho_3 a_3^3 c_3}{3} c_3;$$

$$i_{23} \equiv 0;\ i_{33} \equiv \frac{\rho_3 a_3^3}{3}$$

In the above expressions, ρ is the mass per unit length of the link, and a is its length. The subscripts 1 and 3 are used for the rocker and the coupler. Furthermore, $c_i \equiv \cos\theta_i$ and $s_i \equiv \sin\theta_i$ for the ith link. The 3-dimensional vectors $\mathbf{h}^I$ and the vector of generalized force due to gravity $\boldsymbol{\tau}^{gI}$ for Subsystem-I are given by

$$\mathbf{h}^I \equiv \begin{bmatrix} h_1 \\ h_2 \\ h_3 \end{bmatrix} \text{ and } \boldsymbol{\tau}^{gI} \equiv \begin{bmatrix} \tau_1^{gI} \\ \tau_2^{gI} \\ \tau_3^{gI} \end{bmatrix} \tag{6.3}$$

where

$$h_1 \equiv -\left(\frac{2}{3}\rho_3 a_3^3 s_3(\dot\theta_1 + \dot\theta_2)\dot\theta_3 + \frac{1}{2}\rho_3 a_1 a_3^2 s_2(\dot\theta_2^2 + \dot\theta_3^2 + 2\dot\theta_1\dot\theta_2)\right) c_3$$
$$- \left(\rho_3 a_1 a_3^2 c_2(\dot\theta_1\dot\theta_3 + \dot\theta_2\dot\theta_3)\right) s_3$$
$$h_2 \equiv -\left(\frac{2}{3}\rho_3 a_3^3 s_3(\dot\theta_1 + \dot\theta_2)\dot\theta_3 - \frac{1}{2}\rho_3 a_1 a_3^2 s_2 \dot\theta_1^2\right) c_3$$
$$h_3 \equiv \left(\frac{1}{3}\rho_3 a_3^3 s_3(\dot\theta_1^2 + 2\dot\theta_1\dot\theta_2 + \dot\theta_2^2)\right) c_3 + \frac{1}{2}\rho_3 a_1 a_3^2 c_2 \dot\theta_1^2 s_3$$
$$\tau_1^{gI} \equiv \frac{1}{2}(\rho_1 g a_1^2 c_1 + 2\rho_3 g a_1 a_3 c_1 + \rho_3 g a_3^2 c_{12} c_3)$$
$$\tau_2^{gI} \equiv \frac{1}{2}\rho_3 g a_3^2 c_{12} c_3 \text{ and } \tau_3^{gI} \equiv -\frac{1}{2}\rho_3 g a_3^2 s_{12} s_3 \tag{6.4}$$

Note that, the symbols used in Eqs. (6.1)–(6.4) were already discussed in the previous chapter. Similarly, for Subsystem-II, the terms $\mathbf{I}^{II}$, $\mathbf{h}^{II}$, $\boldsymbol{\tau}^{gII}$ are scalars which are given by

$$\mathbf{I}^{II} \equiv \frac{\rho_4 a_4^3}{3}, \mathbf{h}^{II} \equiv 0, \boldsymbol{\tau}^{gII} \equiv \frac{\rho_4 g a_4^2}{2} c_4 \bar{c}_4 \tag{6.5}$$

where $c_4 \equiv \cos\theta_4$, and $\bar{c}_4 \equiv \cos\alpha_4$. Note that $(\boldsymbol{\tau}^{\lambda})^I = (\mathbf{J}^I)^T \boldsymbol{\lambda}$ and $(\boldsymbol{\tau}^{\lambda})^{II} = (\mathbf{J}^{II})^T \boldsymbol{\lambda}$ in which $\mathbf{J}^I$ and $\mathbf{J}^{II}$ are the Jacobians relating the velocities of the tip, i.e., at joint 5 with the joint rates $\dot{\mathbf{q}}^I = \left[\dot{\theta}_1\ \dot{\theta}_2\ \dot{\theta}_3\right]^T$ and $\dot{\mathbf{q}}^{II} = \dot{\theta}_4$, respectively. The Jacobians $\mathbf{J}^I$ and $\mathbf{J}^{II}$ for Subsystems, I and II, respectively, are given by

$$\mathbf{J}^I \equiv \begin{bmatrix} -a_1 s_1 - a_3 s_{12} c_3 & -a_3 s_{12} c_3 & -a_3 c_{12} s_3 \\ a_1 c_1 + a_3 c_{12} c_3 & a_3 c_{12} c_3 & -a_3 s_{12} s_3 \\ 0 & 0 & -a_3 c_3 \end{bmatrix} \text{ and } \mathbf{J}^{II} \equiv \begin{bmatrix} -a_4 s_4 \\ a_4 c_4 \\ 0 \end{bmatrix} \tag{6.6}$$

The three-dimensional vector $\boldsymbol{\lambda}$ was then solved from the combined dynamic equations of motion and the appropriate constraint equations. Next, the generalized coordinates $\mathbf{q}$ were solved, and forward dynamics simulation was performed.

6.2.1.1 Numerical Simulation

The links of *RRR* spatial four-bar mechanism were assumed to be a slender rod of 40 mm diameter. The lengths of the crank (a_4), coupler (a_3), and rocker (a_1) were taken as 0.5 m, 1.1259 m, 1 m, respectively, and the fixed-base vector was taken as, $a_0 \equiv \left[-1.30\ -0.20\ -0.2036\right]^T$. The masses of the crank (m_4), coupler (m_3) and rocker (m_1) were taken as 7.7 kg, 17.3 kg, and 15.4 kg, respectively.

In this illustration, the deflections due to bending in both Y-direction (u_y) and Z-direction (u_z) were considered for the flexible link, and the axial deflection and the torsional effects were neglected. The initial configuration of the *RRR* mechanism is given in Table 6.1. The mechanism allowed to fall freely due to gravity with no external torque. The results are shown in Figs. 6.2 and 6.3. In order to verify the results, the RSSR mechanism was modeled in the commercial multibody analysis software RecurDyn. This is shown in Fig. 6.4. The link parameters and initial conditions were taken the same as above. The forward dynamics was carried out without any external torque. Figure 6.2 shows the joint angles, whereas Fig. 6.3 shows their corresponding joint rates. The plotted results from the proposed formulation and RecurDyn software are close to each other, which verifies the correctness of the results from the proposed formulation.

Table 6.1 Initial conditions of the RSSR mechanism

	Joint 1	Joint 2	Joint 3	Joint 4
Joint angles (°)	$\theta_1 = 120$	$\theta_2 = -225.72$	$\theta_3 = 10.41$	$\theta_4 = 0$
Joint velocities (rad/s)	$\dot{\theta}_1 = 0$	$\dot{\theta}_2 = 0$	$\dot{\theta}_3 = 0$	$\dot{\theta}_4 = 0$

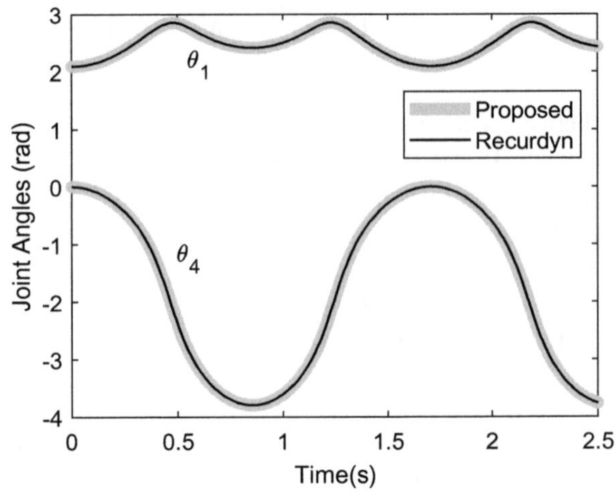

Fig. 6.2 Joint angles of *RRR* spatial mechanism

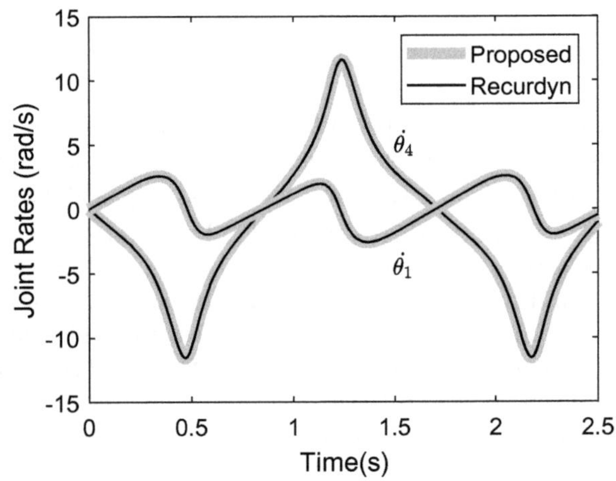

Fig. 6.3 Joint rates of *RRR* spatial mechanism

6.2.2 *RRF Spatial Mechanism*

The dynamic formulation of *RRF* spatial four-bar mechanism was done in which the crank (#4) in Subsystem-II and coupler (#3) in Subsystem-I were treated as rigid. In contrast, the output link, i.e., rocker (#1) in Subsystem-I, was considered flexible. In order to derive the dynamics of the *RRF*-RSSR mechanism, one can either cut joint 2 or joint 5 so that the resulting subsystems are one single-link serial chain with one revolute joint with the base and one two-link serial-chain system with a revolute joint with the base and the spherical joint in between the two links. For the analyses, the mechanism was cut at joint 5, as in Fig. 6.1b. The Subsystem-I was considered to have a two-link flexible–rigid (*FR*) system, and Subsystem-II was with one rigid link (*R*). Modeling of Subsystem-I was already done by considering the equivalent

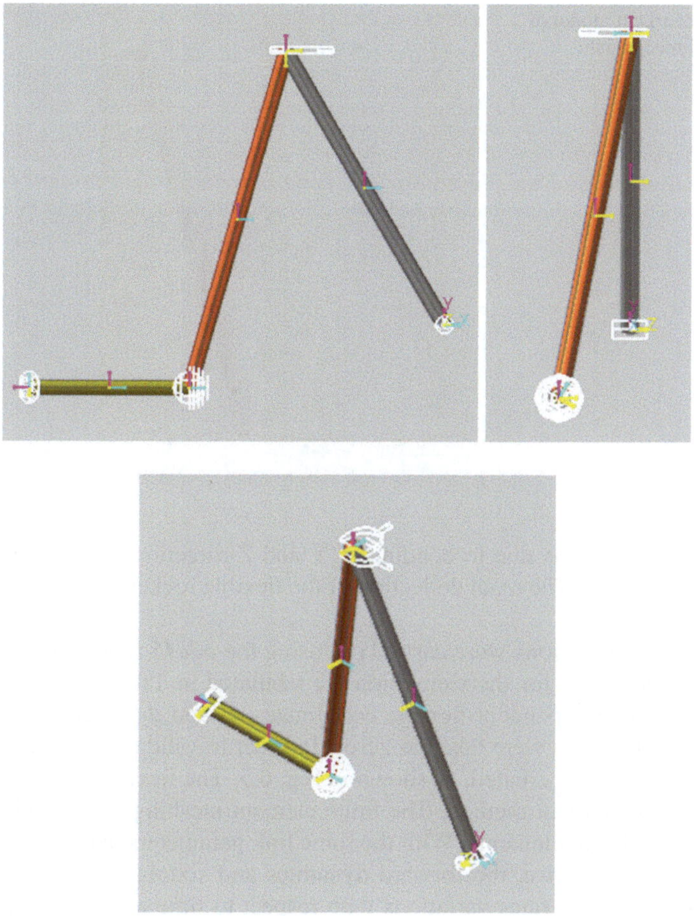

Fig. 6.4 *RRR* spatial mechanism in RecurDyn

three-link serial-chain system, as explained in Sect. 6.2.1. The only difference in the formulation of this section is that the three-link mechanism has the first link flexible (*F*), and the succeeding links are rigid (*R*). Hence, the *FRR* system was modeled using the methodology presented in Sect. 3.1. The EOM for the Subsystem-II are the same as Eq. (6.5).

6.2.2.1 Numerical Simulation

For the *RRF*-RSSR mechanism, the masses and the cross sections of the rigid links, i.e., of the crank and coupler, were taken the same as in Sect. 6.2.1, but the radius and mass of flexible rocker were taken as 15 mm and 5.54 kg, respectively. For simulation purposes, the mechanism was allowed to fall freely under gravity with no external

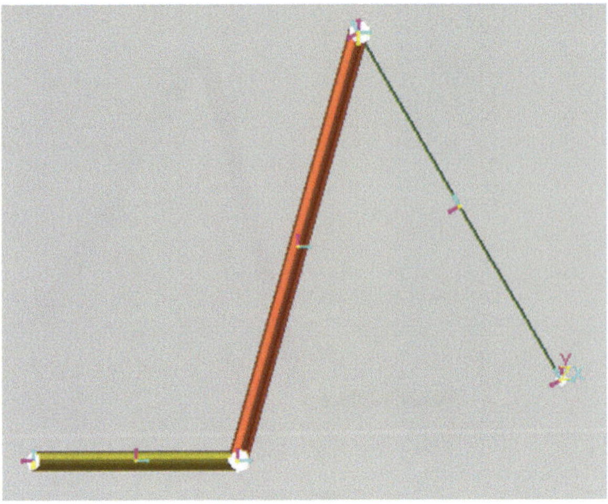

Fig. 6.5 RecurDyn model of *RRF* spatial mechanism

torque. The deflections due to bending in Y and Z-directions were denoted as u_y and u_z, respectively. The axial deflection of the flexible rocker (#1) and its torsional effect were neglected.

Numerical integrations were carried out using the *ode45* function of MATLAB. The initial conditions for the simulation are tabulated in Table 6.1, and the initial values of the time-dependent flexible coordinates d_y and d_z of the flexible rocker and its first derivatives were taken as zeros. In order to validate the results, a model in RecurDyn was also created, as shown in Fig. 6.5. The flexible link was modeled using the finite element method. The finite element meshing of the link was done by considering beam elements. With the same link parameters and the initial conditions as described above, the forward dynamics and simulation were carried out. Joint angles and joint rates variations with respect to time are plotted in Figs. 6.6 and 6.7, respectively. The transverse displacement of the flexible link in Y and Z-directions are also plotted. Figure 6.8 shows the comparison of the plots based on the proposed DeNOC-based formulation and RecurDyn for Y-displacement of the flexible rocker, whereas Fig. 6.9 shows the plot of the Z-displacement of the rocker. The joint angles and the transverse displacements of the flexible output link (rocker) were closely matching with those from RecurDyn software. There is a slight variation in Y-displacement results; this may be due to the complexity introduced in modeling flexible links and the variations in AMM and FEM discretization of the link for the proposed and RecurDyn formulations, respectively.

6.2.3 *FRR Spatial Mechanism*

The dynamic formulation of the *FRR* spatial four-bar mechanism was done in which the coupler (#3) and the output rocker (#1) were treated as rigid, whereas crank (#4)

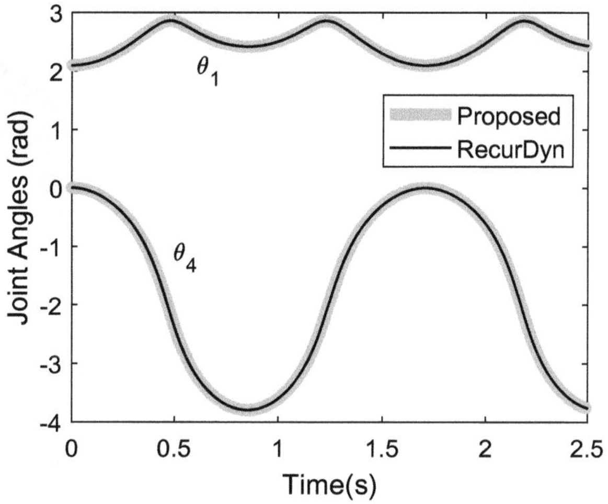

Fig. 6.6 Joint angles of *RRF* spatial mechanism

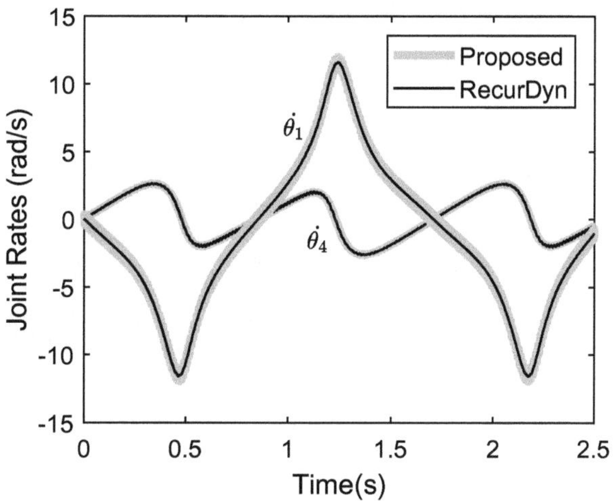

Fig. 6.7 Joint rates of *RRF* spatial mechanism

was considered flexible. In this case, the dynamic modeling was done by cutting the mechanism at joint 5. Subsystem-I has two rigid links (*RR*), whereas Subsystem-II has only one flexible link (*F*). Again Subsystem-I was considered equivalent to the three-link serial system, as explained in Sect. 6.2.1. The matrices associated with the dynamic equations of motion, i.e., Eq. (6.1), are the same as the Eqs. (6.2), (6.3) and (6.4) for Subsystem-I. The equations of motion for Subsystem-II were derived using the methodology presented in Sect. 3.1.

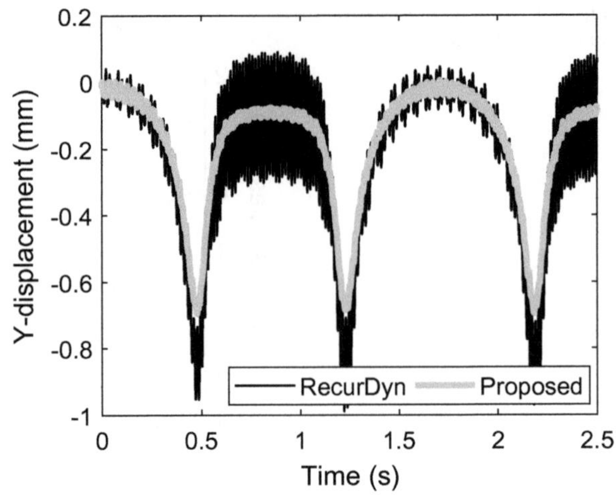

Fig. 6.8 Y-displacement of the flexible rocker for the *RRF* mechanism

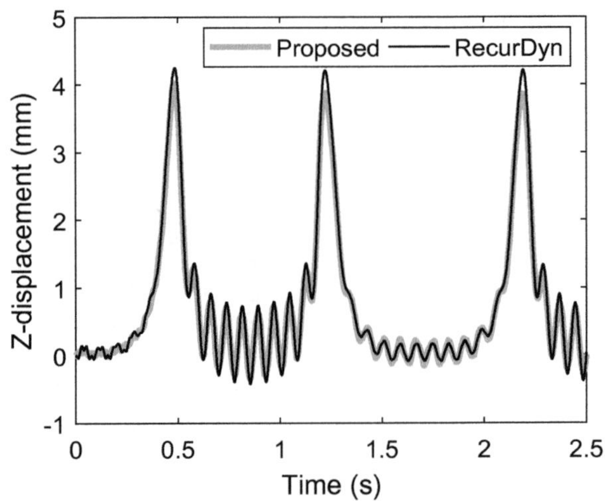

Fig. 6.9 Z-displacement of the flexible rocker for the *RRF* mechanism

6.2.3.1 Numerical Simulation

The spatial four-bar *FRR*-RSSR mechanism was analyzed with the material properties of the crank, coupler, and rocker taken from Sect. 6.2.1.1. But the radius and mass of the flexible rocker (#1) were considered 15 mm and 2.77 kg, respectively. The mechanism was allowed to fall freely under gravity, with no external torque. The deflections for the flexible link due to bending, i.e., u_y and u_z in Y and Z-directions, respectively were determined. The axial deflection and the torsional effect were neglected, mainly due to the simplicity of modeling. Numerical integrations were done using the *ode45* function of MATLAB with initial conditions given in Table 6.1. Note that the initial values of time-dependent flexible coordinates, namely d_y and d_z, and their first derivatives were taken as zeros. The *FRR*-RSSR mechanism was also modeled in RecurDyn to verify the results generated using the proposed formulation. The flexible crank was modeled as beam elements. With the same link parameters and initial conditions, the forward dynamics and simulation were performed.

The plots of joint angles and their joint rates were given in Figs. 6.10 and 6.11, respectively. The transverse displacement of the flexible link along Y and Z-directions are plotted in Fig. 6.12. The joint angles and the transverse displacements of the flexible output link (rocker), obtained from the DeNOC-based approach and the RecurDyn software, are close to each other and thus, verified the results.

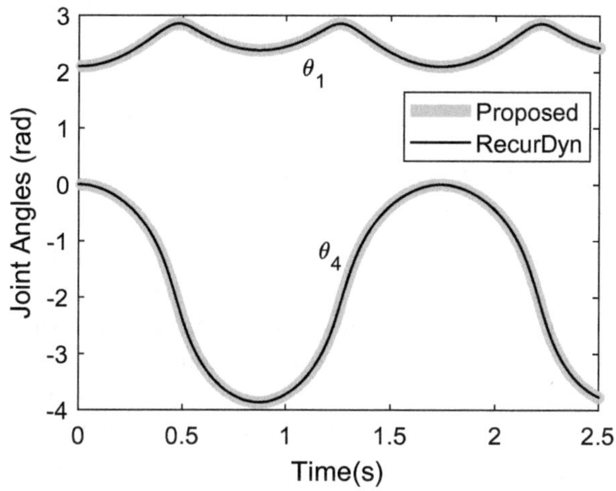

Fig. 6.10 Joint angles of *FRR* spatial mechanism

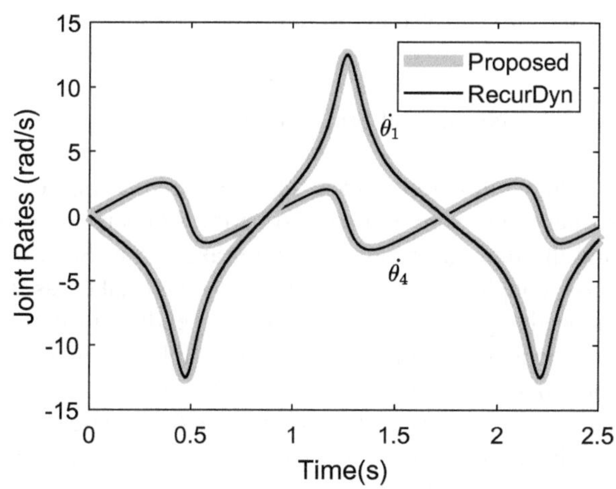

Fig. 6.11 Joint rates of *FRR* spatial mechanism

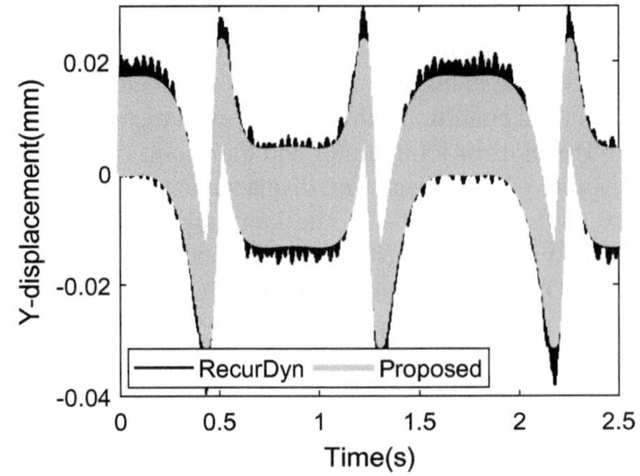

Fig. 6.12 Y-displacement of the flexible crank for *FRR* mechanism

6.2.4 *FFF* Spatial Mechanism

The dynamic formulation of the *FFF* spatial four-bar mechanism was done in which all the links, i.e., crank (#4) in Subsystem-I, coupler (#3), and the output link, i.e., rocker (#1) in Subsystem-II, were considered flexible. The dynamic modeling was done by cutting the mechanism at joint 5. Note that, Subsystem-I has two flexible–flexible links (*FF*) and Subsystem-II has one flexible link (*F*). Again, the spherical joint of the Subsystem-I was initially replaced by a universal joint which is equivalent to two intersecting revolute joints. As a result, Subsystem-I becomes identical to the three-link serial manipulator, where the intermediate link has no length. Hence, treating it as rigid or flexible did not make any difference. The equations of motion for the two subsystems were derived similarly to the method described in Chap. 3.

6.2.4.1 Numerical Simulation

For the simulation purpose, all the three links, i.e., crank (#4), coupler (#3), and rocker (#1), were assumed to be the slender rod of 30 mm diameter. The lengths and the material properties were taken from Sect. 6.2.1. The masses of the crank (m_4), coupler (m_3), and rocker (m_1) were taken as 2.77 kg, 6.24 kg, and 5.54 kg, respectively. The deflection due to bending in both Y (u_y) and Z (u_z) directions was considered for all the flexible links. However, the axial deflections and the torsional effects were neglected. The initial configuration of the *FFF* spatial four-bar mechanism was taken from Table 6.1, whereas the initial values associated with flexible coordinates were taken as zero. Numerical integrations were done using the *ode45* function of MATLAB. The results are shown in Figs. 6.13, 6.14, 6.15, 6.16, 6.17, 6.18, 6.19 and 6.20. To validate the results, the model of the spatial four-bar mechanism (RSSR) was also built in RecurDyn software, as shown in Fig. 6.21.

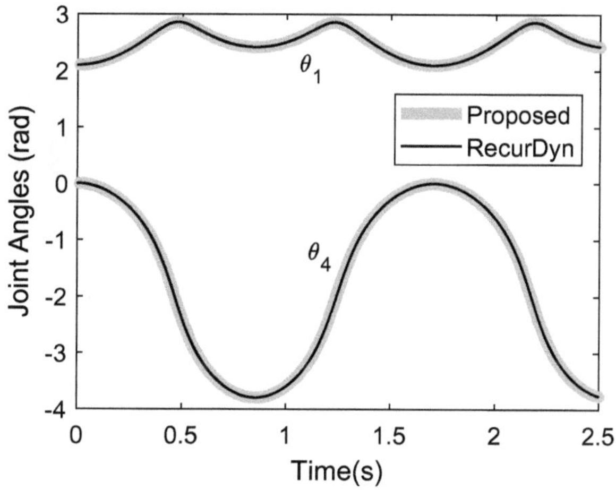

Fig. 6.13 Joint angles of *FFF* spatial mechanism

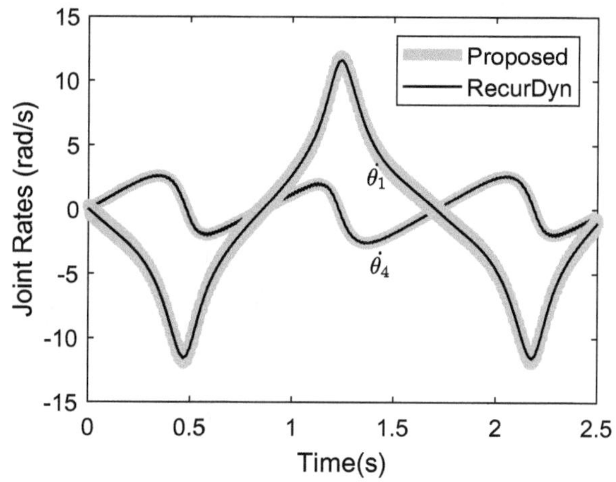

Fig. 6.14 Joint rates of *FFF* spatial mechanism

The link parameters and initial conditions were taken the same as in the DeNOC-based simulation. The simulation results are superimposed in Figs. 6.13, 6.14, 6.15, 6.16, 6.17, 6.18, 6.19 and 6.20 for comparison purposes. The results of both the proposed formulation and RecurDyn in Figs. 6.13 and 6.14 are close to each other, whereas variations are observed in Figs. 6.15, 6.16, 6.17, 6.18, 6.19 and 6.20. Even though the mean values appear to be close, relatively large fluctuations and deviations are observed in Figs. 6.19 and 6.20, respectively. No apparent reason was found except the differences in modeling techniques and numerical solvers employed and many modeling and simulation parameters used by the proposed DeNOC-based formulation in MATLAB and the RecurDyn software, respectively.

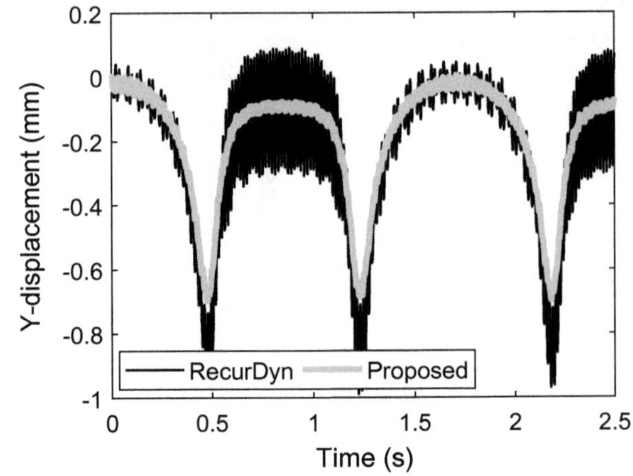

Fig. 6.15 Y-displacement of the flexible rocker for *FFF* mechanism

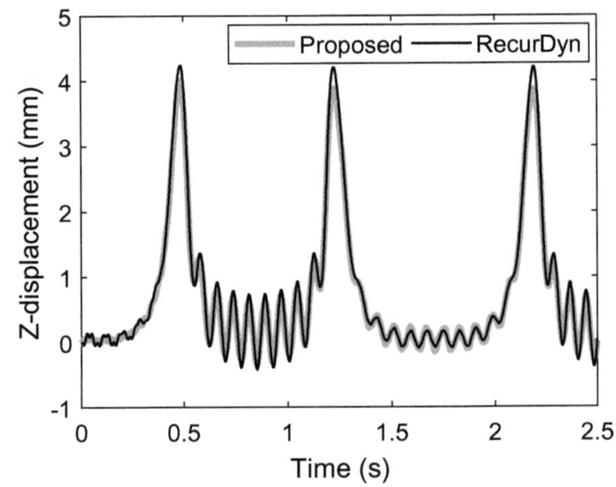

Fig. 6.16 Z-displacement of the flexible rocker for *FFF* mechanism

6.3 Summary

Dynamic modeling and simulation of closed-loop rigid–flexible spatial four-bar mechanism were presented using the DeNOC matrices. This method exploited the recursive formulation of the serial open subsystem by cutting at the appropriate joint of the closed-loop. The flexibility in a link was introduced. Different combinations of the rigid–flexible spatial four-bar mechanism were studied to demonstrate how two modelings of one-link and two-link rigid and rigid–flexible manipulators can solve different types of spatial four-bar mechanisms. Results were verified using the models generated with the help of commercial software, namely RecurDyn, where discretization of the flexible link was done using the finite element method.

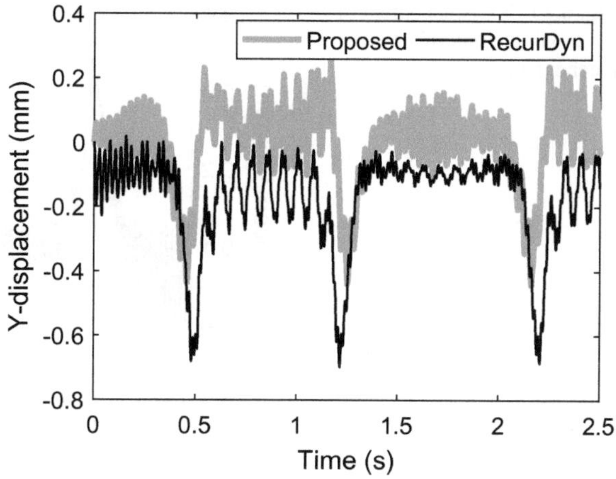

Fig. 6.17 Y-displacement of the flexible coupler of *FFF* mechanism

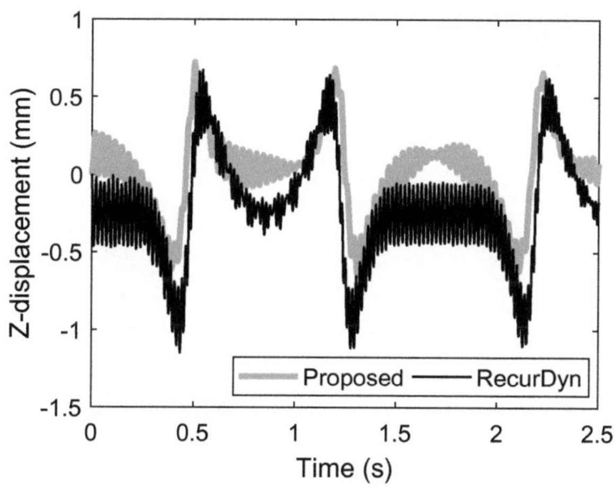

Fig. 6.18 Z-displacement of the flexible coupler of *FFF* mechanism

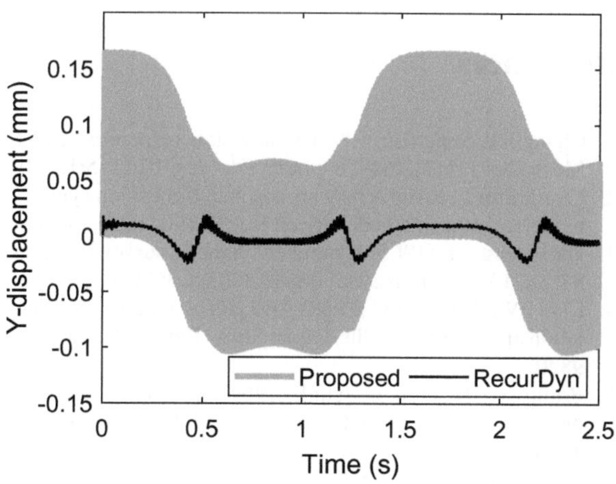

Fig. 6.19 Y-displacement of the flexible crank of *FFF* mechanism

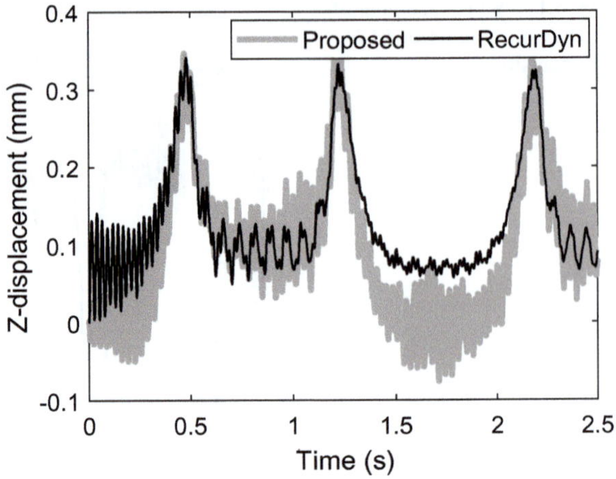

Fig. 6.20 Z-displacement of the flexible crank of *FFF* mechanism

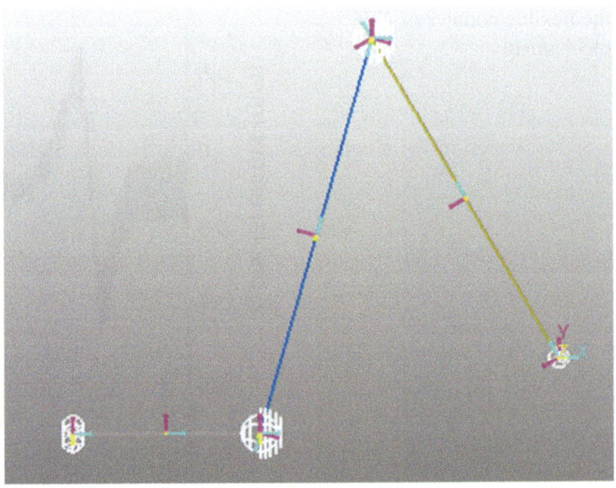

Fig. 6.21 RecurDyn model of *FFF* spatial mechanism

References

1. Chang RY, Sung CK (1994) Steady-state responses of an RSRC-type spatial flexible linkage. J Mech Des 116(1):264–269. https://doi.org/10.1115/1.2919357
2. Chu J, Sun J (2010) A new approach to dimension synthesis of spatial four-bar linkage through numerical atlas method. J Mech Rob 2(4):041004–041014. https://doi.org/10.1115/1.4001774
3. Harrisberger L (1965) A number synthesis survey of three-dimensional mechanisms. J Eng Ind 87(2):213–218. https://doi.org/10.1115/1.3670797
4. Liou FW, Erdman AG (1989) Analysis of a high-speed flexible four-bar linkage: Part I—formulation and solution. J Vib Acoust Stress Reliab Des 111(1):35–41. https://doi.org/10.1115/1.3269820
5. Primrose EJF, Freudenstein F (1969) Spatial motions I—point paths of mechanisms with four or fewer links. J Eng Ind 91(1):103–113. https://doi.org/10.1115/1.3591479
6. Tanık E, Parlaktaş V (2011) A new type of compliant spatial four-bar (RSSR) mechanism. Mech Mach Theor 46(5):593–606. https://doi.org/10.1016/j.mechmachtheory.2011.01.004

Part III
Numerical Stability and Computational Efficiency of Dynamic Algorithm

Chapter 7
Numerical Stability and Efficiency

Simulation, as defined in Chaps. 3 and 4 is a two-step process: (1) forward dynamics, i.e., the computation of the accelerations for the generalized coordinates from the equations of motion of a system at hand for the given actuator forces and torques; and (2) numerical integration of the accelerations computed in step (1) to obtain the corresponding rates and positions. Thus, the nature of simulation results depends as much on the forward dynamics algorithm as on the integrator used in step (2). In fact, it is possible that the simulation results are unstable even when the real system is stable. The role of integrator's algorithm, as well as the forward dynamics scheme, is very crucial in determining the stability characteristics of the simulation results. While working with low-order, linear models, resulting from simpler systems, computers are usually insensitive to numerical difficulties, and the common mathematical software tools like MATLAB, etc., give accurate results. However, the dynamic models of rigid or rigid–flexible robots are a set of highly coupled and nonlinear equations, and the finite-precision arithmetic of a computer is not so forgiving. Hence, it is essential to exercise caution to restrict the cumulative growth of errors through appropriate choice of the forward dynamic algorithm and the numerical integrator. In this chapter, it is assumed that a suitable numerical integrator provides satisfactory results uniformly for a class of systems, and the effect of forward dynamics algorithm on the simulation results will be studied. Note that, to get numerically accurate simulation results for a given forward dynamics problem, using a computer, the following factors are imperative:

1. Condition number of the matrices associated with the dynamic model of the system, mainly the generalized inertia matrix (GIM) of the system at hand derived in Eqs. (2.27) and (3.31) for the rigid and rigid–flexible robotic systems, respectively,
2. Methodology to solve the accelerations for the generalized coordinates, and
3. Software implementation.

P. V. Nandihal et al., *Dynamics of Rigid-Flexible Robots and Multibody Systems*,
Intelligent Systems, Control and Automation: Science and Engineering 100,
https://doi.org/10.1007/978-981-16-2798-9_7

A problem is said to be well-conditioned if small changes in the data cause only small change in its solutions. If small changes in the data have the potential to induce large changes in the solution, the problem is said to be ill-conditioned. A forward dynamics algorithm is numerically stable if it does not introduce any additional sensitivity than that already inherent in the problem due to its physical characteristics. Thus, a stable algorithm cannot be expected to solve an ill-conditioned problem more accurately than the data warrant, but an unstable algorithm can produce poor solutions even to a well-conditioned problem. Similarly, the methodology, e.g., recursive or non-recursive, is also vital. It has been reported in the literature, e.g., [1–3], and others that the conditions for the stability of simulation results are generally met by the recursive forward dynamics algorithms, like the ones presented in Chaps. 3 and 4. Since the order of the algorithm is linear in the number of links and the explicit inversion of the GIM associated with the dynamic model, particularly for rigid systems, is completely avoided, any ill-conditioned problem will not cause significant error amplification in the solution of joint accelerations. However, for the rigid–flexible robots, even a recursive algorithm requires the explicit inversion of a smaller dimension matrix associated with the number of modes of vibration, namely matrix $\hat{I}_i$ of Eq. (3.40). Hence, the algorithm for the rigid–flexible systems is relatively less stable compared to the similar algorithm for rigid systems. This phenomenon is clear from the results of Chap. 5. However, the recursive algorithm for rigid–flexible systems is certainly better than its non-recursive counterpart, where a larger dimension matrix given by Eq. (3.31) needs to be inverted. Obviously, the error amplification effect is more in the latter algorithm and should be avoided.

For comparison purposes, the numerical stability of the recursive forward dynamic algorithms is tested against the conventional algorithms based on Cholesky decomposition [4], where the GIM, $\boldsymbol{I}$ of Eq. (3.31), is first obtained numerically using Eq. (3.32). Then, the GIM is factorized numerically using the Cholesky decomposition, followed by the solution of the accelerations, $\ddot{\boldsymbol{q}}(\ddot{\boldsymbol{\theta}}$ for rigid bodies), using forward and backward substitutions. Such an algorithm is known to have cubic order complexities [5], i.e., the complexities grow with the cube of the number of dynamic equations of motion at hand. Since, the calculations, i.e., $\mathbf{0} = \boldsymbol{\tau} - \mathbf{h}$ of Eq. (3.31) and others are carried out exactly in a manner done for the proposed recursive algorithms. Any difference in the simulation results is attributed to the recursive and non-recursive algorithm.

Note that for the non-recursive algorithms, as described in Sect. 2.2, there exist a number of stabilization methods, namely Baumgarte stabilization and Augmented Lagrangian approach [6]. However, these stabilization techniques modify the dynamic model such that the simulation results obtained do not correspond to the original system but represent some other slightly deviated system [7]. Moreover, the behavior of these stabilization approaches is highly configuration-dependent and does not improve the stability of the system for its near singular configurations [8]. Furthermore, the above stabilization techniques have been used by researchers for only rigid link robotic systems. Literature is comparatively silent on the existence of stabilization methods for flexible link robotic systems. Hence, in order to adopt a uniform comparison process for rigid and flexible robots and avoid the loss in accuracy of simulation results, no stabilization method is used here with any of the algorithms.

Note that the numerical integrator used in both the simulations is the same. The stability aspects of the algorithms are studied based on the criteria presented in Sect. 7.1. Initially, the 6° of freedom (DOF) PUMA robot, Fig. 4.1, with all links rigid, is considered to establish the stability of the proposed recursive forward dynamics algorithm compared to a non-recursive algorithm, as explained above. The robots with rigid–flexible links are then tested, where the three-link Canadarm in planar motion, Fig. 3.22, and the six-link Space Shuttle Remote Manipulator System (SSRMS), Fig. 4.6, in spatial motion are considered.

7.1 Criteria for Numerical Stability

In this section, different criteria used to compare the numerical stability of forward dynamics algorithms are presented. Various criteria based on error blow-up due to zero eigenvalue phenomenon [9], condition number [2, 3], energy conservation [1], quality of accelerations results, etc., are used in the literature. There seems to be no consensus on a unique approach to test the stability of a simulation algorithm. Hence, three different criteria based on the above approaches are adopted, as explained next.

7.1.1 Zero Eigenvalue Phenomenon

In forced simulation, torques/forces required by a robot to traverse the desired maneuver are calculated from the inverse dynamic algorithm of the robot and used as input to the robot joints. It is expected that the simulated joint paths must be the same as the desired maneuver. However, as seen in the forced simulation results of Chap. 4, e.g., Fig. 4.4, the simulation fails after a certain duration of time. This phenomenon of failure in the simulation results is termed by [9] as the zero eigenvalue effect. The effect is very similar to the zero poles of a control system [10], where a little disturbance to the system may cause the system to deviate from its normal path. A similar phenomenon happens with numerical calculations. Numerical errors due to the truncation or otherwise cause the simulations to deviate. When the errors become large beyond the tolerance provided in the numerical integrator, the results start deviating from the stable results. Eventually, the integration stops. Note that the numerical error buildup is more prominent in the following situations:

(I) A rapidly changing function, e.g., the joint accelerations obtained from Eq. (4.31),

(II) Mixture of functions, i.e., differential and algebraic, as in differential algebraic equations (DAE) formulations.

As such, any unwarranted deviation in the simulation results due to large amplification in the errors is an indication of the instability in the results. While comparing two forward dynamics (FD) algorithms, if the error buildup occurs earlier in one algorithm, it can be assumed that it is numerically less stable than the other one, as the deviation from the desired maneuver due to the errors in the latter algorithm occurs later. In the simulation results, the times up to which the simulation results do not deviate sharply are noted down. If this time is larger for one FD algorithm than other one, the former algorithm is considered here as more stable.

7.1.2 Power Difference

This criterion is based on the principle of conservation of energy and power. In the forced simulation, the input power to the robotic system is calculated as, $E_n = \sum_{i=1}^{n} \tau_i \dot{\theta}_i$, where τ_i and $\dot{\theta}_i$ are the torque obtained from the inverse dynamics algorithm and the desired joint rates, respectively, for an all rigid-link system-n being the number of joints. Now, based on the simulation results, the simulated power is computed as, $\widetilde{E}_n = \sum_{i=1}^{n} \tau_i \dot{\tilde{\theta}}_i$, where $\dot{\tilde{\theta}}_i$ is the simulated joint rate. Since no loss due to friction or damping is assumed, desired and simulated powers should be the same. However, if any difference occurs, that must be due to numerical errors. Since the numerical integrator is kept the same in simulations using different forward dynamics (FD) algorithms, any variation in the power difference (PD) is caused by the FD algorithms. Higher the PD, the FD algorithm is considered to produce less numerically stable results.

7.1.3 Acceleration Plots

This criterion is based on error amplification while integrating a rapidly changing quantity, namely the accelerations obtained from the dynamic equations of motion, using an FD algorithm. If the acceleration plots have sharp and high frequency peaks, then the possibility of simulation results for joint velocities and corresponding joint paths to be less stable, as most of the existing numerical integrators cannot handle sharply varying functions beyond a certain limit. Comparing the nature of the plots for the accelerations obtained from two different FD algorithms, their numerical stability can be judged. This is a qualitative comparison.

7.2 Stability and Efficiency for Rigid Robots

The numerical stability of the recursive forward dynamics (FD) algorithm proposed in Sect. 3.3.1 for rigid robots is studied here using the six-link 6-DOF PUMA robot shown in Fig. 4.1. The DH parameters of the robot, along with the mass of each link, hub inertia, etc., are shown in Table 4.1. The numerical integrator used for the simulation is ode45 of MATLAB with the step size of 0.001 s and tolerance of 10^{-8}. Simulation for the input torques/forces required to maintain the trajectory of Eq. (4.1), as shown in Fig. 4.3, is done using two FD algorithms, namely (1) the proposed recursive algorithm given in Chap. 2, Sect. 2.3.1 and (2) a non-recursive algorithm based on the Cholesky decomposition, developed separately for the sake of comparison. The latter type is conventionally used, as in [11, 12], and others. Results of comparisons are explained in the following subsections.

7.2.1 Zero Eigenvalue Phenomenon

Figure 7.1a–f shows the deviation of the simulated joint path obtained using the two FD algorithms mentioned above from the desired joint maneuver, Eq. (4.1). It is clear that the proposed recursive algorithm provides accurate results for all the joints up to 5.4 s compared to the non-recursive algorithm (about 1.5 s for joint 2). The simulation duration beyond which the simulated joint angles deviate from the desired values are shown in Table 7.1 which shows clearly that the proposed recursive FD algorithm is better.

7.2.2 Power Difference

Figure 7.2 shows input power and the powers obtained from the forced simulation of the PUMA robot. Using the two FD algorithms, a simulated joint rate, $\dot{\tilde{\theta}}_i$, is obtained. From Fig. 7.2, it is clear that the simulated power matches with the desired input power for a longer duration of time when $\dot{\tilde{\theta}}_i$ is obtained using the proposed recursive algorithm (about 5.6 s) compared to that obtained using the Cholesky decomposition-based non-recursive algorithm (about 5.0 s). Hence, the proposed recursive FD algorithm provides more stable simulation results.

7.2.3 Acceleration Plots

To investigate the built-up of errors, the nature of the joint accelerations, i.e., $\ddot{\theta}_i$, for $i = 1, \ldots, 6$, are obtained using both the FD algorithms. The joint accelerations for

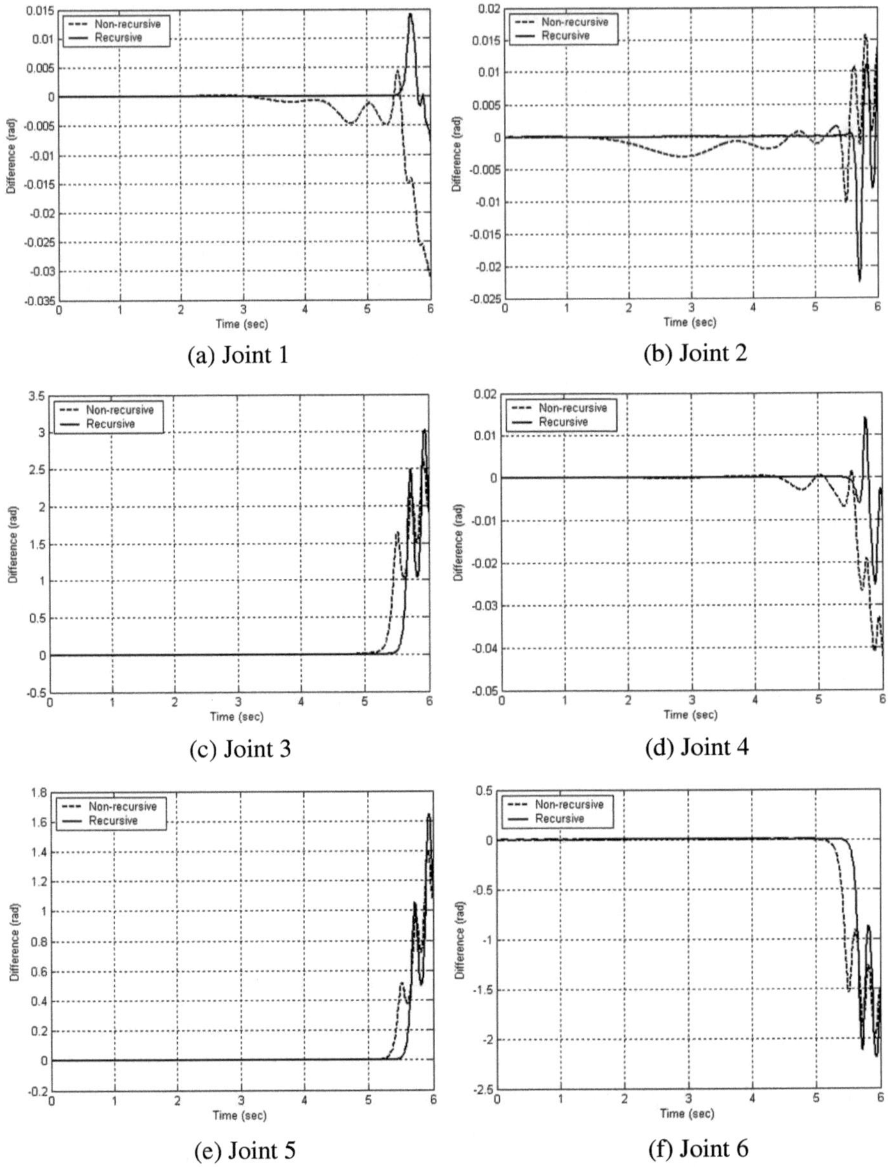

Fig. 7.1 Difference between the desired and simulated results

the recursive algorithm are smoother for a longer duration (about 5.5 s) than those obtained from the conventional algorithm, as shown in Fig. 7.3a, b for a representative joint 3. Similar behaviors were obtained for the other joints also.

The smoother acceleration plot in the case of the proposed algorithm, Fig. 7.3a, is responsible for the stability of the results and also for the computational efficiency

Table 7.1 Acceptable simulation duration

Angle	Recursive (s) (Proposed)	Non-recursive (s)
θ_1	5.4	3.0
θ_2	5.6	1.5
θ_3	5.4	4.8
θ_4	5.4	3.6
θ_5	5.4	5.2
θ_6	5.4	5.2

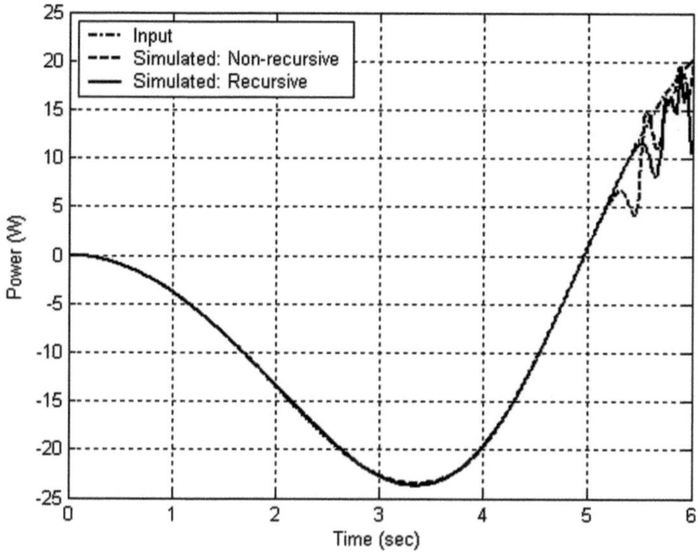

Fig. 7.2 Comparison of powers

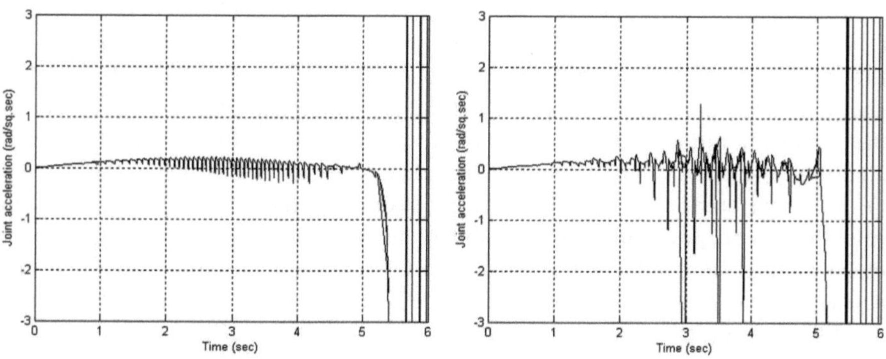

(a) Proposed recursive

(b) Conventional non-recursive

Fig. 7.3 Joint accelerations at joint 3

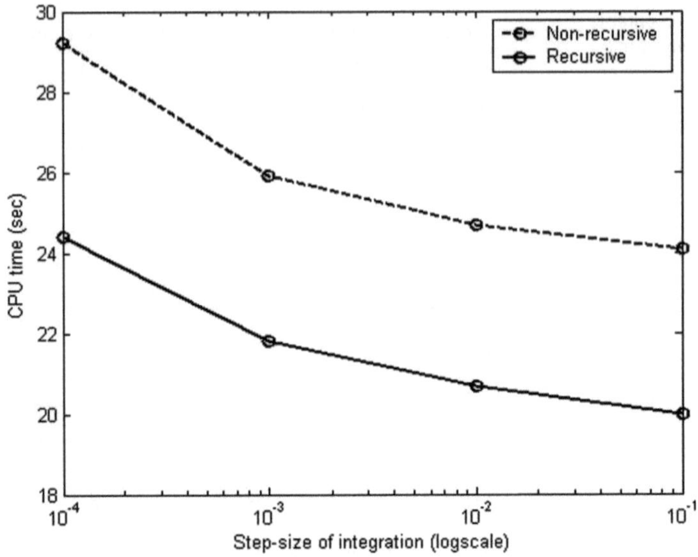

Fig. 7.4 Comparison of CPU times for PUMA robot

of the algorithm, as indicated in Fig. 7.4. From all the aspects of the numerical stability study, it is concluded that the proposed recursive algorithm is more stable than the conventional non-recursive approach.

7.2.4 Computational Complexity

As a verification of the computational efficiency, claimed in Chap. 1, Table 1.1, the CPU times were taken by a 3.5 GHz Pentium PC for the simulations using the two FD algorithms are compared in Fig. 7.4. The CPU times were obtained using ‘tic’ and ‘toc’ commands of the MATLAB at the beginning and end of the program for the simulation duration of 6 s. It is clear from Fig. 7.4 that the proposed recursive algorithm is efficient than its non-recursive counterpart. This is due to the fact that the joint accelerations, $\ddot{\theta}_i$, for $i = 1, \ldots, 6$, obtained from the recursive algorithm are smoother. Hence, even if the FD computations for the non-recursive algorithm are much smaller, as in Table 1.1, it apparently takes more iterations before the results converge to a solution. As a result, the CPU time is more for the non-recursive algorithms. In order to see the effect of the step sizes on the CPU times taken by the proposed and the non-recursive algorithms, step sizes of 0.1, 0.01, 0.001, and 0.0001 were used in ‘ode45’. No significant effect was observed.

7.3 Stability and Efficiency for Rigid–Flexible Robots

The scheme adopted for investigating the numerical stability of a forward dynamics (FD) algorithm for rigid–flexible robots, given in Sect. 3.4, is the same as that adopted in Sect. 7.2 for the rigid robots. The stability of the proposed algorithm is demonstrated based on the comparisons of simulation results obtained using the proposed recursive and non-recursive algorithms. To study the phenomenon of numerical stability of the proposed algorithm for rigid–flexible robots, two examples, i.e., (1) The three-link Canadarm in planar motion and (2) the six-link Space Shuttle Remote Manipulator System (SSRMS) in spatial motion, are undertaken.

7.3.1 Three-Link Planar Canadarm

Figure 3.22 shows the three-link Canadarm whose first and second links are flexible. The architecture and the DH parameters are shown in Table 3.5. Forced simulation, as done in Sect. 3.10.2, is performed. The initial conditions are the same as explained in Sect. 3.10.2. The stability of the simulation results obtained using the two algorithms is then investigated using the three criteria listed above in Sect. 7.1.

7.3.1.1 Zero Eigenvalue Phenomena

The simulation results obtained using the two FD algorithms are compared with the desired trajectory of Eq. (4.1). Figures 7.5a–c shows the deviation of the simulated joint paths from the desired paths. Note that, unlike the plots in Fig. 7.1, where the differences are zero until the simulation results are considered acceptable, there are significant differences in the plots of Fig. 7.5. This is due to the fact that the input torques to the FD algorithms of the rigid–flexible system are calculated for the rigid-link systems. Hence, there are obvious variations in the joint angles. However, the non-recursive algorithm fails to provide any result after about 5 s, whereas the proposed one continues to give acceptable results. Hence, the proposed algorithm can be considered numerically more stable.

7.3.1.2 Power Difference

The power difference (PD) for the three-link Canadarm is now investigated for the two FD algorithms. Since a flexible link is an energy-dissipating system, assuming no losses due to friction and damping, the output energy is equal to the input energy supplied to the system minus the energy dissipated due to vibrations. Accordingly, the output power of the system is calculated as,

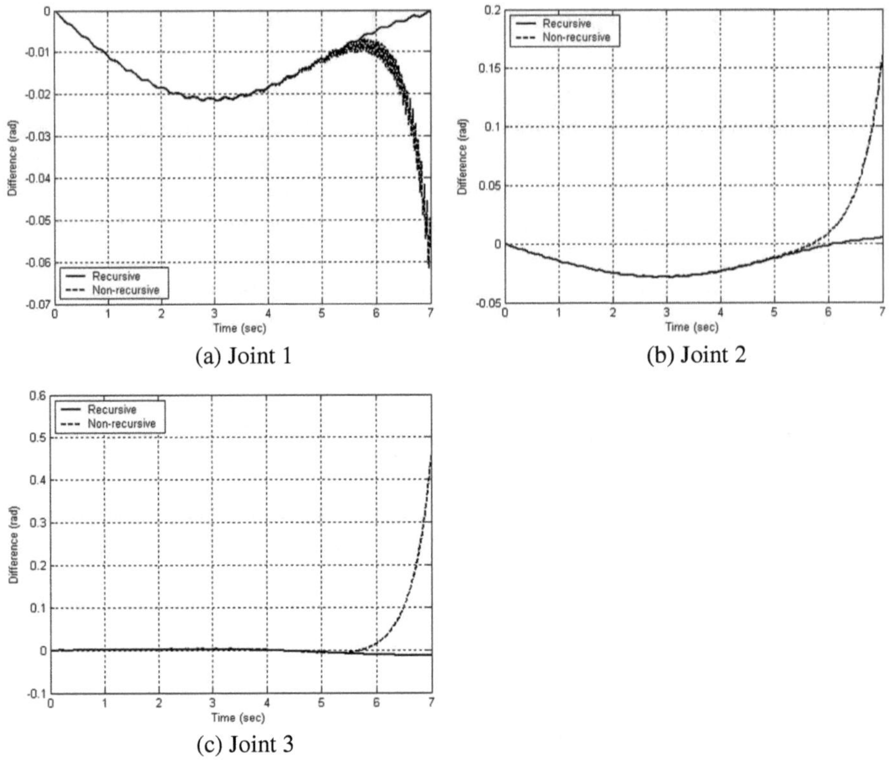

Fig. 7.5 Comparison of simulated joints

$$\widetilde{E}_n = \sum_{i=1}^{n} \tau_i \dot{\tilde{\theta}}_i \tag{7.1}$$

where τ_i and $\dot{\tilde{\theta}}_i$ are the torque obtained from the inverse dynamics algorithm and the simulated joint rates, respectively, and n is the number of joints. The output power is calculated from both the proposed and conventional algorithms. The results are plotted in Fig. 7.6a, b. The PD for the non-recursive seems to be increasing to an unacceptable level after 3.5 s, whereas the proposed recursive algorithm continues to provide acceptable PD, even after 7 s. This shows that the proposed algorithm is more stable.

7.3.1.3 Acceleration Plots

To investigate the built-up of the errors, the nature of the joint accelerations, $\ddot{\theta}_i$, for $i = 1,\ldots,3$, is obtained using both the FD algorithms. As shown in Fig. 7.7a, b for joint

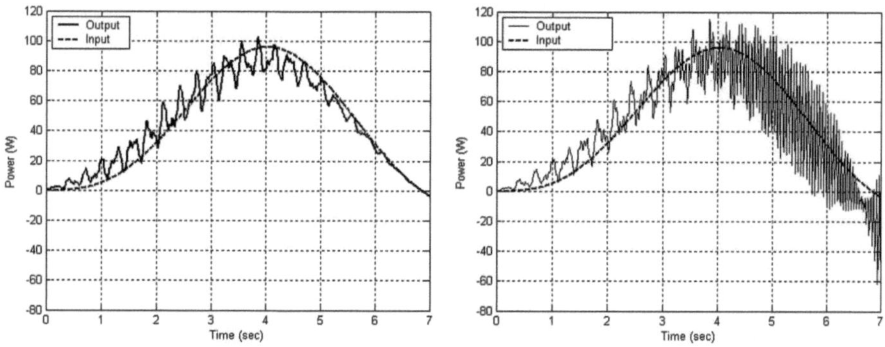

(a) Recursive (Proposed) (b) Non-recursive

Fig. 7.6 Comparison of desired and simulated powers

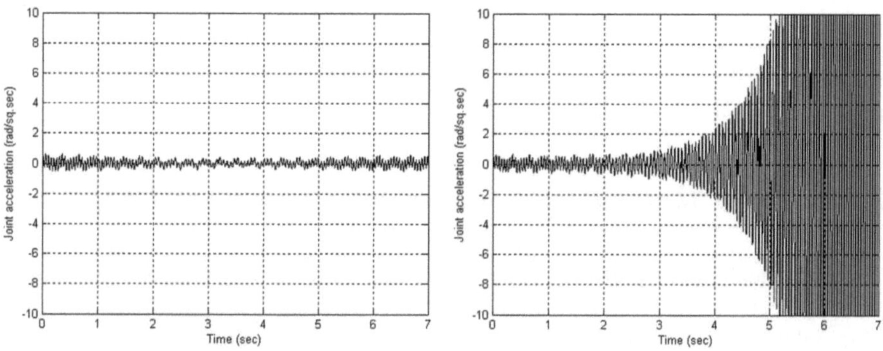

(a) Recursive (Proposed) (b) Non-recursive

Fig. 7.7 Joint accelerations at joint 1

1, the joint acceleration curve, $\ddot{\theta}_1$, for the proposed recursive algorithm, is smoother for a longer duration of time while the same from the non-recursive algorithm is not, which shows sharp peaks after 3.5 s. Similar behaviors are observed for $\ddot{\theta}_2$ and $\ddot{\theta}_3$ also. From the error plots and study of numerical stability from other aspects, it is concluded that the proposed FD algorithm, as given in Sect. 3.4.1, is numerically more stable.

7.3.1.4 Computational Efficiency

The CPU time taken by the 3.5 GHz Pentium PC used for the simulations, as mentioned in Sect. 7.2.4, using the two FD algorithms, are now compared for a simulation duration of 5 s. This is because the simulations results are stable for both algorithms till 5 s, as clear from Figs. 7.5, 7.6, and 7.7. It is clear from Fig. 7.8 that the proposed recursive algorithm is efficient than its counterpart.

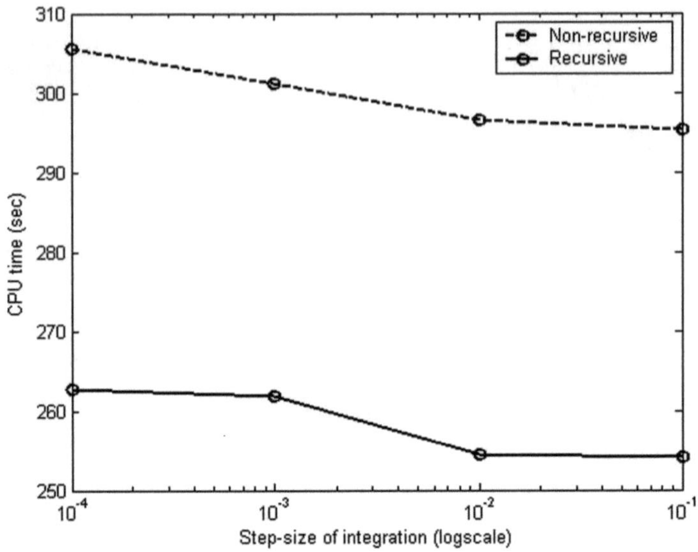

Fig. 7.8 Comparison of CPU times

7.3.2 Space Shuttle Remote Manipulator System (SSRMS)

The numerical stability of the proposed recursive forward dynamics algorithm, as given in Sect. 3.4.1, is now investigated for the spatial six-link Space Shuttle Remote Manipulator System (SSRMS) robot whose second and third links are flexible, and the rest are rigid. Figure 4.6 shows the SSRMS, whose architecture and DH parameters are shown in Table 4.2. Simulation is performed for the initial conditions mentioned in Sect. 4.2.

7.3.2.1 Zero Eigenvalue Phenomenon

The effect of the numerical instability is very pronounced for the six-link spatial SSRMS, where the simulation based on the non-recursive algorithm totally fails after about 2.5 s. On the other hand, the simulation using the proposed recursive algorithm continues up to 7 s, albeit the error blows up. The simulation results obtained using the two algorithms are compared with the desired trajectory of Eq. (4.2). The difference between the simulated joint path obtained from the two FD algorithms and the desired joint maneuver are plotted in Fig. 7.9 for joint 4. While up to about 1.1 s results from both the algorithm are the same, the non-recursive results tend to blow-up slowly. Hence, the proposed algorithm is proven to be more stable.

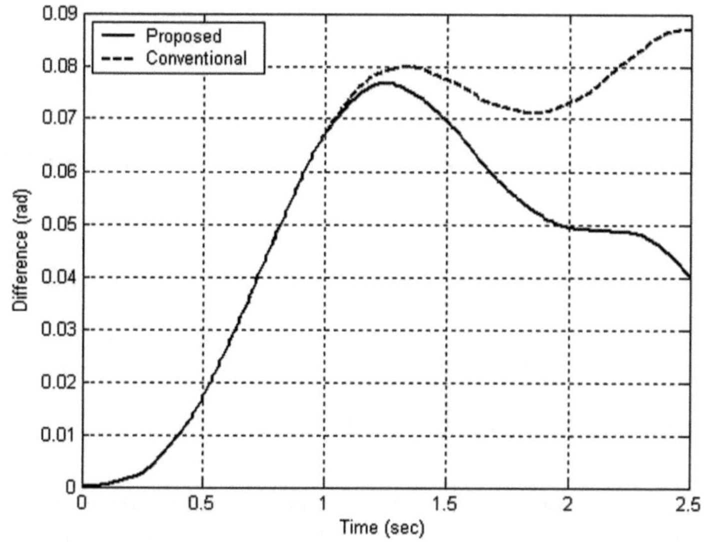

Fig. 7.9 Error deviation for joint 4 of SSRMS

7.3.2.2 Power Difference

The input and simulated powers for the two algorithms are shown in Fig. 7.10. For the non-recursive algorithm, results become unstable, and the simulation stops soon

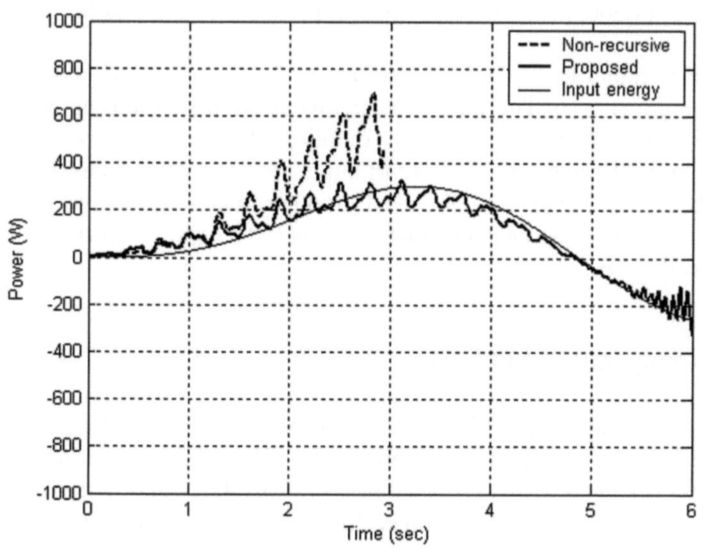

Fig. 7.10 Comparison of desired and simulated powers

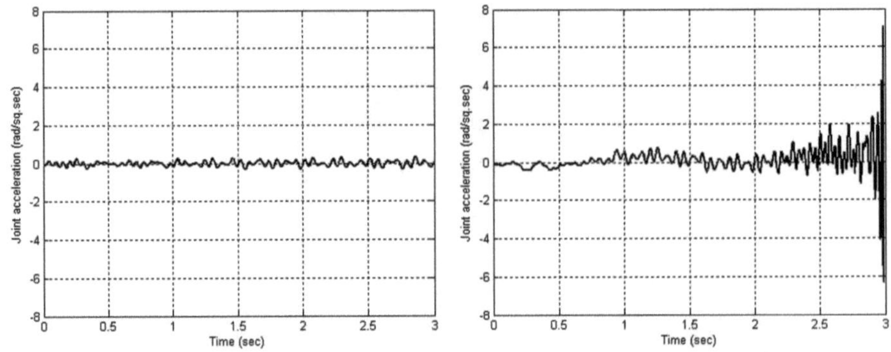

(a) Recursive (Proposed) (b) Non-recursive

Fig. 7.11 Joint accelerations for joint 6 of SSRMS

after 2.5 s. However, for the recursive algorithm, the output power almost matches with the input one even up to 6 s. Moreover, the drift in the non-recursive algorithm increases after about 1.1 s before it fails after 2.5 s.

The drift for the recursive algorithm is much less, even up to 6 s. Hence, the numerical stability of the latter algorithm is established.

7.3.2.3 Acceleration Plots

To investigate the built-up of the errors, the nature of the joint accelerations, $\ddot{\theta}_i$, for $i = 1, \ldots, 3$, is obtained using both the proposed recursive and the non-recursive algorithms.

Here, the results for the joint acceleration, $\ddot{\theta}_6$, are presented and compared as shown in Fig. 7.11a, b. The joint acceleration curve from the proposed algorithm shows smoother behavior for a longer duration of time, while the joint acceleration curve from the non-recursive algorithm is non-smooth and shows sharp peaks, Fig. 7.11b. Similar behavior is observed for other joint accelerations as well. From all the three studies, it is concluded that the proposed recursive algorithm, Sect. 3.4.1, is numerically more stable.

7.3.2.4 Computational Efficiency

The CPU times taken by the 3.5 GHz Pentium PC using the two algorithms are given in Fig. 6.12 for the simulation duration of 2.5 s. The reason for taking a simulation time of 2.5 s is similar to that provided in Sect. 7.3.1.4. It can be seen from Fig. 7.12 that the proposed recursive algorithm is efficient than the conventional non-recursive one.

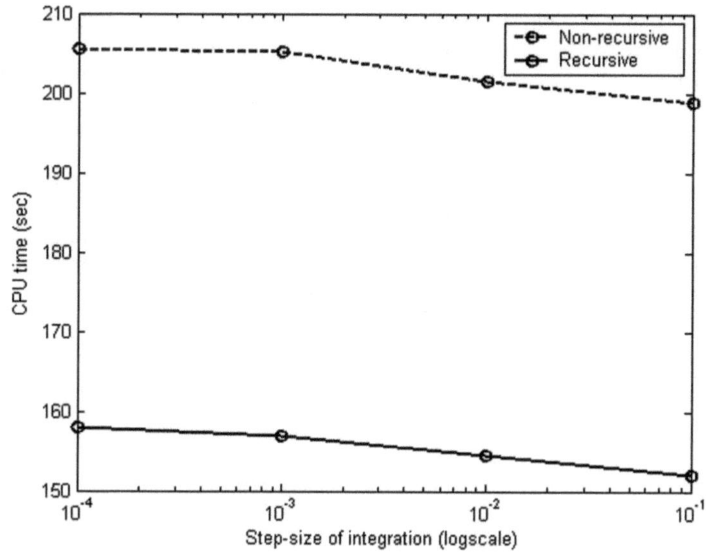

Fig. 7.12 Comparison of CPU times

7.4 Summary

This chapter analyzes the numerical stability aspects and the computational efficiencies of the proposed forward dynamics algorithms for the rigid and rigid–flexible serial robotic systems. The proposed algorithms are shown to be numerically more stable than a non-recursive algorithm based on the Cholesky decomposition. First, the importance of the study of numerical stability characteristics is underlined, and various criteria for testing it are listed. Then, using a series of numerical experiments, their behaviors are compared. Examples of the spatial six-link rigid PUMA robot, the planar three-link rigid–flexible Canadarm, and the spatial six-link rigid–flexible Space Shuttle Remote Manipulator System are considered to demonstrate the stability and efficiency of the proposed algorithms.

References

1. Sharf I, Damaren C (1992) Simulation of flexible-link manipulators: basis functions and nonlinear terms in the motion equations, pp 1956–1962
2. Cloutier BP, Pai DK, Ascher UM (1995) Formulation stiffness of forward dynamics algorithms and implications for robot simulation, pp 2816–2822
3. Ascher UM, Pai DK, Cloutier BP (1997) Forward dynamics, elimination methods, and formulation stiffness in robot simulation. Int J Rob Res 16(6):749–758
4. Strang G (1980) Linear Algebra and its applications. Harcourt, Brace, Jovanovich Pub., Florida
5. Stewart GW (1973) Introduction to matrix computations. Academic Press

6. Neto MA, Ambrósio J (2003) Stabilization methods for the integration of DAE in the presence of redundant constraints. Multibody Syst Dynam 10(1):81–105
7. Schiehlen W (2005) Recent developments in multibody dynamics. J Mechl Sci Tech 19(1):227–326
8. Nikravesh PE (1988) Computer-aided analysis of mechanical systems. Prentice-Hall Englewood Cliffs, New Jersey
9. Saha SK, Schiehlen WO (2001) Recursive kinematics and dynamics for parallel structured closed-loop multibody systems. Mech Struct Mach 29(2):143–175
10. Ogata K (1995) Modern Control Engineering. Prentice-Hall of India, New Delhi
11. Luh JYS, Walker MW, Paul RPC (1980) On-line computational scheme for mechanical manipulators. J Dyn Syst Meas Control Trans ASME 102(2):69–76
12. Cyril X (1988) Dynamics of flexible-link manipulators. McGill University, Canada

Part IV
Experimental Study of Flexible System

Chapter 8
Experimental Results

Dynamics of multilink flexible manipulators are highly nonlinear, configuration-dependent, and computationally complex. Moreover, real systems always involve external factors such as joint damping, structural damping in the flexible links, and the robot's working environment, which cannot be modeled without reasonable assumptions. As a result, the dynamic equations of the rigid and rigid–flexible robots, like those presented in Chaps. 2 and 3, represent only idealized and approximate models of the actual systems. Furthermore, attempts by various researchers [1–3] to theoretically estimate the damping factors of a system have been only partially successful. As such, it becomes necessary to complement the theoretical models with the experimental data for estimation of the damping factor and its incorporation into the dynamic model of the systems for accurate dynamic analysis of the system.

Consequently, a series of experiments was conducted on a single flexible link and a two flexible links robotic arms. The experimental results were used to estimate the damping, which was then incorporated into the dynamic model. The resulting simulation results for the modified model are then compared with those available from the experiments. Hence, the objectives of the experiments conducted are (i) to incorporate damping into the model and (ii) to validate the theoretical model. The chapter is divided into two parts. In the first part, the modus operandi adopted for the estimation of the damping factors for the system under study and the corresponding modifications in the dynamic model are presented. In the second part, the scheme of experiments, details of the experimental setup and equipment used are presented.

8.1 Damping in Dynamic Model

In this section, modifications in the dynamic equations of motion for the rigid and rigid–flexible robotic systems due to damping are presented. The inclusion of

P. V. Nandihal et al., *Dynamics of Rigid-Flexible Robots and Multibody Systems*,
Intelligent Systems, Control and Automation: Science and Engineering 100,
https://doi.org/10.1007/978-981-16-2798-9_8

damping factors into the dynamic model is important for any realistic simulation and meaningful comparison of the simulation results with the experimental ones. While damping in a rigid robotic system results mainly due to joint assembly, damping in a flexible link is of two types: joint damping and structural damping. The decay in the amplitude of the oscillations of a link is due to friction at its joint, which is referred to as joint damping. In contrast, the decay in the amplitude of vibrations of a flexible link is due to its structural stiffness, which is called as structural damping. Both the types of damping characteristics are directly proportional to the first derivative (rate) of the associated generalized coordinates [4]. For a rigid link, structural damping is absent as the generalized coordinates associated with the vibrations of the link do not exist. Dynamic equations of motion for an undamped, n-link rigid–flexible serial robot are given by Eq. (3.31), which is reproduced here as,

$$\mathbf{I}\ddot{\mathbf{q}} = \boldsymbol{\varphi} \tag{8.1}$$

where $\mathbf{I}$ is the $\overline{n} \times \overline{n}$ generalized inertia matrix (GIM) and $\boldsymbol{\varphi}$ is the $\overline{n}$-dimensional vector corresponding to the vector of external generalized torques, vector of convective inertia, Coriolis, and those associated with the potential, strain, and dissipation energy terms. Moreover, $\mathbf{q}$ is the generalized coordinates associated with the joint angles or displacements and those of the vibration amplitudes for the flexible links. Note that $\overline{n}$ is the degree of freedom (DOF) of the rigid–flexible robot given by, $\overline{n} \equiv n + \sum_{i=1}^{n_f} (3m_i + \overline{m}_i)$, where $n \equiv n_r + n_f$, is the total number of links–n_r and n_f being the number of rigid and flexible links, respectively. Moreover, m_i and $\overline{m}_i$ are the number of modes of vibration considered to model the link deflections in bending and torsion, respectively. Note that, the number 3 in the formula for $\overline{n}$ corresponds to the X, Y, and Z components of the spatial 3-dimensional deflection of ith flexible link, and its corresponding generalized coordinates in bending, $\mathbf{d}_i^x$, $\mathbf{d}_i^y$, and $\mathbf{d}_i^z$, defined after Eq. (3.4c). In this chapter, experiments are conducted on the links which by virtue of its geometry, have vibrations only about the axes of the joints. For example, a beam as shown in Fig. 8.1, with a rectangular cross section of very large length and a large height/width ratio, will have significant vibrations only its length. Following the frame notations, as followed in this book and given in Appendix A, the vibrations of the beam shown in Fig. 8.1 will be significant only about Z_i-axis, which is the joint axis.

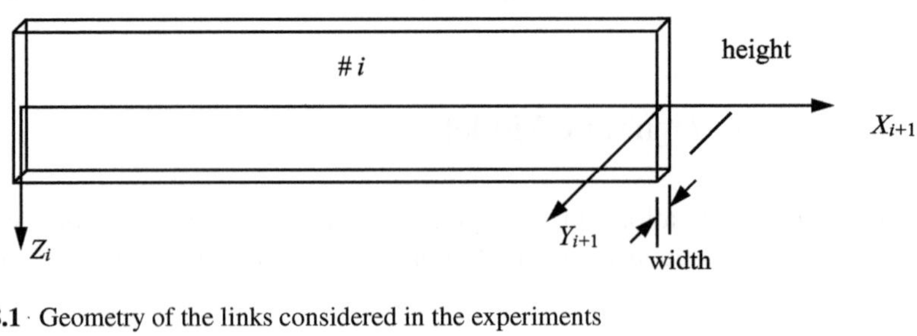

Fig. 8.1 Geometry of the links considered in the experiments

In the series of experiments presented in this chapter, the links are of similar geometry, Fig. 8.1, and no vibrations are considered about X_i and Y_i axes. Consequently, the links are assumed to be vibrating in m_i number of modes about Z_i-axis and the only degree of freedom associated with the vibrations in bending is for $j = 1, \ldots, m_i$ is $\mathbf{d}_i^z$. Thus, for the systems at hand in this chapter, $\overline{n}$ is given by $\overline{n} \equiv n + \sum_{i=1}^{n_f} (m_i + \overline{m}_i)$. Now, considering the damping in the system, Eq. (8.1) is modified as [4],

$$\mathbf{I}\ddot{\mathbf{q}} = \boldsymbol{\varphi} + \mathbf{Z}\dot{\mathbf{q}} \tag{8.2}$$

where $\dot{\mathbf{q}}$ is the $\overline{n}$-dimensional vector of joint rates defined as

$$\dot{\mathbf{q}} \equiv \left[\dot{q}_1^T \cdots \dot{q}_n^T\right]^T, \quad \text{and} \quad \dot{\mathbf{q}}_i \equiv \left[\dot{\theta}_i \; \dot{d}_i^T \; \dot{c}_i^T\right]^T$$

in which $\dot{\theta}_i$ is the rate of the rotational or translational displacement of the ith joint, θ_i, depending on its type, i.e., revolute or prismatic, respectively, and vectors $\dot{\mathbf{d}}_i$ and $\dot{\mathbf{c}}_i$ are the rates of generalized coordinates $\boldsymbol{d}_i$ and $\boldsymbol{c}_i$ as defined after Eq. (3.1a). The $\overline{n} \times \overline{n}$ damping coefficient matrix, $\mathbf{Z}$, is accordingly represented as

$$\mathbf{Z} = \begin{bmatrix} \mathbf{Z}_1 & \cdots & \mathbf{O} \\ \vdots & \ddots & \\ \mathbf{O} & & \mathbf{Z}_n \end{bmatrix} \tag{8.3a}$$

where Z_i, for $i = 1, \ldots, n$, is the $(1 + m_i + \overline{m}_i)$-dimensional matrix given by

$$\mathbf{Z}_i = \begin{bmatrix} \kappa_i & \mathbf{0} \\ \mathbf{0} & \boldsymbol{\zeta}_i \end{bmatrix} \tag{8.3b}$$

in which the scalar κ_i represents the damping coefficient at the joint and the $(m_i + \overline{m}_i) \times (m_i + \overline{m}_i)$-dimensional diagonal matrix, $\boldsymbol{\zeta}_i$, corresponds to the structural damping of vibration of the link in m_i modes in bending and $\overline{m}_i$ modes in torsion. Moreover, for rigid links, the damping coefficient matrix associated with link, i, namely $\mathbf{Z}_i$, reduces to scalar κ_i. The joint damping coefficient κ_i and the elements of the structural damping coefficients matrix $\boldsymbol{\zeta}_i$ are determined here from the experimental data. The methodologies adopted for the determination of the joint damping and structural damping coefficients are presented next.

8.2 Damping Coefficients

In this section, methodologies to determine the joint and structural damping coefficients are presented.

8.2.1 Joint Damping

The joint damping coefficient, κ_i, for a robotic system can be given by [4]

$$\kappa_i = \xi_i \overline{\kappa}_i \tag{8.4a}$$

where ξ_i is the damping ratio and $\overline{\kappa}_i$ is the critical damping factor associated with the ith joint. The damping ratio, ξ_i, is determined by measuring the rate of logarithmic decay of the oscillations of the link and is given by

$$\xi_i = \frac{1}{2\pi(k-1)} \log_e\left(\frac{x_1}{x_k}\right) \tag{8.4b}$$

in which x_i and x_p are, respectively, the amplitudes in 1st and kth cycles of the oscillation of the link. The critical damping factor associated with the ith joint, $\overline{\kappa}_i$, is obtained as

$$\overline{\kappa}_i = 4\pi\rho_i a_i \eta_f \tag{8.4c}$$

where ρ_i is the mass per unit length, a_i is the length—one of the four DH parameters—and η_f is the natural frequency of oscillation of the ith link.

8.2.1.1 An Illustration: A Single Rigid Link

The method of determination of the joint damping coefficient, κ_i, as presented above, is now illustrated using an example of a single rigid link falling freely under gravity. The only scalar equation of motion represented in the form of Eq. (8.2) is given by

$$\left(\frac{\rho a^3}{3} + I_h\right)\ddot{\theta} = \tau + \kappa\dot{\theta} \tag{8.5}$$

where ρ, a, θ, I_h, and τ are, respectively, the mass per unit length, link length, joint angle, inertia due to the hub at the root of link, and the external torque due to gravity, whereas κ is the associated joint damping coefficient of the system, that will be determined from the experimental data and using Eq. (8.4a–c). The scheme of the experiment conducted is shown in Fig. 8.2. The physical parameters of the

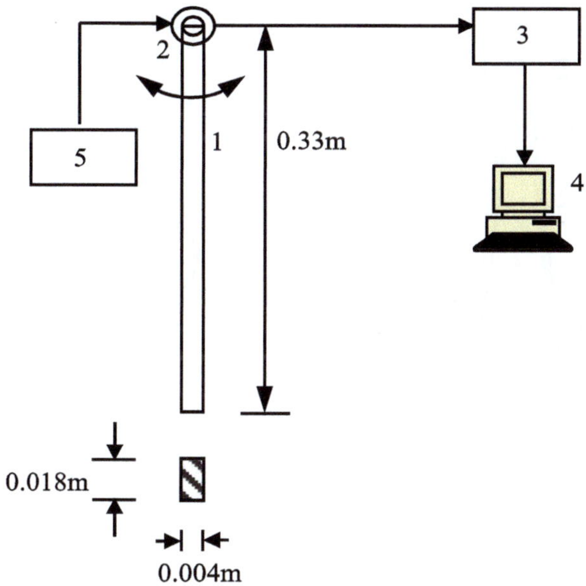

1. Rigid link; 2. Wire-pot; 3. Pico CRO; 4. Computer; 5. Power supply

Fig. 8.2 Experimental setup for the single rigid link

link and the setup are shown in Table 8.1. The angular displacement of the link is measured using the wire-pot potentiometer, 2 of Fig. 8.2, mounted at the joint. The link is allowed to fall freely under gravity from the horizontal position. The corresponding angular displacement of the link, as measured by the potentiometer, is shown in Fig. 8.3. The amplitudes of the 1st and 2nd oscillations are x_1 and x_2, as indicated in the figure. The ratio of the successive amplitudes of oscillations is obtained from it and used in Eq. (8.4b) to obtain the damping ratio, $\xi_1 = 0.0142$, whereas the critical damping factor associated with the joint, namely, $\overline{\kappa}_1$, is obtained from Eq. (8.4c) using the physical parameters of the link given in Table 8.1a and

Table 8.1 Experiment with single rigid link (Fig. 8.2)

(a) Physical parameters of the link, 1

Material	Length, a (m)	Cross section (m^2)	Mass (Kg)	Mass/length, ρ (Kg/m)	Hub inertia (Kgm^2)
Carbon steel	0.33	0.018×0.004	0.180	0.5455	2.95×10^{-5}

(b) Specifications of the equipment

Equipment	Specifications
Potentiometer, 2	10 KΩ, ± 0.25% linearity, wire-wound pot
Data acquisition system, 3 and 4	PICO CRO with Pentium IV Intel processor computer
Power supply, 5	Regulated 5 V, 1A, DC

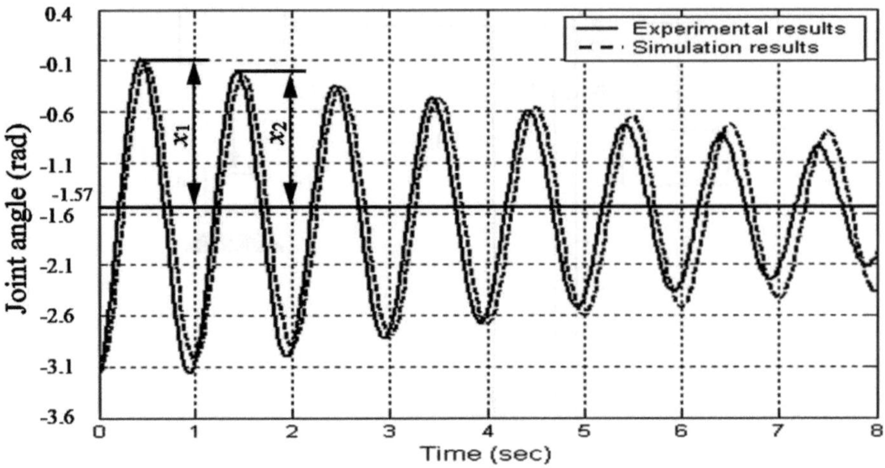

Fig. 8.3 Experimental and simulation results for a single rigid link

its natural frequency, η_f, as obtained from the experimental results. The value of $\overline{\kappa}_1 = 2.1716$ kg/s. The joint damping coefficient, κ_1, is then obtained from Eq. (8.4a) as 0.0308 kg/s for the single rigid link under study. The joint damping coefficient, κ_1, is next incorporated into Eq. (8.5), and the simulation for the same initial conditions is repeated. The resulting angular variation is plotted in Fig. 8.3 and compared with those obtained from the experiments. It is seen that the experimental and simulation results match closely, thus, verifying the modeling technique of the joint damping in the dynamic model, i.e., Eq. (8.5). The frequency of oscillation is also obtained from the Fourier transform of the experimental response for the joint angles. This is obtained as 0.96 Hz, while the frequency of oscillation of the link, as obtained from the simulation results is 1.0 Hz, which is a good match. The difference in the experimental and simulation results is, however, increasing in the latter part. This is mainly due to the joint damping, given by Eq. (8.4b), as linear. In reality, it is not so, rather hyperbolic [4] that remains linear during the initial cycles only. Modeling the hyperbolic decay of oscillation amplitudes is quite complex. Hence, a linear model was implemented, which provided fairly accurate results.

8.2.2 *Structural Damping*

The vibration amplitude of a flexible link is a combination of several modes. In order to determine the structural damping, it is required to separate out the decay in amplitudes of vibration due to each mode. This decoupling is essential because the structural damping coefficient of a beam is different in each of its modes. For a beam, vibrating in m_i modes of vibrations in bending and $\overline{m}_i$ modes of vibrations in torsion, the associated structural damping coefficients are given by $(m_i + \overline{m}_i)$ diagonal elements of the $(m_i + \overline{m}_i) \times (m_i + \overline{m}_i)$ diagonal matrix, $\boldsymbol{\zeta}_i$ of Eq. (8.3b). Similar to the joint damping, Eq. (8.4a) the matrix, $\boldsymbol{\zeta}_i$, for structural damping coefficients is defined as:

$$\boldsymbol{\zeta}_i = \tilde{\xi}_i \overline{\zeta}_i \tag{8.6a}$$

where $\tilde{\xi}_i$ is the $(m_i + \overline{m}_i) \times (m_i + \overline{m}_i)$-dimensional diagonal matrix of damping ratios associated with the m_i modes of vibrations in bending and $\overline{m}_i$ modes of vibrations in torsion of the ith link, and $\overline{\zeta}_i$ is the $(m_i + \overline{m}_i) \times (m_i + \overline{m}_i)$-dimensional matrix of critical damping factors associated with the respective modes. The matrices $\tilde{\xi}_i$ and $\overline{\zeta}_i$ are thus defined as,

$$\tilde{\xi}_i = \begin{bmatrix} \tilde{\xi}_{i1} & \cdots & & 0 \\ \vdots & \ddots & & \\ \vdots & & \tilde{\xi}_{ij} & \\ 0 & & & \tilde{\xi}_{iC} \end{bmatrix} \quad \overline{\zeta}_i = \begin{bmatrix} \overline{\zeta}_{i1} & \cdots & & 0 \\ \vdots & \ddots & & \\ \vdots & & \overline{\zeta}_{ij} & \\ 0 & & & \overline{\zeta}_{iC} \end{bmatrix} \tag{8.6b}$$

where $\tilde{\xi}_{ij}$, for $j = 1, \ldots, m_i$, is the damping ratio of the ith link in its jth mode, referred here as the modal damping ratio, and $\tilde{\xi}_{iC}$ is the $\overline{m}_i \times \overline{m}_i$ damping ratio matrix associated with the vibration of the ith link in torsion. In order to isolate the decay in the amplitude of vibration of link in each mode and to determine the corresponding damping ratios $\tilde{\xi}_{ij}$, the method of evolving spectra (McIntyre and Woodhouse 1988) is adopted here, whereas the critical damping factor, $\overline{\zeta}_{ij}$, for $j = 1, \ldots, m_i$, is obtained from the modal analysis of the link [4]. In this chapter, for the comparison of experimental and simulation results, the torsional vibration is neglected due to the types of beams selected in the experimental setup. Hence, the damping ratio matrix, $\tilde{\xi}_{iC}$, vanishes.

8.2.2.1 Method of Evolving Spectra

The method also referred as waterfall method (McIntyre and Woodhouse 1988), is based on a time window selected for the amplitude response of the link. Fast Fourier transform (FFT) of the response in this time window is performed. The natural frequencies of the link, known in advance from a real-time analyzer, are identified as the sharp peaks in the FFT response. Thus, the corresponding amplitude of the link vibration in each mode is obtained. The time window is then shifted forward, and the FFT of the response in the shifted time window is performed again. The amplitude corresponding to each mode is then noted for this shifted position of the time window. The process is repeated several times by shifting the time window. The amplitudes of the FFT curves corresponding to a particular mode in each time window are then plotted. It is found that the decay curve for the amplitudes corresponding to a mode is logarithmic, which provides the damping ratio as,

$$\tilde{\xi}_{ij} = \frac{1}{2\pi\eta_j \Delta t_k}\log_e\left(\frac{x_{j1}}{x_{jk}}\right) \tag{8.6c}$$

where η_j is the natural frequency of vibration of ith link in its jth mode, and x_{j1} and x_{jk} are the amplitudes of the vibrations in the 1st and k^{th} time windows of the response. Moreover, Δt_k is the shift in time for the kth time window with respect to the first-time window. These steps will be further illustrated in Sect. 8.2.2.3.

8.2.2.2 Critical Damping

The matrix of critical damping factors, $\overline{\zeta}_i$ of Eq. (8.6b), is obtained by the modal analysis of the link [4], which is given by

$$\overline{\zeta}_i = 2\sqrt{\widetilde{\mathbf{K}}_{di}\widetilde{\mathbf{M}}_{di}} \tag{8.7a}$$

where the $(m_i + \overline{m}_i) \times (m_i + \overline{m}_i)$-dimensional diagonal matrices, $\widetilde{\mathbf{K}}_{di}$ and $\widetilde{\mathbf{M}}_{di}$, are, respectively, the modal stiffness and modal mass matrices of the ith link. For a link vibrating in space in all three directions, namely X, Y, and Z, in m_i modes, $\widetilde{\mathbf{K}}_{di}$ and $\widetilde{\mathbf{M}}_{di}$ are the $(3m_i + \overline{m}_i) \times (3m_i + \overline{m}_i)$ matrices, which can be written using the modal analysis as

$$\widetilde{\mathbf{K}}_{di} \equiv \int_0^{a_i} E_i I_i \begin{bmatrix} \overline{\mathbf{S}}_i^x & & & \\ & \overline{\mathbf{S}}_i^y & & \\ & & \overline{\mathbf{S}}_i^z & \\ & & & \mathbf{O}_{\overline{m}_i} \end{bmatrix}^{\mathrm{T}} \begin{bmatrix} \overline{\mathbf{S}}_i^x & & & \\ & \overline{\mathbf{S}}_i^y & & \\ & & \overline{\mathbf{S}}_i^z & \\ & & & \mathbf{O}_{\overline{m}_i} \end{bmatrix} d\overline{a}_{i+1} \tag{8.7b}$$

$$\widetilde{\mathbf{M}}_{di} \equiv \int_0^{a_i} \rho_i \begin{bmatrix} \mathbf{S}_i^x & & & \\ & \mathbf{S}_i^y & & \\ & & \mathbf{S}_i^z & \\ & & & \mathbf{O}_{\overline{m}_i} \end{bmatrix} \begin{bmatrix} \mathbf{S}_i^x & & & \\ & \mathbf{S}_i^y & & \\ & & \mathbf{S}_i^z & \\ & & & \mathbf{O}_{\overline{m}_i} \end{bmatrix} d\overline{a}_{i+1} \tag{8.7c}$$

where $E_i I_i$ and ρ_i are, respectively, the flexure stiffness and mass per unit length of the ith link, and $\mathbf{O}_{\overline{m}_i}$ is the $\overline{m}_i \times \overline{m}_i$ zero matrix due to the absence of torsional vibration. However, the $m_i \times m_i$-dimensional matrices, $\overline{\mathbf{S}}_i^x$, $\overline{\mathbf{S}}_i^y$, and $\overline{\mathbf{S}}_i^z$; and $\mathbf{S}_i^x$, $\mathbf{S}_i^y$, and $\mathbf{S}_i^z$, are the matrices associated with the shape functions for the vibrations about $\widehat{X}_{i+1}$-, $\widehat{Y}_{i+1}$-, and $\widehat{Z}_{i+1}$- axis, respectively. Thus,

$$\overline{\mathbf{S}}_i^x \equiv \begin{bmatrix} \frac{\partial s_{i,1}^x}{\partial \overline{a}_i} & & O \\ & \ddots & \\ O & & \frac{\partial s_{i,m_i}^x}{\partial \overline{a}_i} \end{bmatrix}; \mathbf{S}_i^x \equiv \begin{bmatrix} s_{i,1}^x & & O \\ & \ddots & \\ O & & s_{i,m_i}^x \end{bmatrix} \tag{8.7d}$$

where for, $j = 1, \ldots, m_i, s_{i,j}^x$ is the shape function for the link in its jth mode. The $m_i \times m_i$-dimensional matrices $\mathbf{S}_i^y$, $\mathbf{S}_i^z$, $\overline{\mathbf{S}}_i^y$, and $\overline{\mathbf{S}}_i^z$ are defined similarly for the shape functions associated with vibrations about $\widehat{Y}_{i+1}$ and $\widehat{Z}_{i+1}$ axes. In this chapter, the beams are considered to vibrate about the joint axis only, as presented in Sect. 7.1. Hence, the vibrations about $\widehat{X}_{i+1}$, $\widehat{Y}_{i+1}$ axes are neglected, i.e., the matrices, $\mathbf{S}_i^x$, $\mathbf{S}_i^y$, $\overline{\mathbf{S}}_i^x$, and $\overline{\mathbf{S}}_i^y$ vanish, and $\widetilde{\mathbf{K}}_{di}$ and $\widetilde{\mathbf{M}}_{di}$ become $(m_i + \overline{m}_i) \times (m_i + \overline{m}_i)$-dimensional matrices that are expressed as

$$\widetilde{\mathbf{K}}_{di} \equiv \int_0^{a_i} E_i I_i \begin{bmatrix} \overline{\mathbf{S}}_i^z & \\ & \mathbf{O}_{\overline{m}_i} \end{bmatrix}^{\mathrm{T}} \begin{bmatrix} \overline{\mathbf{S}}_i^z & \\ & \mathbf{O}_{\overline{m}_i} \end{bmatrix} d\overline{a}_{i+1} \tag{8.7e}$$

$$\widetilde{\mathbf{M}}_{di} \equiv \int_0^{a_i} \rho_i \begin{bmatrix} \mathbf{S}_i^z & \\ & \mathbf{O}_{\overline{m}_i} \end{bmatrix}^{T} \begin{bmatrix} \mathbf{S}_i^z & \\ & \mathbf{O}_{\overline{m}_i} \end{bmatrix} d\overline{a}_{i+1} \tag{8.7f}$$

where $m_i \times m_i$-dimensional matrices $\overline{\mathbf{S}}_i^z$ and $\mathbf{S}_i^z$ are defined similar to Eq. (8.7d).

8.2.2.3 An Illustration: A Single Flexible Link

The method of determining structural damping coefficient, as presented above, is now illustrated using an example of a single flexible link. Only the link vibrations without any joint motion are considered, i.e., the flexible link is clamped at one end, as shown in Fig. 8.4. The beam is kept in such a way that it vibrates in the horizontal plane. For

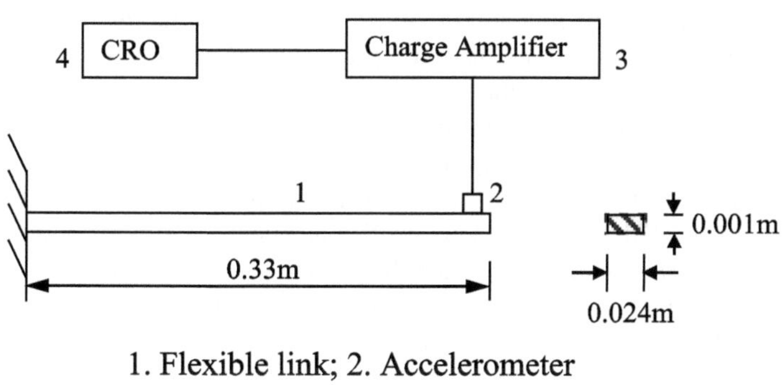

Fig. 8.4 Determination of structural damping coefficients

the purpose of modeling, only first two modes are considered. Correspondingly, the determination of the structural damping coefficient in the first two modes is shown.

For a single flexible link vibrating in its first two modes, equations of motion, Eq. (8.1), are expressed as,

$$\rho\begin{bmatrix} \frac{a^3}{3}+k_{11}d_1^2+k_{12}d_2^2 & k_{21} & k_{22} \\ & k_{11} & \\ sym & & k_{12} \end{bmatrix}\begin{bmatrix} \ddot{\theta} \\ \ddot{d}_1 \\ \ddot{d}_2 \end{bmatrix}+\rho\begin{bmatrix} 2\dot{\theta}\left(k_{11}\dot{d}_1^2+k_{12}\dot{d}_2^2\right) \\ -k_{11}d_1\dot{\theta}^2 \\ -k_{12}d_2\dot{\theta}^2 \end{bmatrix}$$
$$=\begin{bmatrix} \tau_\theta \\ \tau_{d_1} \\ \tau_{d_2} \end{bmatrix}+\begin{bmatrix} 0 \\ \tau_{d_1}^s \\ \tau_{d_2}^s \end{bmatrix}+\begin{bmatrix} \kappa & 0 & 0 \\ & \zeta_1 & 0 \\ sym & & \zeta_2 \end{bmatrix}\begin{bmatrix} \dot{\theta} \\ \dot{d}_1 \\ \dot{d}_2 \end{bmatrix} \tag{8.8a}$$

where ρ is the mass per unit length, a is the link length, θ is the joint angle, d_1 and d_2 are the generalized coordinates of the link vibrating in 1st and 2nd modes, respectively. Moreover, the terms, τ_θ, τ_{d_1} and τ_{d_2} are, respectively, the corresponding generalized forces corresponding the generalized coordinates, θ, d_1 and d_2. The terms $\tau_{d_1}^s$ and $\tau_{d_2}^s$ are the terms due to strain energy. The expressions for $\tau_{d_1}^s$ and $\tau_{d_2}^s$ are given after Eq. (3.20e–f), whereas k_{1j} and k_{2j}, for $j = 1, 2$, are the constants associated with the shape functions, i.e.,

$$k_{1j} \equiv \int_0^a s_{ij}^2 d\bar{a} \quad \text{and} \quad k_{2j} \equiv \int_0^a s_{ij} d\bar{a} \tag{8.8b}$$

in which s_{ij} is the shape function of the link in jth mode, as given by Eq. (3.4a). Furthermore, κ is the joint damping coefficient, and ζ_j, for $j = 1, 2$, is the associated structural damping coefficient of the link which will be determined here from the experimental data. The scheme of the experiment is shown in Fig. 8.4. Corresponding to the two modes of vibrations, the structural damping coefficients are given by, $\zeta_j = \xi_j\bar{\zeta}_j$, for $j = 1,2$, in which $\tilde{\xi}_j$ and $\bar{\zeta}_j$ are, respectively, the structural damping ratio and critical damping factor associated with vibration in the jth mode. The physical parameters of the link and the setup are presented in Table 8.2.

The physical dimensions and the mass properties of the link are so selected that the flexibility characteristics of the link are prominent. Moreover, the link is clamped horizontally, and an accelerometer is mounted on its end, as indicated in Fig. 8.4. The output of the accelerometer is amplified using a charge amplifier, and the readings are taken on a CRO. To determine the structural damping coefficients of the link, the tip of the link is deflected by a known displacement (0.015 m) and released to vibrate in its natural condition. The amplitude decay curve of the link, as obtained from the accelerometer, is shown in Fig. 8.5a. The structural damping coefficient, ζ_j, is then estimated by first determining the structural damping ratio $\tilde{\xi}_j$ using the method of evolving spectra explained in Sect. 8.2.2.1. A time window of 0.05 s is selected. Two such positions of the time window, namely the 1st and 4th, are shown in Fig. 8.5a. FFT of the responses in different time windows are plotted to

Table 8.2 Experiment with single flexible link (Fig. 8.4)

(a) Physical parameters of the link

Material	Length, a_i (m)	Cross section (m^2)	Mass (Kg)	Mass/length, ρ_i (kg/m)	Flexure stiffness (E_iI_i) (Nm^2)
Spring steel	0.33	0.024 × 0.001	0.060	0.1818	0.4

(b) Specifications of the equipment

Equipment	Specifications
Accelerometer, 2	20 g; Diaphragm type
Charge amplifier, 3	Bruel and Kjaer, 2635; Lower-frequency cut-off: 2 Hz; Higher-frequency cut-off: 30 Hz; Calibration factor: 100 mV/mm; Charge sensitivity: 1.003 pC/ms^{-2}; Voltage sensitivity: 0.88 mV/ms^{-2}
Data acquisition system, 4	Agilent, 54621oscilloscope; 60 MHz, 200 MSa/s

get Fig. 8.5b. Figure 8.5b is called the waterfall plot. For clarity of representation, FFT responses in only four locations of the time window are shown. Since the peaks in the FFT curves are located corresponding to the natural frequencies of the link, the amplitudes of the successive peaks on a particular natural frequency are now noted. Since the determination of structural damping coefficients is performed for the first and second modes, amplitudes of successive peaks, namely x_1 and x_2 for the 1st mode, as shown in Fig. 8.5b, are noted. Then, using Eq. (8.6c), the structural damping ratio $\tilde{\xi}_1$ is calculated. Similarly, the ratio for the 2nd mode, $\tilde{\xi}_2$, is obtained. Alternatively, the amplitude decay of the peaks of the FFT curves for the 1st and 2nd natural frequencies are plotted against time on a semi-logarithmic scale, as shown in Fig. 8.5c. The slope of the curve divided by the corresponding natural frequency of the mode gives the structural damping ratio for the particular mode at hand. For the link understudy, the latter methodology, i.e., Fig. 8.5c, is used, which results in $\tilde{\xi}_1 = 0.1980$ and $\tilde{\xi}_2 = 0.0428$. Next, the critical damping ratios, $\overline{\zeta}_j$, for $j = 1, 2$, are calculated using the link length a, the shape functions, and Eq. (8.7). The values are $\overline{\zeta}_1 = 6.9 \times 10^{-2}$ kg/s and $\overline{\zeta}_2 = 6.5 \times 10^{-3}$ kg/s. Finally, the structural damping coefficient, ζ_j, is obtained for $j = 1, 2$, using $\zeta_j = \tilde{\xi}_j \overline{\zeta}_j$, i.e., as $\zeta_1 = 1.37 \times 10^{-2}$ kg/s and $\zeta_2 = 2.8 \times 10^{-3}$ kg/s. They are then incorporated into Eq. (8.8a). Simulation is then performed with the same initial conditions as that of Fig. 8.4.

The initial conditions are taken as follows: Since the tip of the link is initially deflected by a known amount, $\boldsymbol{u}$ of Eq. (3.1d), i.e.,

$$\boldsymbol{u} \equiv \begin{bmatrix} \overline{s}_1^z & 0 \\ 0 & \overline{s}_2^z \end{bmatrix} \begin{bmatrix} d_1 \\ d_2 \end{bmatrix}$$

where for $j = 1, 2$; $\overline{s}_j^z$ is the shape function associated with the bending of the link in jth mode and evaluated at a, i.e., the tip of the link. These values are known.

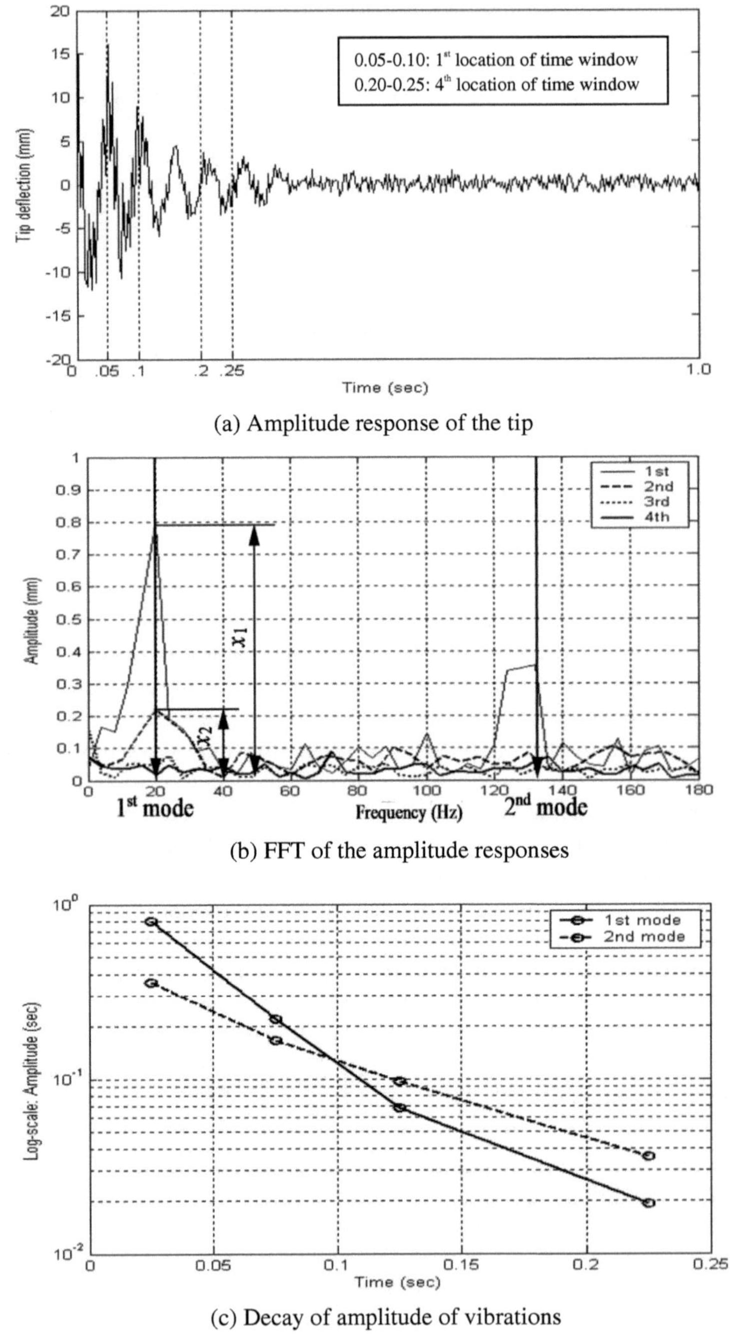

Fig. 8.5 Determination of structural damping ratio

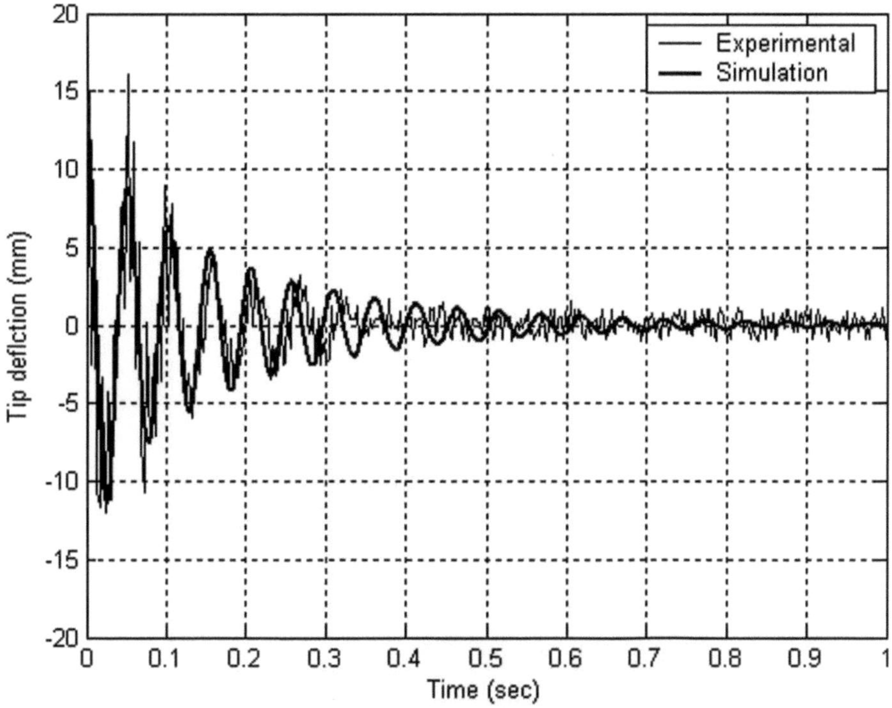

Fig. 8.6 Simulation results for clamped flexible link

Assuming $d_2 = 0$, d_1 is evaluated. Now, in order to use Eq. (8.8a) for the simulation, θ is always zero, i.e., $\theta = 0$ since one end of the link is clamped. Accordingly, $\dot{\theta} = \ddot{\theta} = 0$. Also, initially, $\dot{d}_1 = \dot{d}_2 = 0$. Simulation results for the tip deflection of the link thus obtained are then compared with those obtained experimentally, as in Fig. 8.6. It is clear from the plots that the experimental and simulation results match closely, thus verifying the values for the structural damping coefficients as correct.

8.3 A Robotic Arm with Single Flexible Link

In this section, results of the experiments conducted on a single flexible link arm are reported. The experimental results are compared with those obtained through simulations that are based on the dynamic equations of motion derived in Chap. 3. The flexible link used as a moving arm is the same as used in Sect. 8.2.2.3. The scheme of the experiment and its photograph are shown in Figs. 8.7a, b. The physical parameters of the link and the experimental setup are shown in Table 8.3. The joint displacement of the link is measured using the potentiometer placed at the joint, whose inertia is included in the dynamic model as hub inertia. The deflection at the

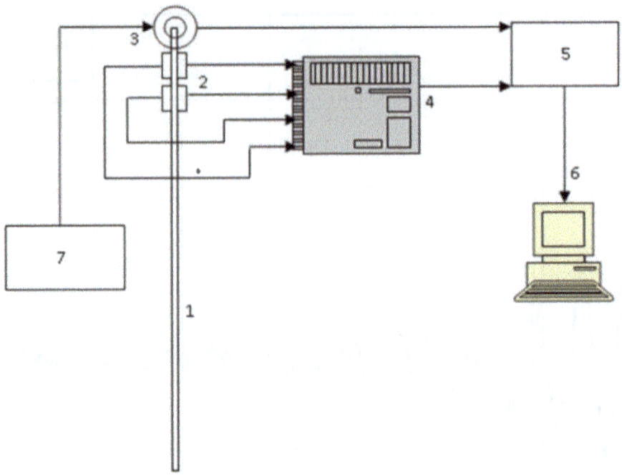

(a) Schematic

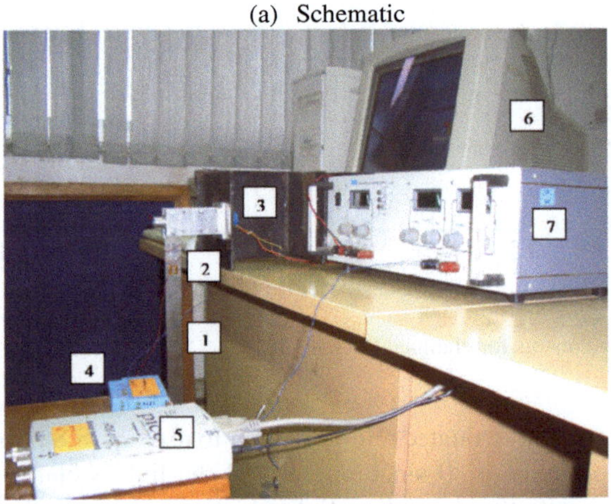

(b) Photograph

1. Flexible link; 2. Strain-gauges; 3. Potentiometer; 4. Amplifier; 5. PICO-CRO; 6. Computer; 7. Power supply

Fig. 8.7 Experimental setup for single flexible link arm

tip of the link is measured using a full strain gauge Wheatstone bridge mounted at the root of the link. Two strain gauges are mounted on each side of the link so that the proper conditioning of the readings is obtained, giving the deflection direction correctly. The strain gauge signals are amplified using an amplifier, and the output is taken on the PICO CRO attached to the computer. The characteristic details of the equipment are shown in Table 8.3b.

Table 8.3 Single flexible link arm (Fig. 8.7)

(a) Physical parameters of the link, 1

Material	Length, a (m)	Cross section (m^2)	Mass (kg)	Mass/ Length, ρ (kg/m)	Flexure stiffness, EI (Nm^2)	Hub-inertia, I_h (kg/m^2)
Spring steel	0.33	0.024 × 0.001	0.060	0.1818	0.4	2×10^{-5}

(b) Specifications of the equipment used

Equipment	Specifications
Amplifier, 4	ADAM-3016, DIN rail-mounted
Strain gauges, 2	350 Ω, 100% gain
Computer, 6	3.5 GHz Pentium IV desktop PC
Potentiometer, 3	Same as Table 7.1b
Data acquisition system, 5 and 6	
Power Supply, 7	

8.3.1 *Calibration*

In order to use the strain gauges to find the link deflection at its tip, which is quite economical compared to vision and other systems, it needs to be calibrated. First, the output of the strain gauge bridge, i.e., the strain induced in the strain gauges due to the link bending, is recorded for a predetermined deflection of the link tip.

Figures 8.8a, b show the schematic and the actual experimental setup for this calibration, in which the link is clamped at one end, as shown in Fig. 8.8a, and its tip is deflected by putting a known load on the load cup. The deflection is measured using the dial gauge, and the corresponding change in the output voltage is recorded on a Sanwa DMM, PC5000digital multimeter of 0.01 Ω resistance and 0.01 mV resolution. Figure 8.9 shows the calibration curve for the strain gauge bridge outputs. It can be seen that the bridge exhibits linear characteristics in the range of deflection, which is given by the equation written in Fig. 8.9 itself.

8.3.2 *Free Fall*

In this subsection, results of the free-fall experiments with the single flexible link arm are presented which are also compared with those obtained from the simulation results. To obtain the results, the link is allowed to fall freely under gravity from the horizontal position, as per Fig. 8.7, with no initial deflection and no external torque applied on it. The angular displacement of the joint was measured by the potentiometer and the tip deflection of the link using the strain gauges. The results are shown in Fig. 8.10a, b. To compare the experimental results with those from the

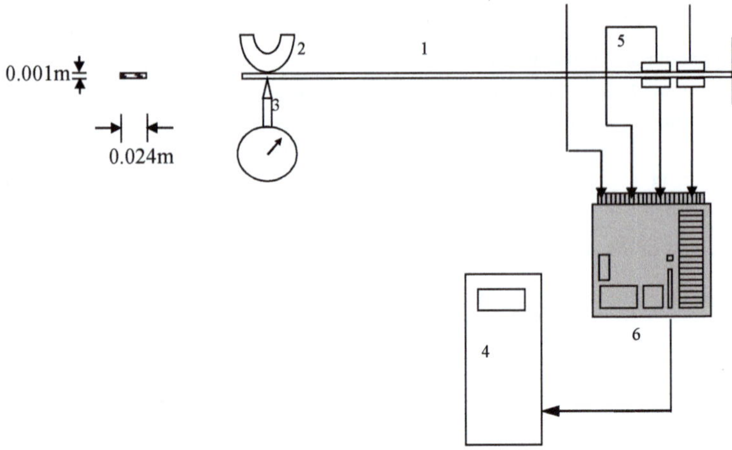

(a) Schematic

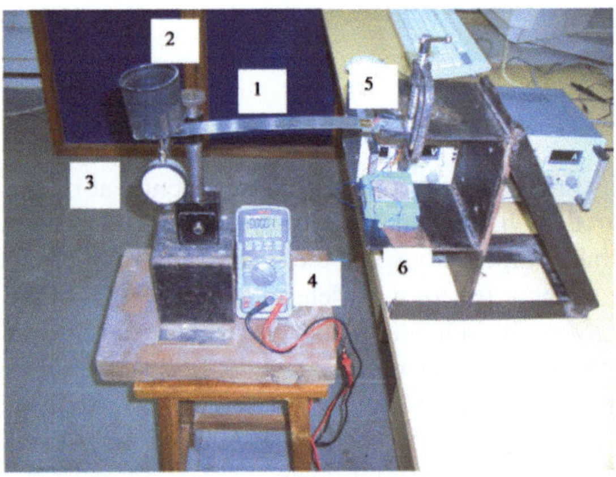

(b) Photograph
1. Flexible link; 2. Load-cup; 3. Dial-gage; 4. Multimeter; 5. Strain-gage bridge; 6. Amplifier

Fig. 8.8 Calibration of the strain gauge bridge

simulations, joint damping coefficient and structural damping were first calculated, respectively as $\kappa = 0.0106\,\text{kg/s}$ and $\zeta_1 = 1.37 \times 10^{-2}\,\text{kg/s}$, $\zeta_2 = 2.8 \times 10^{-3}\,\text{kg/s}$ using the scheme presented in Sect. 8.1. In this subsection, the same arm, as considered in the previous subsection, was subjected to a specified torque excitation at its joint. For this, the experimental setup is shown in Fig. 8.11. The specifications of the setup are shown in Table 8.4. The torque is applied on the link by the servomotor. The control system on the servomotor is designed to apply its maximum torque (0.011 Nm) while moving the link from any random initial position to the reference position of the setup. Moreover, as soon as the link reaches its final position, which

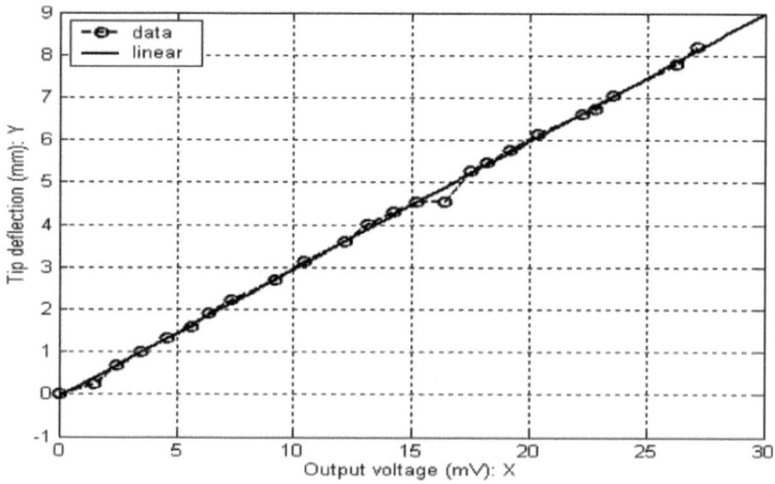

o: Reading points; y=0.3x-0.086

Fig. 8.9 Calibration curve for the strain gauge bridge

is the reference position of the link for the control circuit, the motor stops applying torque, and brakes are applied on the link to bring it to the rest. The angular rotation of the link is measured using the potentiometer mounted on the servomotor. The deflection of the link is measured using two load cells; one mounted on each side of the link. The output of the load cells and the potentiometer are taken on a CRO having an interface with the computer.

The damping factors were then incorporated into Eq. (8.8a) and the free-fall simulation results were obtained using the proposed forward dynamics algorithm. Initially, conditions for simulation were taken as, $\theta = d_1 = d_2 = 0; \dot{\theta} = \dot{d}_1 = \dot{d}_2 = 0$. Figures 8.10a, b show the comparison of the experimental and simulation results for the joint angle and tip deflection of the link, respectively. The fast Fourier transformation, i.e., FFT, of the experimental results, Figs. 8.10a, b show an oscillation frequency of 1.1 Hz and the tip vibration frequency of 20.5 Hz., respectively. From the simulation results, the joint oscillation and tip vibration frequencies are 1.1 Hz is 21 Hz, respectively. Also, the amplitudes of joint oscillations and tip vibrations also match closely, particularly during the first three–four cycles. The mismatch in the peak height values is attributed to the fact that the real beam vibrates in all possible modes of vibration, whereas the simulation considered only for first two modes of vibration. The worsening of results in the latter part is mainly due to linear decay considerations of the damping factors given by Eq. (8.4b, c).

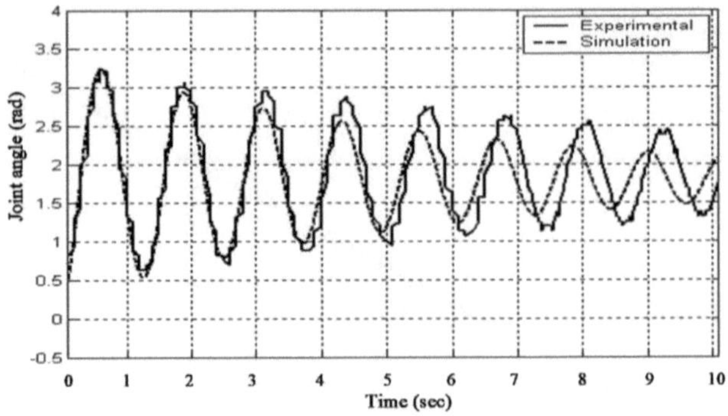

(a) Joint angle

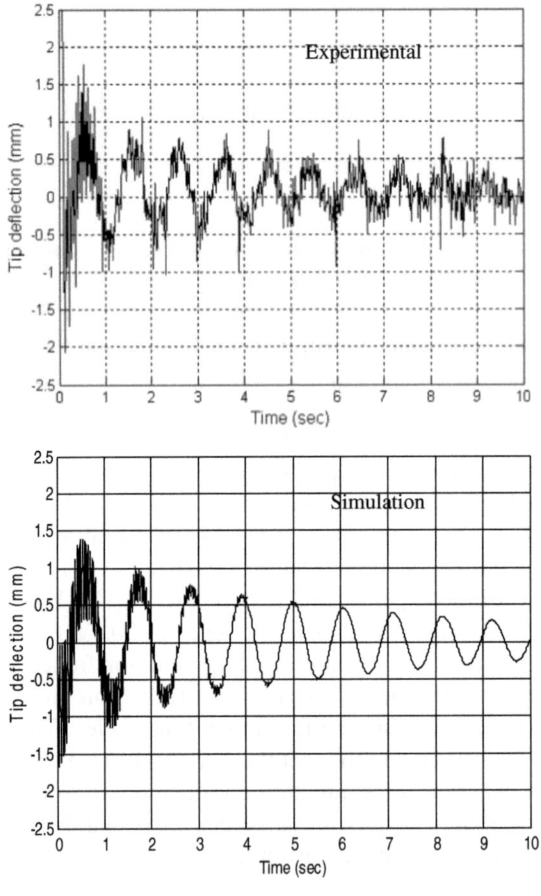

(b) Tip deflection

Fig. 8.10 Free-fall results for the single flexible link arm

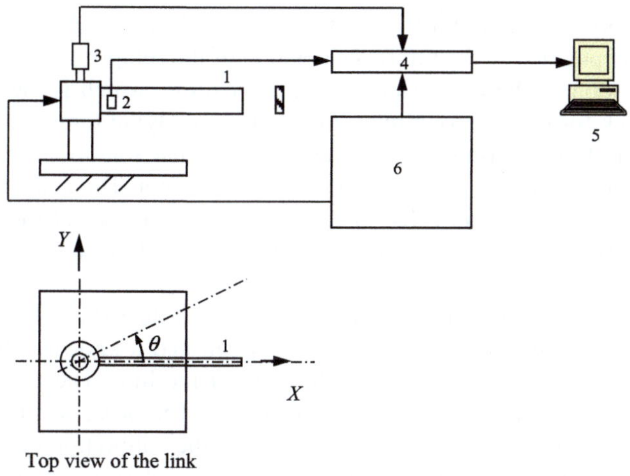

1. Flexible link; 2. Load-cell; 3. Servomotor and potentiometer; 4. PICO CRO; 5. Computer; 6. Power supply

Fig. 8.11 Setup for the forced experiment with single flexible link arm

Table 8.4 Setup for the forced experiment with single flexible link arm (Fig. 8.11)

(a) Physical parameters of the link, 1

Material	Length, a (m)	Cross section (m^2)	Mass (kg)	Mass/length, ρ (kg/m)	Flexure rigidity, EI (Nm^2)	Hub-inertia (kg/m^2)
Spring steel	0.33	0.024×0.001	0.060	0.1818	0.4	3×10^{-5}

(b) Specifications of the equipment

Equipment	Specifications
Actuator, 3	M5945G-11 Standard Delux digital coreless servomotor operating voltage range: 4.8–6.0 V
Potentiometer, 3	Wire-wound
Load-cell, 2	350Ω, activation voltage 5 V
Data acquisition system, 4 and 5	Same as Table 7.1b
Power supply, 6	

8.3.3 Forced Response

In this subsection, the same arm, as considered in the previous subsection, was subjected to a specified torque excitation at its joint. For this, the experimental setup is shown in Fig. 8.11. The specifications of the setup are shown in Table 8.4. The torque is applied on the link by the servomotor. The control system on the servomotor is designed to apply its maximum torque (0.011 Nm) while moving the link from any random initial position to the reference position of the setup.

Moreover, as soon as the link reaches its final position, which is the reference position of the link for the control circuit, motor stops applying torque, and brakes are applied on the link to bring it to the rest. The angular rotation of the link is measured using the potentiometer mounted on the servomotor. The deflection of the link is measured using two load cells; one mounted on each side of the link. The output of the load cells and the potentiometer are taken on a CRO having an interface with the computer.

To generate the experimental results, first, the output of the load cells bridge was calibrated. To do this, the link was locked at its joint, and a known deflection in the lateral direction was imparted at the other end of the link. The output of the load cells bridge for the given deflection was recorded. The process is repeated by changing the deflection of the link. Finally, to obtain the calibration curve for the load cells, the link deflections are plotted against the output voltages, as shown in Fig. 8.12a. A series of experiments was then conducted for different initial angular displacements, namely $\theta = -80°, -60°, -40°$, as per Fig. 8.11. Simulation results were then also obtained after incorporating the joint and structural damping, where the initial conditions were taken as, $\theta = -40°, d_1 = d_2 = 0; \dot{\theta} = \dot{d}_1 = \dot{d}_2 = 0$. The comparison of the results is shown in Figs. 8.12b, c. The results corresponding to other initial conditions are similar and hence not reported. Since the end conditions of the link remain the same as those in Sect. 8.2.2.3, the same structural damping coefficients, i.e., $\zeta_1 = 1.37 \times 10^{-2}$ kg/s, $\zeta_2 = 2.8 \times 10^{-3}$ kg/s, were taken.

Note in Fig. 8.12b that both the joint angles have an increasing trend. However, their numerical values do not match. This is apparently due to reasons cited for the robotic arm with two flexible links in Sect. 8.4. On the other hand, the tip deflections have very good match, as evident from Fig. 8.12c.

8.4 A Robot Arm with Two Flexible Links

In the previous section, a comprehensive experimental validation of the proposed dynamic modeling for a flexible link under different conditions has been presented. This section presents the experimental validation of the proposed dynamic modeling for a two flexible link robotic arm, as desired in Chap. 3. The two flexible link robotic arm represents a system of multiple flexible links coupled together. Hence, the main objective behind presenting the experimental results and their comparison with the simulation results for a two flexible link robotic arm is to validate the model for a representative serial multiple-link robot. Moreover, the methodology for the estimation of damping and its incorporation in the dynamic model is also tested for a multiple flexible link robot. The schematic diagram of the experimental setup for conducting the experiments on two flexible link arms and its photograph are shown in Fig. 8.13. Some of the closer views are also shown in Fig. 8.14. Characteristic details

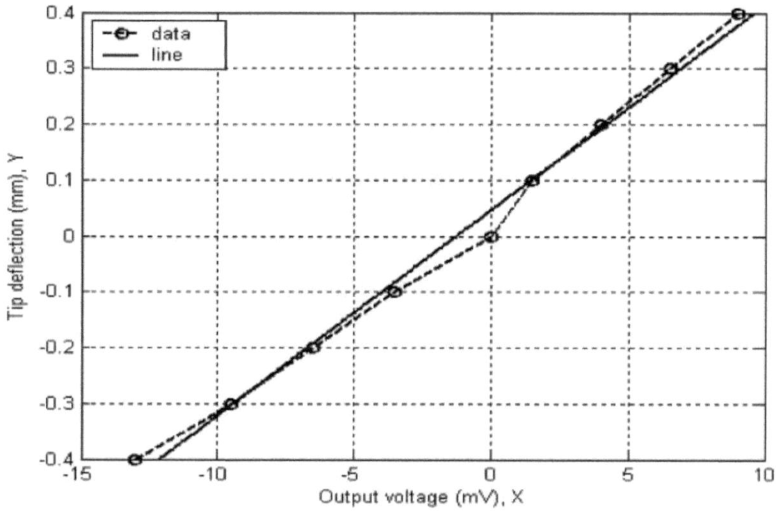

o: Reading points; y=0.037x+0.04

(a) Calibration curve

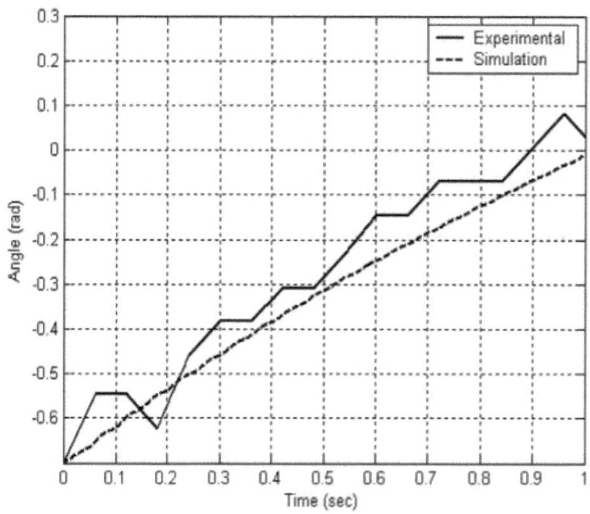

(b) Joint angle

Fig. 8.12 Forced response of the single flexible link arm

of the equipment used are shown in Table 8.4b, whereas the physical parameters of the links are shown in Table 8.5.

Like forced single flexible link arm experiments, torques are applied by the servomotors at the two joints. For the simulation, the effect of servomotors, potentiometers, and the joint fixtures at each joint is taken into account as hub inertias on the link.

Fig. 8.12 (continued)

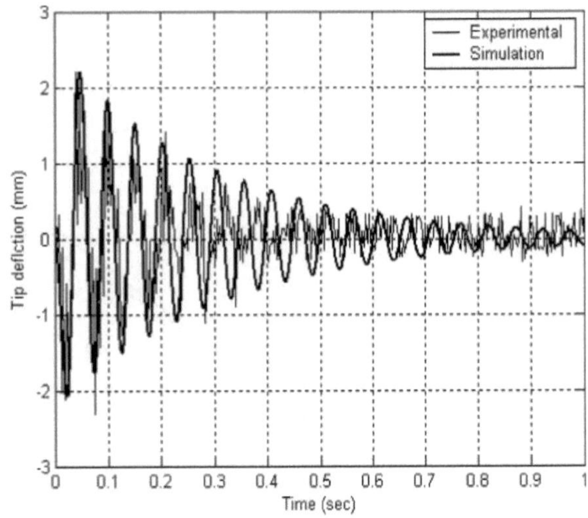

(c) Tip deflection

Moreover, for the purpose of obtaining the experimental results, the potentiometer and load-cell at each joint are calibrated, as plotted in Fig. 8.15a, b. The characteristics of the potentiometer and the load cells are linear, as seen in Fig. 8.15a, b. The forced response of the arm is then obtained by applying a torque of 0.011 Nm on the 1st joint and 0.007 Nm on the 2nd joint. A series of experiments determining the forced responses of the arms, for different initial configurations of the arm, namely $\theta_1 = 80°, \theta_2 = -80°$; $\theta_1 = 60°, \theta_2 = -60°$; and $\theta_1 = 40°, \theta_2 = -40°$ were obtained. The results are compared for $\theta_1 = 40°, \theta_2 = -40°$ with the simulation results, as shown in Figs. 8.15c, d. Behaviors for other initial conditions are similar. The joint angles were measured with the potentiometers, whereas the deflections were measured with the load cells. Note that the determination of the damping was done based on the methodology presented in Sect. 8.2.2, whose plots are shown in Fig. 8.16. The corresponding values are $\zeta_{11} = 0.0922$ kg/s, $\zeta_{12} = 0.0708$ kg/s, $\zeta_{21} = 0.0807$ kg/s and $\zeta_{22} = 0.1348$ kg/s. Again, for simulation only, the first two modes were considered for the vibration of each link. Since the link is brought to rest through the feedback control systems of the servomotors, the natural joint damping coefficients were taken as zeros, i.e., $\kappa_1 = \kappa_2 = 0$. Initial conditions for the simulation were taken as, $\theta_1 = 40°, \theta_2 = -40°$, and, for $i = 1, 2$, $d_{i1} = d_{i2} = 0$; $\dot{\theta}_1 = \dot{\theta}_2 = \dot{d}_{i1} = \dot{d}_{i2} = 0$.

The simulation results for the joint angle variations and the tip deflections are shown in Figs. 8.15c, d, which are compared with those obtained experimentally. The results for the joint angles are not matching closely, as seen in Fig. 8.15c. For joint 1, experimental and simulation results follow a similar pattern even though the values are different. For joint 2, the results match only up to 0.3 s but deviate afterward. The reasons for such variation are mainly attributed to:

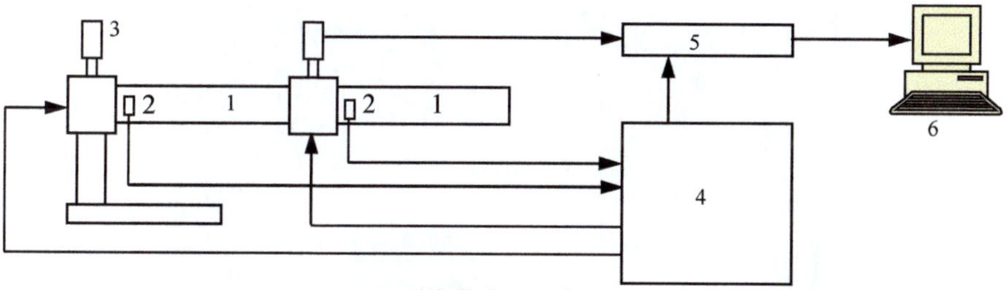

(a) Schematic

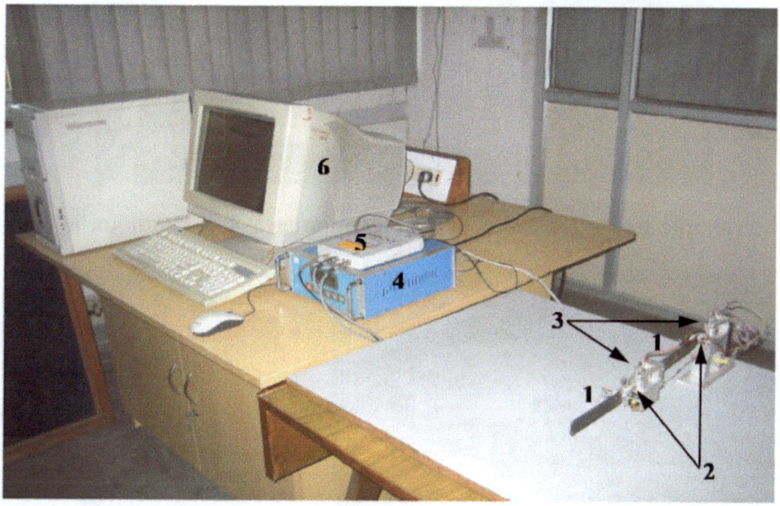

(b) Photograph

1. Two flexible-links; 2. Load cells; 3. Motor and pot; 4. Power supply and amplifier; 5. Pico-CRO; 6. Computer

Fig. 8.13 Setup for a robotic arm with two flexible link

1. It was difficult to estimate the exact torque applied at the joints, as there are in-built controllers in the commercially built setup.
2. Due to the complex shapes of the servomotors and the joint fixtures, it was difficult to measure the hub inertia at the joint of each link accurately.
3. The servomotor assembly at joint 2, which forms the payload for link 1, was also difficult to estimate because of its complex geometry.

The tip deflection results, however, match closely. For example, the frequency of vibrations of the tip of the first link obtained from the experimental results is 11.75 Hz, whereas the same from the simulation results is 11.5 Hz, which is close. For link 2, the frequencies of tip vibrations from the experiment and simulation are 13.5 Hz and 13.0 Hz, respectively. The amplitudes of vibrations are also closely matching, as seen in Fig. 8.15d.

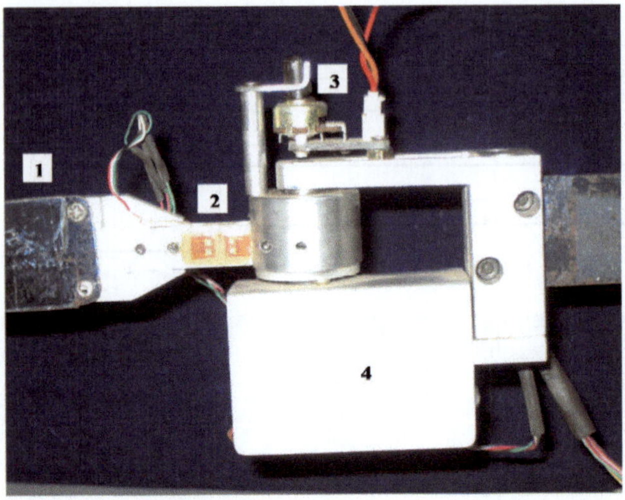

1. Two flexible-links; 2. Load cells; 3. Potentiometer; 4. Servomotor assembly

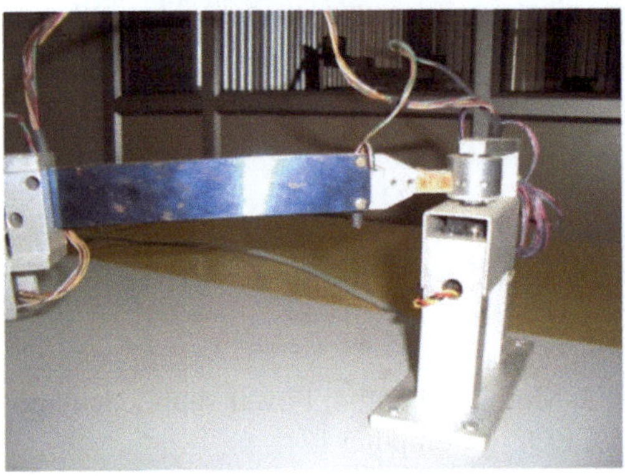

(a) Load-cells mounted at the root

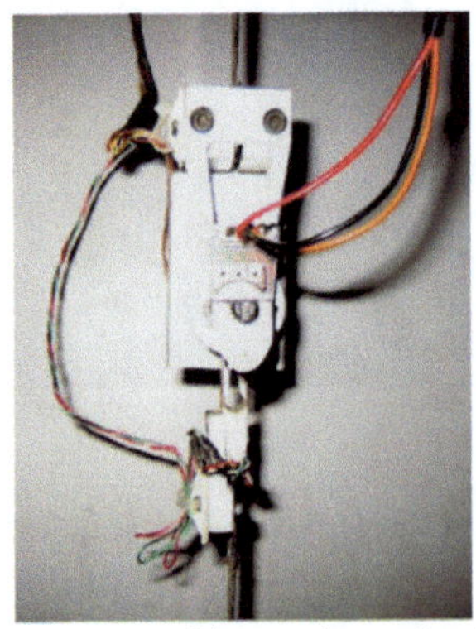

(c) Servomotor with pot (top view)

Fig. 8.14 Closer views of the two flexible link setup

Table 8.5 Physical parameters of the links

i	Material	Length, a_i (m)	Cross section (m^2)	Mass (kg)	Mass/length, ρ_i (kg/m)	Hub Inertia (kgm^2)	Flexure rigidity, E_iI_i (Nm^2)	Payload (kg)
1	Spring steel	0.24	0.024 × 0.0005	0.020	0.0833	3×10^{-5}	0.05	0.19
2	Spring steel	0.19	0.024 × 0.0005	0.017	0.1053	3×10^{-5}	0.05	0

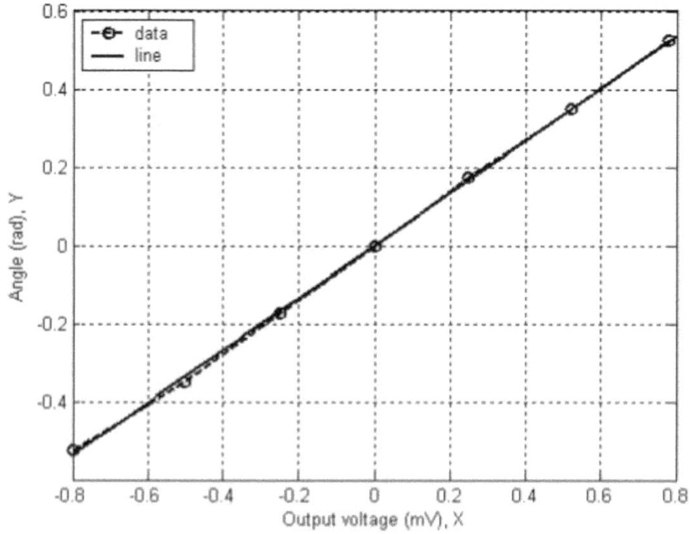

o: Reading points; y=0.667x
Joint 1

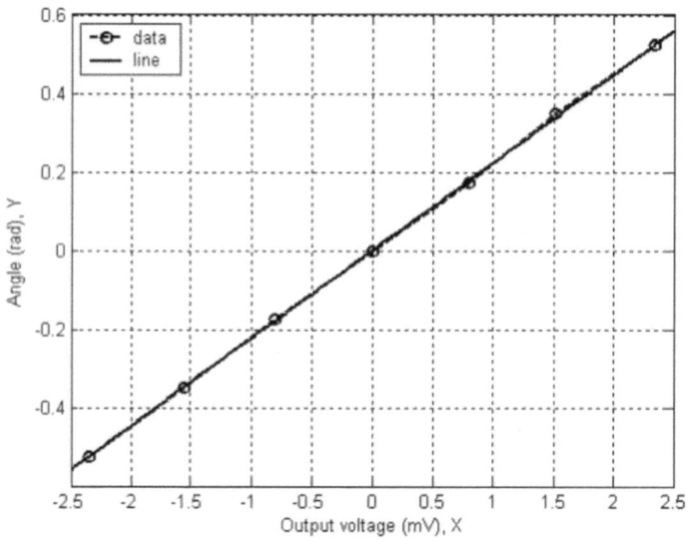

o: Reading points; y=0.22x
Joint 2
(a) Calibration curves for pots

Fig. 8.15 Results for two flexible link robotic arm

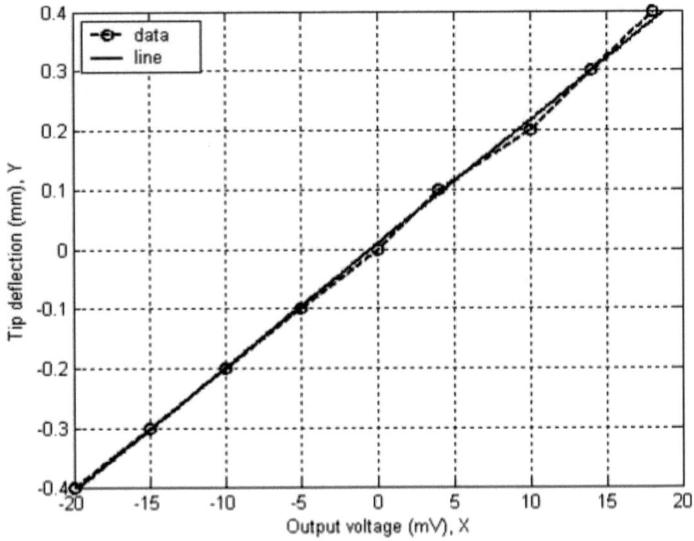

o: Reading points; y=0.02x+0.013
Joint 1

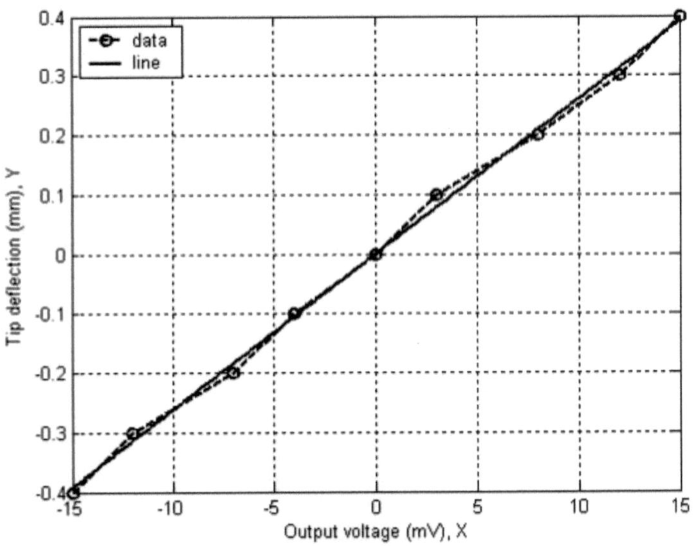

o: Reading points; y=0.025x+0.01
Joint 2
(b) Calibration curves for load cells

Fig. 8.15 (continued)

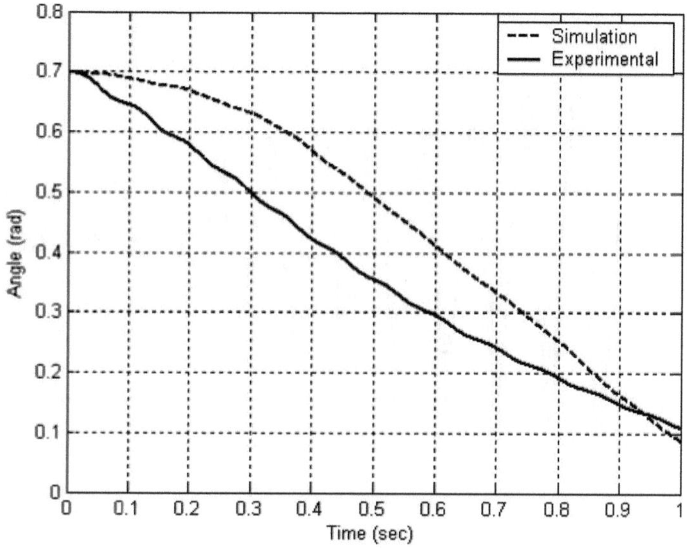

Joint 1

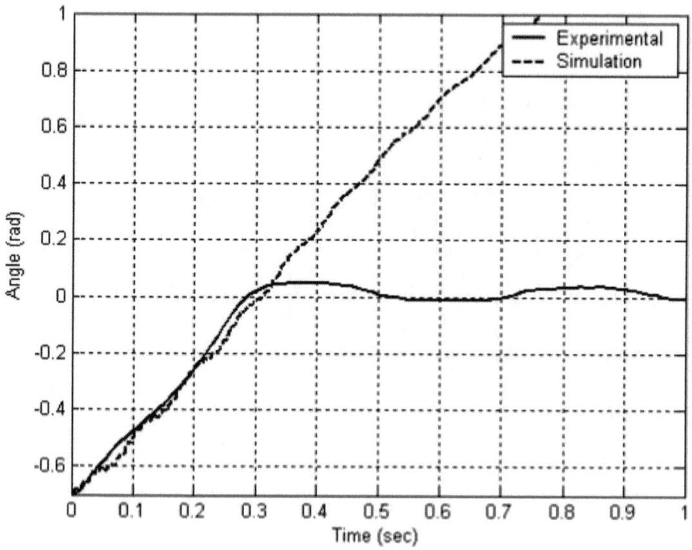

Joint 2

(c) Joint angles

Fig. 8.15 (continued)

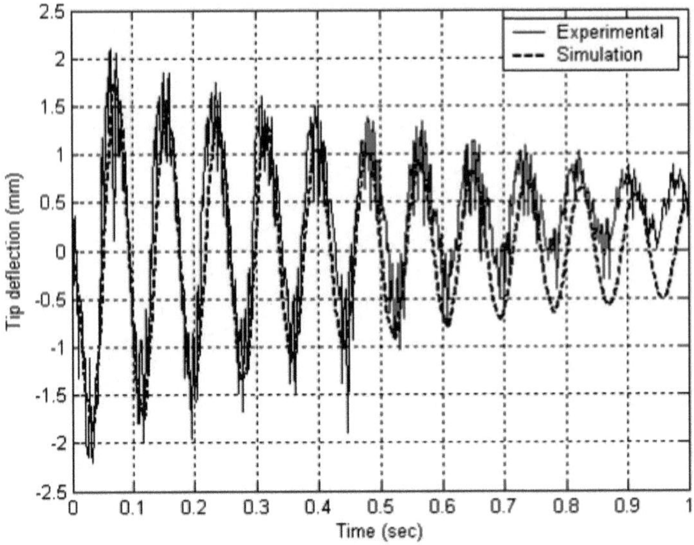

Link 1

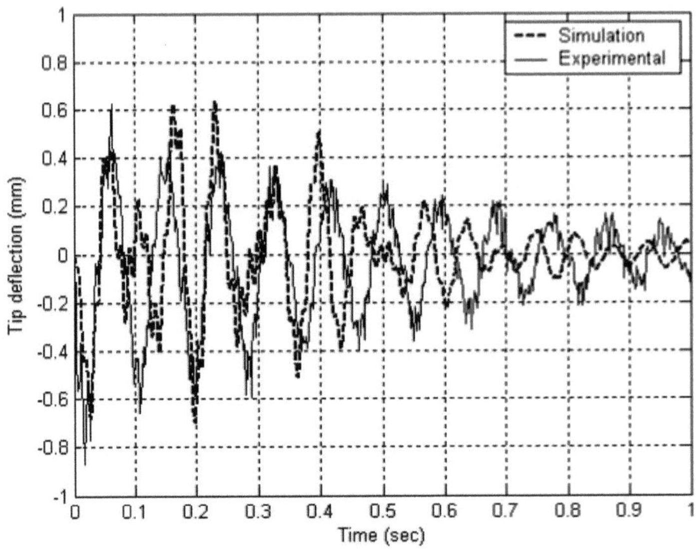

Link 2
(d) Tip deflection of the links

Fig. 8.15 (continued)

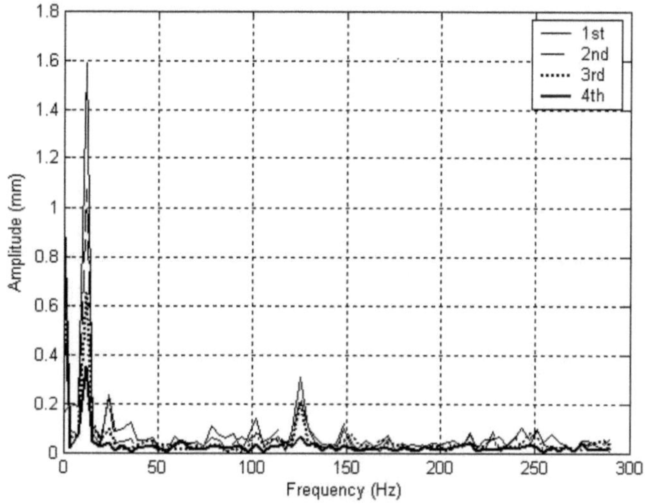

Link 1

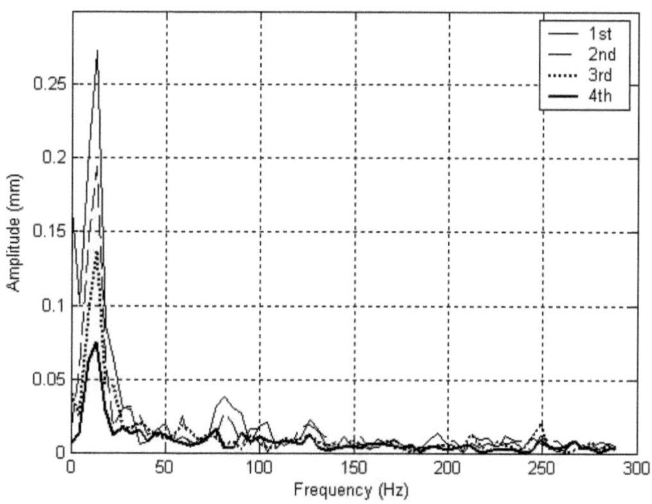

Link 2

(a) Fourier transforms of the amplitude response in time window at four different locations

Fig. 8.16 Determination of structural damping ratios for two flexible link arm

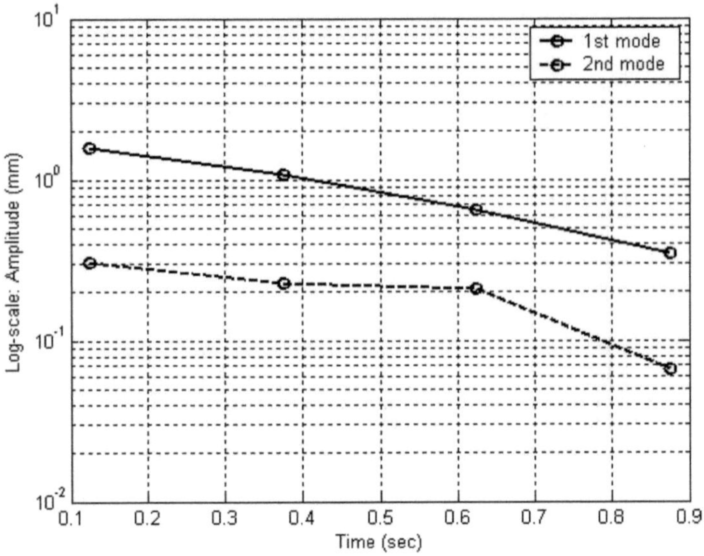

Link 1

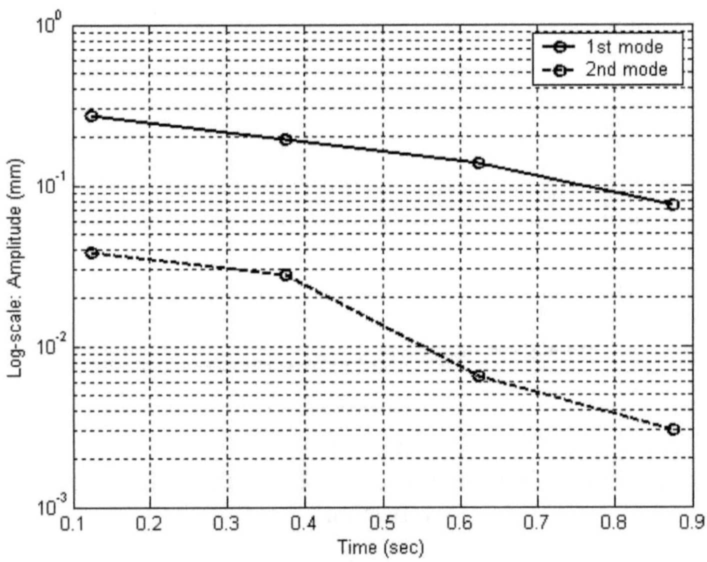

Link 2

(b) Decay of amplitude of vibrations corresponding to 1st and 2nd modes

Fig. 8.16 (continued)

8.5 Summary

In this chapter, experimental verification of the dynamic model and the forward dynamics algorithm proposed in Chap. 3 is presented. Results of a series of experiments conducted on a real single flexible link and two flexible link robotic arms are presented. Since a real robotic system always involves factors such as joint damping and structural damping, modifications are introduced in the dynamic equations to incorporate them. The modified dynamic equations are given by Eq. (8.2). As the damping factors cannot be accurately modeled because of its dependency on the external environmental factors, the data from the experimental results were used to evaluate the damping factors in the models. Later, experiments on a commercially built two flexible link robotic arm are explained and the results for the joint displacements and tip deflections are presented.

References

1. Lankarani HM, Nikravesh PE (1990) A contact force model with hysteresis damping for impact analysis of multibody systems. ASME J Mechanical Design 112:369–376. https://doi.org/10.1115/DETC1989-0104
2. Ambrosio JAC, Rayn P (1997) Elastodynamics of multibody systems using generalized inertial coordinates and structural damping. J Mechanical Structural Mechanics 25(2):201–219. https://doi.org/10.1080/08905459708905287
3. Austin EM, Inman DJ (2000) Errors in damping predictions due to kinematic assumptions for sandwich beams. Proc of Int Conf on Structural Dynamics–Recent Advances, pp 515–526
4. Thompson WT (1988) Theory of vibration with applications. Prentice Hall, London

Appendix A
Denavit and Hartenberg Parameters

The Denavit and Hartenberg (DH) parameters (Denavit and Hartenberg 1955) are a systematic method to define the relative position and orientation of the consecutive links in a multibody robotic system. The definitions help in computing the coordinate transformations between the frames attached to the links. Note, however, that the DH parameters can be assigned differently for the same system, as in Yoshikawa (1998), Angeles (2003). The DH parameters that are used in this book are explained next. Referring to Fig. 2.1, the serial robot manipulator under study consists of $(n + 1)$ bodies or links, namely the fixed base and the bodies numbered as #1, …, #n. All the bodies are coupled by n joints, numbered as 1, …, n. As shown in Fig. 2.1, the ith joint couples the $(i\text{-}1)^{st}$ link for the $(i + 1)^{st}$ frame, i.e., X_{i+1}, Y_{i+1}, Z_{i+1}, it is clearly shown in Fig. 2.2 that it is attached to the ith link.

Now, for the first n frames, the DH parameters are defined according to the following rules:

Referring to Fig. A.1

1. Z_i is the axis of the ith joint. Its positive direction can be chosen arbitrarily.
2. X_i is defined as the common perpendicular to Z_{i-1} and Z_i, directed from the former to the latter. The origin of the ith frame, O_i, is the point where X_i intersects Z_i. If these two axes intersect, the positive direction of X_i is chosen arbitrarily. And the origin, O_i, coincides with the origin of the $(i-1)^{st}$ frame, i.e., O_{i-1}.
3. The distance between Z_i and Z_{i+1} is defined as a_i, which is a non-negative number.
4. The Z_i-coordinate of the intersection of X_{i+1} axis with Z_i, which is shown in Fig. A.1 as the distance between O_i and $O_i^{'}$ is defined as b_i. This can be either positive or negative. For a prismatic joint, b_i is a variable.
5. The angle between Z_i and Z_{i+1} is defined as α_i and is measured about the positive direction of X_{i+1}.
6. The angle between X_i and X_{i+1} is defined as θ_i and is measured about the positive direction of Z_i. For a revolute joint, θ_i is a variable.

P. V. Nandihal et al., *Dynamics of Rigid-Flexible Robots and Multibody Systems*, Intelligent Systems, Control and Automation: Science and Engineering 100,
https://doi.org/10.1007/978-981-16-2798-9

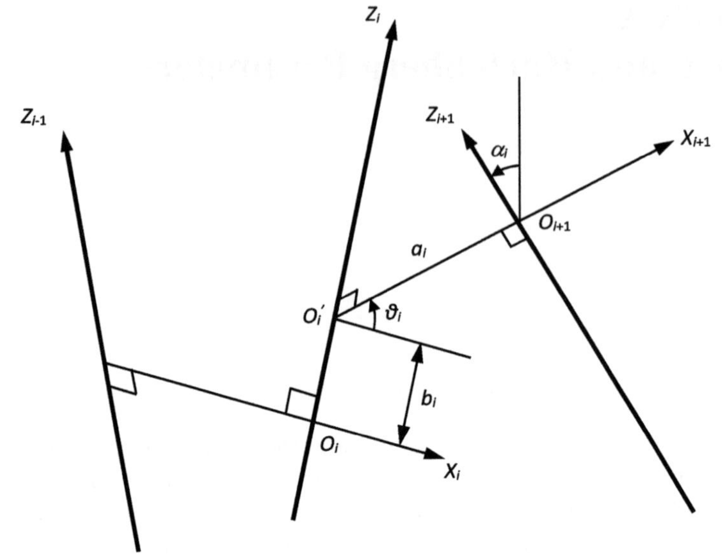

Fig. A.1 Definition of DH parameters

Since no $(n + 1)^{st}$ link exists, the above definitions do not apply to the $(n + 1)^{st}$ frame, and its axes can be chosen at will.

Appendix B
Derivation of Eq. (2.16)

Using the expressions for the total kinetic energy, $T = \sum_{i=1}^{n} T_i$, where T_i is given by Eq. (2.12), partial differentiations with respect to the jth generalized speed and coordinate, i.e., $\dot{\theta}_j$ and θ_j, respectively, as required in Eq. (2.15a) are obtained as

$$\frac{\partial T}{\partial \dot{\theta}_j} = \sum_{i=1}^{n}\left[\int_0^{b_i} \rho_i \dot{\mathbf{r}}_i^T \frac{\partial \dot{\mathbf{r}}_i}{\partial \dot{\theta}_j} d\bar{b}_i + \int_0^{a_i} \rho_i \dot{\tilde{\mathbf{r}}}_i^T \frac{\partial \dot{\tilde{\mathbf{r}}}_i}{\partial \dot{\theta}_j} d\bar{a}_i + m_{pi} \dot{\mathbf{r}}_{pi}^T \frac{\partial \dot{\mathbf{r}}_{pi}}{\partial \dot{\theta}_j} + (\mathbf{I}_{hi}\boldsymbol{\omega}_i)^T \frac{\partial \boldsymbol{\omega}_i}{\partial \dot{\theta}_j}\right] \tag{B.1a}$$

$$\frac{\partial T}{\partial \theta_j} = \sum_{i=1}^{n}\left[\int_0^{b_i} \rho_i \dot{\mathbf{r}}_i^T \frac{\partial \dot{\mathbf{r}}_i}{\partial \theta_j} d\bar{b}_i + \int_0^{a_i} \rho_i \dot{\tilde{\mathbf{r}}}_i^T \frac{\partial \dot{\tilde{\mathbf{r}}}_i}{\partial \theta_j} d\bar{a}_i + m_{pi} \dot{\mathbf{r}}_{pi}^T \frac{\partial \dot{\mathbf{r}}_{pi}}{\partial \theta_j} + (\mathbf{I}_{hi}\boldsymbol{\omega}_i)^T \frac{\partial \boldsymbol{\omega}_i}{\partial \theta_j}\right] \tag{B.1b}$$

Next, $\frac{d}{dt}\left(\frac{\partial T}{\partial \dot{\theta}_j}\right)$, is derived from Eq. (B.1a), as

$$\begin{aligned}
\frac{d}{dt}\left(\frac{\partial T}{\partial \dot{\theta}_j}\right) = \sum_{i=1}^{n}\Bigg[&\int_0^{b_i} \rho_i \left\{\ddot{\mathbf{r}}_i^T \frac{\partial \dot{\mathbf{r}}_i}{\partial \dot{\theta}_j} + \dot{\mathbf{r}}_i^T \frac{d}{dt}\left(\frac{\partial \dot{\mathbf{r}}_i}{\partial \dot{\theta}_j}\right)\right\} d\bar{b}_i \\
&+ \int_0^{a_i} \rho_i \left\{\ddot{\tilde{\mathbf{r}}}_i^T \frac{\partial \dot{\tilde{\mathbf{r}}}_i}{\partial \dot{\theta}_j} + \dot{\tilde{\mathbf{r}}}_i^T \frac{d}{dt}\left(\frac{\partial \dot{\tilde{\mathbf{r}}}_i}{\partial \dot{\theta}_j}\right)\right\} d\bar{a}_i \\
&+ m_{pi}\left\{\ddot{\mathbf{r}}_{pi}^T \frac{\partial \dot{\mathbf{r}}_{pi}}{\partial \dot{\theta}_j} + \dot{\mathbf{r}}_{pi}^T \frac{d}{dt}\left(\frac{\partial \dot{\mathbf{r}}_{pi}}{\partial \dot{\theta}_j}\right)\right\} \\
&+ \left\{(\mathbf{I}_{hi}\dot{\boldsymbol{\omega}}_i + \boldsymbol{\omega}_i \times \mathbf{I}_{hi}\boldsymbol{\omega}_i)^T \frac{\partial \boldsymbol{\omega}_i}{\partial \dot{\theta}_j} + (\mathbf{I}_{hi}\boldsymbol{\omega}_i)^T \frac{d}{dt}\left(\frac{\partial \boldsymbol{\omega}_i}{\partial \dot{\theta}_j}\right)\right\}\Bigg]
\end{aligned} \tag{B.2}$$

As shown in Stejskal and Valasek (1996) and others, it is evident that

P. V. Nandihal et al., *Dynamics of Rigid-Flexible Robots and Multibody Systems*, Intelligent Systems, Control and Automation: Science and Engineering 100,
https://doi.org/10.1007/978-981-16-2798-9

$$\frac{\partial \dot{\mathbf{r}}_i}{\partial \dot{\theta}_j} = \frac{\partial \mathbf{r}_i}{\partial \theta_j};\ \frac{\partial \dot{\tilde{\mathbf{r}}}_i}{\partial \dot{\theta}_j} = \frac{\partial \tilde{\mathbf{r}}_i}{\partial \theta_j} \text{ and } \frac{\partial \dot{\mathbf{r}}_{pi}}{\partial \dot{\theta}_j} = \frac{\partial \mathbf{r}_{pi}}{\partial \theta_j} \tag{B.3}$$

Hence, the parts of the 2nd, 4th, and 6th terms on the right-hand side of Eq. (B.2) are given by

$$\frac{\mathrm{d}}{\mathrm{d}t}\left(\frac{\partial \dot{\mathbf{r}}_i}{\partial \dot{\theta}_j}\right) = \frac{\mathrm{d}}{\mathrm{d}t}\left(\frac{\partial \mathbf{r}_i}{\partial \theta_j}\right) = \frac{\partial \dot{\mathbf{r}}_i}{\partial \theta_j};\ \frac{\mathrm{d}}{\mathrm{d}t}\left(\frac{\partial \dot{\tilde{\mathbf{r}}}_i}{\partial \dot{\theta}_j}\right) = \frac{\mathrm{d}}{\mathrm{d}t}\left(\frac{\partial \tilde{\mathbf{r}}_i}{\partial \theta_j}\right) = \frac{\partial \dot{\tilde{\mathbf{r}}}_i}{\partial \theta_j} \text{ and}$$
$$\frac{\mathrm{d}}{\mathrm{d}t}\left(\frac{\partial \dot{\mathbf{r}}_{pi}}{\partial \dot{\theta}_j}\right) = \frac{\mathrm{d}}{\mathrm{d}t}\left(\frac{\partial \mathbf{r}_{pi}}{\partial \theta_j}\right) = \frac{\partial \dot{\mathbf{r}}_{pi}}{\partial \theta_j} \tag{B.4}$$

Similar to Eq. (B.4) it can also be shown that

$$\frac{d}{dt}\left(\frac{\partial \boldsymbol{\omega}_i}{\partial \dot{\theta}_j}\right) = \frac{\partial \boldsymbol{\omega}_i}{\partial \theta_j} \tag{B.5}$$

Substituting Eqs. (B.4) and (B.5) into Eq. (B.2), and using the resulting expressions, along with Eq. (B.1b), the left-hand side of Eq. (2.15a) yields

$$\sum_{i=1}^{n}\left[\int_0^{b_i} \rho_i \ddot{\mathbf{r}}_i^T \frac{\partial \dot{\mathbf{r}}_i}{\partial \dot{\theta}_j} \mathrm{d}\bar{b}_i + \int_0^{a_i} \rho_i \ddot{\tilde{\mathbf{r}}}_i^T \frac{\partial \dot{\tilde{\mathbf{r}}}_i}{\partial \dot{\theta}_j} \mathrm{d}\bar{a}_i + m_{pi} \ddot{\mathbf{r}}_{pi}^T \frac{\partial \dot{\mathbf{r}}_{pi}}{\partial \dot{\theta}_j} + (\mathbf{I}_{hi}\dot{\boldsymbol{\omega}}_i + \boldsymbol{\omega}_i \times \mathbf{I}_{hi}\boldsymbol{\omega}_i)\frac{\partial \boldsymbol{\omega}_i}{\partial \dot{\theta}_j}\right] = \tau_{\mathrm{j}} \tag{B.6}$$

Next, $\ddot{\mathbf{r}}_i$, $\ddot{\tilde{\mathbf{r}}}_i$, and $\ddot{\mathbf{r}}_{pi}$, are obtained from the time differentiation of expressions of Eq. (2.11) as,

$$\ddot{\mathbf{r}}_i = \dot{\mathbf{v}}_i + \dot{\boldsymbol{\omega}}_i \times \bar{\mathbf{b}}_i + \boldsymbol{\omega}_i \times (\boldsymbol{\omega}_i \times \bar{\mathbf{b}}_i) \tag{B.7a}$$

$$\ddot{\tilde{\mathbf{r}}}_i = \dot{\mathbf{v}}_i + \dot{\boldsymbol{\omega}}_i \times \bar{\mathbf{r}}_i + \boldsymbol{\omega}_i \times (\boldsymbol{\omega}_i \times \bar{\mathbf{r}}_i) \tag{B.7b}$$

$$\ddot{\tilde{\mathbf{r}}}_{pi} = \dot{\mathbf{v}}_i + \dot{\boldsymbol{\omega}}_i \times \bar{\mathbf{r}}_{pi} + \boldsymbol{\omega}_i \times \left(\boldsymbol{\omega}_i \times \bar{\mathbf{r}}_{pi}\right) \tag{B.7c}$$

Substituting $\dot{\mathbf{r}}_i$, ,$\dot{\tilde{\mathbf{r}}}_i$ and $\dot{\mathbf{r}}_{pi}$ from Eq. (2.13) into Eq. (B.6), one obtains

$$\sum_{i=1}^{n}\left[\int_0^{b_i} \rho_i \left\{\ddot{\mathbf{r}}_i^T \frac{\partial \mathbf{v}_i}{\partial \dot{\theta}_j} + \ddot{\mathbf{r}}_i^T \frac{\partial(\boldsymbol{\omega}_i \times \bar{b}_i \mathbf{z}_i)}{\partial \dot{\theta}_j}\right\} \mathrm{d}\bar{b}_i + \int_0^{a_i} \rho_i \left\{\ddot{\tilde{\mathbf{r}}}_i^T \frac{\partial \mathbf{v}_i}{\partial \dot{\theta}_j} + \ddot{\tilde{\mathbf{r}}}_i^T \frac{\partial[\boldsymbol{\omega}_i \times \bar{\mathbf{r}}_i]}{\partial \dot{\theta}_j}\right\} \mathrm{d}\bar{a}_i\right.$$

$$+m_{pi}\left\{\ddot{\mathbf{r}}_{pi}^T\frac{\partial \mathbf{v}_i}{\partial \dot{\theta}_j}+\ddot{\mathbf{r}}_{pi}^T\frac{\partial[\boldsymbol{\omega}_i\times\bar{\mathbf{r}}_{pi}]}{\partial \dot{\theta}_j}\right\}+(\mathbf{I}_{hi}\dot{\boldsymbol{\omega}}_i+\boldsymbol{\omega}_i\times\mathbf{I}_{hi}\boldsymbol{\omega}_i)\frac{\partial \boldsymbol{\omega}_i}{\partial \dot{\theta}_j}\Bigg]=\boldsymbol{\tau}_j \tag{B.8}$$

Using the vector triple product rule (Strang 1980), $\mathbf{a}^T(\mathbf{b}\times\mathbf{c})=(\mathbf{c}\times\mathbf{a})^T\mathbf{b}$—**a**, **b** and **c** are any 3-dimensional Cartesian vectors—one can show that

$$\ddot{\mathbf{r}}_i^T\frac{\partial(\boldsymbol{\omega}_i\times\bar{b}_i\mathbf{z}_i)}{\partial\dot{\theta}_j}=\bar{b}_i\ddot{\mathbf{r}}_i^T\left[\frac{\partial\boldsymbol{\omega}_i}{\partial\dot{\theta}_j}\times\mathbf{z}_i+\boldsymbol{\omega}_\mathbf{i}\times\frac{\partial\mathbf{z}_i}{\partial\dot{\theta}_j}\right]=\bar{b}_i(\mathbf{z}_i\times\ddot{\mathbf{r}}_i)^T\frac{\partial\boldsymbol{\omega}_i}{\partial\dot{\theta}_j} \tag{B.9a}$$

$$\ddot{\tilde{\mathbf{r}}}_i^T\frac{\partial(\boldsymbol{\omega}_i\times\bar{\mathbf{r}}_i)}{\partial\dot{\theta}_j}=\ddot{\tilde{\mathbf{r}}}_i^T\left[\frac{\partial\boldsymbol{\omega}_i}{\partial\dot{\theta}_j}\times\bar{\mathbf{r}}_i+\boldsymbol{\omega}_i\times\frac{\partial\bar{\mathbf{r}}_i}{\partial\dot{\theta}_j}\right]=[\bar{\mathbf{r}}_i\times\ddot{\tilde{\mathbf{r}}}_i]^T\frac{\partial\boldsymbol{\omega}_i}{\partial\dot{\theta}_j} \tag{B.9b}$$

$$\ddot{\mathbf{r}}_{pi}^T\frac{\partial\left(\boldsymbol{\omega}_i\times\bar{\mathbf{r}}_{pi}\right)}{\partial\dot{\theta}_j}=\ddot{\mathbf{r}}_{pi}^T\left[\frac{\partial\boldsymbol{\omega}_i}{\partial\dot{\theta}_j}\times\bar{\mathbf{r}}_{pi}+\boldsymbol{\omega}_i\times\frac{\partial\bar{\mathbf{r}}_{pi}}{\partial\dot{\theta}_j}\right]=\left(\bar{\mathbf{r}}_{pi}\times\ddot{\mathbf{r}}_{pi}\right)^T\frac{\partial\boldsymbol{\omega}_i}{\partial\dot{\theta}_j} \tag{B.9c}$$

In Eqs. (B.9a–c), $\partial\mathbf{z}_i/\partial\dot{\theta}_j=0$, $\partial\bar{\mathbf{r}}_i/\partial\dot{\theta}_j=0$, and $\partial\bar{\mathbf{r}}_{pi}/\partial\dot{\theta}_j=0$ as $\mathbf{z}_i$ $\bar{\mathbf{r}}_i$ and $\bar{\mathbf{r}}_{pi}$ are functions of $\boldsymbol{\theta}_j$'s only, not $\dot{\boldsymbol{\theta}}_j$'s. Hence, using Eqs. (B.9a–c), (B.8) is rewritten as

$$\sum_{i=1}^{n}\Bigg[\int_0^{b_i}\rho_i\left\{\ddot{\mathbf{r}}_i^T\frac{\partial\mathbf{v}_i}{\partial\dot{\theta}_j}+\bar{b}_i(\mathbf{z}_i\times\ddot{\mathbf{r}}_i)^T\frac{\partial\boldsymbol{\omega}_i}{\partial\dot{\theta}_j}\right\}\mathrm{d}\bar{b}_i+\int_0^{a_i}\rho_i\left\{\ddot{\tilde{\mathbf{r}}}_i^T\frac{\partial\mathbf{v}_i}{\partial\dot{\theta}_j}+(\bar{\mathbf{r}}_i\times\ddot{\tilde{\mathbf{r}}}_i)^T\frac{\partial\boldsymbol{\omega}_i}{\partial\dot{\theta}_j}\right\}\mathrm{d}\bar{a}_i$$
$$+m_{pi}\left\{\ddot{\mathbf{r}}_{pi}^T\frac{\partial\mathbf{v}_i}{\partial\dot{\theta}_j}+(\bar{\mathbf{r}}_{pi}\times\ddot{\mathbf{r}}_{pi})^T\frac{\partial\boldsymbol{\omega}_i}{\partial\dot{\theta}_j}\right\}+(\mathbf{I}_{hi}\dot{\boldsymbol{\omega}}_i+\boldsymbol{\omega}_i\times\mathbf{I}_{hi}\boldsymbol{\omega}_i)\frac{\partial\boldsymbol{\omega}_i}{\partial\dot{\theta}_j}\Bigg]=\boldsymbol{\tau}_j \tag{B.10}$$

Now, introducing the following definitions

$$\mathbf{w}_i^*\equiv\begin{bmatrix}\int_0^{b_i}\rho_i\ddot{\mathbf{r}}_i\mathrm{d}\bar{b}_i+\int_0^{a_i}\rho_i\ddot{\tilde{\mathbf{r}}}_i\mathrm{d}\bar{a}_i+m_{pi}\ddot{\mathbf{r}}_{pi}\\ \int_0^{b_i}\rho_i\bar{b}(\mathbf{z}_i\times\ddot{\mathbf{r}}_i)\mathrm{d}\bar{b}_i+\int_0^{a_i}\rho_i\left(\bar{\mathbf{r}}_i\times\ddot{\tilde{\mathbf{r}}}_i\right)\mathrm{d}\bar{a}_i\\ +m_{pi}\left(\bar{\mathbf{r}}_{pi}\times\ddot{\mathbf{r}}_{pi}\right)+(\mathbf{I}_{hi}\dot{\omega}_i+\omega_i\times\mathbf{I}_{hi}\omega_i)\end{bmatrix} \tag{B.11}$$

Equation (B.10) is rewritten in compact form as

$$\left[\left(\frac{\partial\mathbf{t}_1}{\partial\dot{\theta}_j}\right)^T\cdots\left(\frac{\partial\mathbf{t}_n}{\partial\dot{\theta}_j}\right)^T\right]\begin{bmatrix}\mathbf{w}_1^*\\ \vdots\\ \mathbf{w}_n^*\end{bmatrix}=\tau_j \tag{B.12}$$

which is the desired equations of motion (Eq. 2.16) to obtain the generalized force corresponding to the jth generalized coordinate.

Appendix C
Computational Counts for Rigid Robots

In order to compute the joint accelerations, $\ddot{\boldsymbol{\theta}}$ of Eq. (2.27), using the recursive algorithm presented in Sect. 2.31, the following inputs are required: For $i = 1, \ldots, n$

(i) Constant Denavit–Hartenberg (DH) parameters (Denavit and Hartenberg 1955; Bhangale 2004) of the system under study, i.e., α_i, a_i, b_i (for revolute joints) and θ_i (for prismatic joints);
(ii) Time history of the variable DH parameter, i.e., θ_i, for the revolute joint, and b_i, for the prismatic joint, and the initial values for their respective first time derivatives.
(iii) Mass per unit length of each body, ρ_i.
(iv) Time history of input generalized force, $\boldsymbol{\tau}$.

The count is performed in terms of the number of divisions/multiplications (M), subtractions/additions (A), and trigonometric operations (T). Each step is accompanied by the M, A, T, counts that are given inside {and}. To illustrate some of the computations of the intermediate steps, the in-between computations are also given after the colon ':'. Moreover, a vector or a matrix represented in frame i is denoted by $[\cdot]_i$. The calculation sequence is as follows:

C.1 Find 3-dimensional vector $[\mathbf{a}_i]_i$ and $[\mathbf{a}_i]_{i+1}$. For $I = 1, \ldots, N$,

$$c_i \equiv \cos\theta_i,\ s_i \equiv \sin\theta_i \qquad \{2\mathrm{T}(n)\}$$
$$[\mathbf{a}_i]_i = \begin{bmatrix} a_i c_i & a_i s_i & b_i \end{bmatrix}^T \qquad \{2\mathrm{M}(n)\}$$

and

$$\bar{c}_i \equiv \cos\alpha_i, \bar{s}_i \equiv \sin\alpha_i \qquad \{\text{nil}\}$$
$$[\mathbf{a}_i]_{i+1} \equiv \begin{bmatrix} a_i & b_i\bar{s}_i & b_i\bar{c}_i \end{bmatrix}^T \qquad \{2\mathrm{M}(n)\}$$

P. V. Nandihal et al., *Dynamics of Rigid-Flexible Robots and Multibody Systems*, Intelligent Systems, Control and Automation: Science and Engineering 100,
https://doi.org/10.1007/978-981-16-2798-9

in which the DH parameters, a_i, b_i, α_i and θ_i are known from the inputs, and $\bar{c}_i$ and $\bar{s}_i$ would be done offline. Hence, their computations are not counted.

C.2 Find 3-dimensional unit vectors, $[\mathbf{x}_{i+1}]_i$ and $[\mathbf{z}_i]_i$. For $I = 1,\ldots, N$;

$$[\mathbf{x}_{i+1}]_i = \mathbf{Q}_i[\mathbf{x}_{i+1}]_{i+1} \qquad \{\text{nil}\}$$

where $[\mathbf{x}_{i+1}]_{i+1} \equiv [1\ 0\ 0]^T$, and $[\mathbf{z}_i]_i \equiv [0\ 0\ 1]^T$. Vectors The $\mathbf{x}_{i+1}$ and $\mathbf{z}_i$ are, respectively, the unit vectors along X_{i+1} and Z_i axes, and $[\mathbf{Q}_i]_i$ is the 3×3 rotation matrix that defines the orientation of the $(i+1)^{\text{st}}$ frame with respect to the ith frame, i.e.,

$$\mathbf{Q}_i \equiv \begin{bmatrix} c_i & -s_i\bar{c}_i & s_i\bar{s}_i \\ s_i & c_i\bar{c}_i & -c_i\bar{s}_i \\ 0 & \bar{s}_i & \bar{c}_i \end{bmatrix} \qquad \{4\text{M}(n)\}$$

Hence, $[\mathbf{x}_{i+1}]_i$ is nothing but the first column of $\mathbf{Q}_i$ which does not require any computations.

C.3 Find the 6×6 mass matrix, $[\mathbf{M}_i]_i$, which is denoted as

$$\mathbf{M}_i \equiv \begin{bmatrix} \mathbf{M}_{ivv} & \mathbf{M}_{ivw} \\ \text{sym} & \mathbf{M}_{iww} \end{bmatrix}$$

where $\mathbf{M}_{ivv}$, $\mathbf{M}_{ivw}$, and $\mathbf{M}_{iww}$ are the 3×3 submatrices whose expressions are available from Eq. (3.19a) and 'sym' denotes the symmetric elements of the matrix. The nomenclature of $\mathbf{M}_{ivv}$, $\mathbf{M}_{ivw}$, and $\mathbf{M}_{iww}$ is done based upon the associated components of the twist vector, $\mathbf{v}$ and $\mathbf{w}$, Eq. (3.1). For example, $\mathbf{M}_{ivv}$ is the mass matrix associated with the linear velocity, $\mathbf{v}$. The computations are given, for $i = 1,\ldots, n$, as

$$[\mathbf{M}_{ivv}]_i = \rho_i(a_i + b_i)\mathbf{1} \qquad \{\text{nil}\}$$

where $\rho_i(a_i + b_i)$ is constant and need not be counted. Then, the matrix $[\mathbf{M}_{ivw}]_i$ is obtained as

$$[\mathbf{M}_{ivw}]_i = C_{1i}[\mathbf{z}_i]_i \times \mathbf{1} + C_{2i}[\mathbf{x}_{i+1}]_i \times \mathbf{1} \qquad \{2\text{M}(n)\}$$

where $C_{1i} \equiv -\rho_i b_i\left(\frac{b_i}{2} + a_i\right)$ and $C_{2i} \equiv -\rho_i \frac{a_i^2}{2}$, which are constants and computed offline. Hence,

$$[\mathbf{M}_{iww}]_i = C_{3i}[\mathbf{z}_i]_i^T[\mathbf{z}_i] + C_{4i}\left[\mathbf{x}_{i+1}\right]_i^T\left[\mathbf{x}_{i+1}\right]_i + C_{5i}\left([\mathbf{z}_i]_i^T\left[\mathbf{x}_{i+1}\right]_i + \left[\mathbf{x}_{i+1}\right]_i^T[\mathbf{z}_i]_i\right)$$

{7M(n) 2A(n)}

In which $C_{3i} \equiv \rho_i b_i^2\left(\frac{b_i}{3} + a_i\right)$, $C_{4i} \equiv \rho_i \frac{a_i^3}{3}$ and $C_{5i} \equiv \rho_i a_i^2 b_i$, are also the constants that are computed offline. As an illustration of the computations involved in $[\mathbf{M}_{iww}]_i$, the stepwise details of the computations are shown below:

(i) $$C_{3i}[\mathbf{z}_i]_i^T[\mathbf{z}_i]_i = C_{3i}\begin{bmatrix} 1 & 0 & 0 \\ 0 & 1 & 0 \\ 0 & 0 & 0 \end{bmatrix} = \begin{bmatrix} C_{3i} & 0 & 0 \\ 0 & C_{3i} & 0 \\ 0 & 0 & 0 \end{bmatrix} \quad : \text{nil}$$

(ii) $$C_{4i}\left[\mathbf{x}_{i+1}\right]_i^T\left[\mathbf{x}_{i+1}\right]_i = C_{4i}\begin{bmatrix} s_i^2 & -s_i c_i & 0 \\ -s_i c_i & c_i^2 & 0 \\ 0 & 0 & 1 \end{bmatrix} = \begin{bmatrix} C_{4i}s_i^2 & -C_{4i}s_i c_i & 0 \\ & C_{4i}c_i^2 & 0 \\ \text{sym} & & C_{4i} \end{bmatrix} \quad : 5\text{M}$$

where 'sym' stands for symmetric.

(iii) $$C_{5i}\left([\mathbf{z}_i]_i^T\left[\mathbf{x}_{i+1}\right]_i + \left[\mathbf{x}_{i+1}\right]_i^T[\mathbf{z}_i]_i\right) = C_{5i}\left(\begin{bmatrix} 0 & 0 & -c_i \\ 0 & 0 & -s_i \\ 0 & 0 & 0 \end{bmatrix} + \begin{bmatrix} 0 & 0 & 0 \\ 0 & 0 & 0 \\ -c_i & -s_i & 0 \end{bmatrix}\right)$$
$$= \begin{bmatrix} 0 & 0 & -C_{5i}c_i \\ & 0 & -C_{5i}s_i \\ \text{sym} & & 0 \end{bmatrix} \quad : 2\text{M}$$

$[\mathbf{M}_{iww}]_i$ is then obtained by summation of (i—iii), i.e.,

$$[\mathbf{M}_{iww}]_i = \begin{bmatrix} C_{3i} + C_{4i}s_i^2 & -C_{4i}s_i c_i & -C_{5i}c_i \\ & C_{3i} + C_{4i}c_i^2 & -C_{5i}s_i \\ \text{sym} & & C_{4i} \end{bmatrix} \quad : 2\text{A}$$

Thus, the computation of $[\mathbf{M}_{iww}]_i$ involves 7 M and 2A. Since the terms C_{3i}, C_{4i} and C_{5i} will be computed offline, no count is reported.

C.4 Find the n-dimensional vector, φ, from Eq. (3.29): For $I = 1, \ldots, N$

$$\boldsymbol{\varphi} = \boldsymbol{\tau} - \bar{\boldsymbol{\tau}}, \text{ where } \bar{\boldsymbol{\tau}} = \mathbf{N}_d^T\mathbf{N}_l^T\left[\mathbf{M}(\mathbf{N}_l\dot{\mathbf{N}}_d + \dot{\mathbf{N}}_l\mathbf{N}_d)\dot{\boldsymbol{\theta}}\right] \qquad \{1\text{A(n)}\}$$

where the vector, $\bar{\boldsymbol{\tau}}$, is expected to be known, which can be evaluated using any recursive inverse dynamics algorithm (Walker and Orin 1982; Saha 1999a) while $\ddot{\boldsymbol{\theta}}_i = 0$, and $\boldsymbol{\tau} \equiv \begin{bmatrix} \tau_1 & \tau_2 & \cdots & \tau_n \end{bmatrix}^T$ is known from input.

C.5 Find the 6 × 6 matrix, $\left[\hat{\mathbf{M}}_i\right]_i$. For the purpose of showing the computational details,

$\hat{\mathbf{M}}_i$ is written similar to $\mathbf{M}_i$ in Step C.3 as

$$\hat{\mathbf{M}}_i \equiv \begin{bmatrix} \hat{\mathbf{M}}_{ivv} & \hat{\mathbf{M}}_{ivw} \\ \text{sym} & \hat{\mathbf{M}}_{iww} \end{bmatrix}$$

where the 3 × 3 submatrices, $\hat{\mathbf{M}}_{ivv}$ and $\hat{\mathbf{M}}_{iww}$ are also symmetric. For $i = n$.

$$\left[\hat{\mathbf{M}}_{nvv}\right]_n = [\mathbf{M}_{nvv}]_n : \text{nil}; \left[\hat{\mathbf{M}}_{nvw}\right]_n = [\mathbf{M}_{nvw}]_n : \text{nil}; \left[\hat{\mathbf{M}}_{nww}\right]_n = [\mathbf{M}_{nww}]_n : \text{nil}$$

For clarity of representation, the computation, $\left(\left[\hat{\mathbf{M}}_{i+1,ww}\right]_i\right)[\mathbf{a}_i]_i \times \mathbf{1}$, which occur more than once in the computation of $\left[\hat{\mathbf{M}}_i\right]_i$ is represented as $[\mathbf{K}_i]_i$ here. Thus for $i = n-1,\ldots,2$,

$$\left[\hat{\mathbf{M}}_{i+1,ww}\right]_i \equiv \mathbf{Q}_i\left[\hat{\mathbf{M}}_{i+1,ww}\right]_{i+1}\mathbf{Q}_i^T : 16\text{M}17\text{A}; \rightarrow [\mathbf{K}_i]_i \equiv -()[\mathbf{a}_i]_i \times \mathbf{1} : 18\text{M}\,9\text{A};$$
$$\rightarrow ([\mathbf{a}_i]_i \times \mathbf{1})() : 6\text{M};$$
$$\rightarrow \left[\hat{\mathbf{M}}_{i,ww}\right]_i = [\mathbf{M}_{iww}]_i + \mathbf{Q}_i\left\{\left[\bar{\mathbf{M}}_{i+1,ww}\right]_{i+1} + \left[\bar{\mathbf{M}}_{i+1,vw}\right]_{i+1}([\mathbf{a}_i]_{i+1} \times \mathbf{1})\right.$$
$$\left. + ([\mathbf{a}_i]_{i+1} \times \mathbf{1})^T \left[\bar{\mathbf{M}}_{i+1,vw}\right]_{i+1}^T\right\}\mathbf{Q}_i^T + () \quad : 40\text{M}\,56\text{A};$$
$$\rightarrow \left[\hat{\mathbf{M}}_{i,vw}\right]_i = [\mathbf{M}_{ivw}]_i + [\mathbf{K}_i]_i \quad : 6\text{A};$$
$$\rightarrow \left[\hat{\mathbf{M}}_{i,vv}\right]_{i\,i} = [\mathbf{M}_{ivv}]_i + \left[\hat{\mathbf{M}}_{i+1,ww}\right]_i \quad : 3\text{A}$$

{80M 91A(n − 1)}

In the above steps, the notation, ( ), represents the expression computed just before it. Moreover, the frame transformation of the symmetric matrices requiring two 3 × 3 matrix multiplication is done only with 16 M 17A (Bhangale 2004), instead of straightforward 2(27 M 18A) = 54 M 36A—27 M 18A are the values corresponding to the two 3 × 3 matrix multiplication. This is due to the structure of orthogonal matrix $\mathbf{Q}_i$, and the symmetric elements of $\left[\hat{\mathbf{M}}_{iww}\right]_i$. The details are shown in Bhangale (2004). Similarly, a vector transformation from one to another frame, e.g., $[\mathbf{x}]_i = \mathbf{Q}_i[\mathbf{x}]_{i+1}$, can be done efficiently using only 8 M 4A instead of 9 M 6A, where $\mathbf{x}$ is any 3-dimensional vector.

C.6 Find the 6-dimensional vector, $\left[\hat{\boldsymbol{\psi}}_i\right]_i$, which is expressed as

$$\hat{\boldsymbol{\psi}}_i \equiv \left[\hat{\boldsymbol{\psi}}_{i1}^T \; \hat{\boldsymbol{\psi}}_{i2}^T \right]^T$$

where $\hat{\boldsymbol{\psi}}_{i1}$ and $\hat{\boldsymbol{\psi}}_{i2}$ are the 3-dimensional vectors. Now, for $i = n, \ldots, 1$

$$\left[\hat{\boldsymbol{\psi}}_{i1}\right]_i = \left[\hat{\mathbf{M}}_{ivw}\right]_i [\mathbf{z}_i]_i : \text{nil}; \left[\hat{\boldsymbol{\psi}}_{i2}\right]_i = \left[\hat{\mathbf{M}}_{iww}\right]_i [\mathbf{z}_i]_i : \text{nil for revolute joint} \quad \{\text{nil}\}$$

$$\left[\hat{\boldsymbol{\psi}}_{i1}\right]_i = \left[\hat{\mathbf{M}}_{ivv}\right]_i [\mathbf{z}_i]_i : \text{nil}; \left[\hat{\boldsymbol{\psi}}_{i2}\right]_i = \left[\hat{\mathbf{M}}_{ivw}\right]_i^T [\mathbf{z}_i]_i : \text{nil for prismatic joint} \quad \{\text{nil}\}$$

where $[\mathbf{z}_i]_i = [0\ 0\ 1]^T$.

C.7 Find the scaler $\hat{\imath}_i$: For $I = n, \ldots, 1$,

$$\hat{\imath}_i = [\mathbf{z}_i]_i^T \left[\hat{\boldsymbol{\psi}}_{i2}\right]_i \text{ for revolute joint; } \hat{\imath}_i = [\mathbf{z}_i]_i^T \left[\hat{\boldsymbol{\psi}}_{i1}\right]_i \text{ for prismatic joint;} \quad \{\text{nil}\}$$

C.8 Find the 6-dimensional vector, $[\boldsymbol{\psi}_i]_i$: For $I = n, \ldots, 1$

$$[\boldsymbol{\psi}_{i1}]_i = \left[\hat{\boldsymbol{\psi}}_{i1}\right]_i / \hat{\imath}_{i\,i} : 2\text{M}; \quad [\boldsymbol{\psi}_{i2}]_i = \left[\hat{\boldsymbol{\psi}}_{i2}\right]_i / \hat{\imath}_i : 3\text{M} \qquad \{5\text{M(n)}\}$$

C.9 Find the 6 × 6 matrix, $[\bar{\mathbf{M}}_i]_i$: A definition similar to $\left[\hat{\mathbf{M}}_i\right]_i$ is also used here as

$$\bar{\mathbf{M}}_i \equiv \begin{bmatrix} \bar{\mathbf{M}}_{ivv} & \bar{\mathbf{M}}_{ivw} \\ \text{sym} & \bar{\mathbf{M}}_{iww} \end{bmatrix}$$

where the 3 × 3 submatrices, $\bar{\mathbf{M}}_{ivv}$ and $\bar{\mathbf{M}}_{iww}$, are symmetric.

(a) For $i = n$, $[\bar{\mathbf{M}}_n]_n$ {16M14A}

$$\left[\hat{\boldsymbol{\psi}}_{n1}\right]_n [\boldsymbol{\psi}_{n1}]_n^T : 4\text{M}; \rightarrow [\bar{\mathbf{M}}_{n11}]_n = \left[\hat{\mathbf{M}}_{nvv}\right]_n - () : 2\text{A};$$

$$\left[\hat{\boldsymbol{\psi}}_{n2}\right]_n [\boldsymbol{\psi}_{n1}]_n^T : 6\text{M}; \rightarrow [\bar{\mathbf{M}}_{nwv}]_n = \left[\hat{\mathbf{M}}_{nwv}\right]_n - () : 6\text{A};$$

$$\left[\hat{\boldsymbol{\psi}}_{n2}\right]_n [\boldsymbol{\psi}_{n2}]_n^T : 6\text{M}; \rightarrow [\bar{\mathbf{M}}_{nww}]_n = \left[\hat{\mathbf{M}}_{nww}\right]_n - () : 6\text{A}$$

(b) For $i = n-1, \ldots, 2$, $[\bar{\mathbf{M}}_i]_i$ {16M18A(n − 1)}

$$\left[\hat{\boldsymbol{\psi}}_{i1}\right]_i\left[\boldsymbol{\psi}_{i1}\right]_i^T : 4\text{M}; \rightarrow \left[\bar{\mathbf{M}}_{ivv}\right]_i = \left[\hat{\mathbf{M}}_{ivv}\right]_i - () : 6\text{A};$$
$$\left[\hat{\boldsymbol{\psi}}_{i2}\right]_i\left[\boldsymbol{\psi}_{i1}\right]_i^T : 6\text{M}; \rightarrow \left[\bar{\mathbf{M}}_{iwv}\right]_i = \left[\hat{\mathbf{M}}_{iwv}\right]_i - () : 6\text{A};$$
$$\left[\hat{\boldsymbol{\psi}}_{i2}\right]_i\left[\boldsymbol{\psi}_{i2}\right]_i^T : 6\text{M}; \rightarrow \left[\bar{\mathbf{M}}_{iww}\right]_i = \left[\hat{\mathbf{M}}_{iww}\right]_i - () : 6\text{A}$$

C.10 Find the scaler $\hat{\varphi}_i$:

(a) For $i = n, \hat{\varphi}_n = \hat{\varphi}_n$ {nil}
(b) For $i = n - 1, \ldots, 1$: {14M 8A (n – 1)}

$$[\bar{\mathbf{a}}_i]_{i+1} = [\mathbf{a}_i]_{i+1} \times \left[\boldsymbol{\eta}_{i+1,1}\right]_{i+1} : 6\text{M};$$
$$\left[\tilde{\mathbf{a}}_i\right]_i = \mathbf{Q}_i\left(\left[\boldsymbol{\eta}_{i+1,2}\right]_{i+1} + [\bar{\mathbf{a}}_i]_{i+1}\right) : 8\text{M}7\text{A};$$
$$\hat{\varphi}_n = \varphi_n - [\mathbf{z}_i]_i^T\left[\tilde{\mathbf{a}}_i\right]_i : 1\text{A}$$

C.11 Find the 6-dimensional vector $\left[\boldsymbol{\eta}_i\right]_i$. The vector $\boldsymbol{\eta}_i$ is Defined as

$$\boldsymbol{\eta}_i \equiv \left[\boldsymbol{\eta}_{i1}^T \; \boldsymbol{\eta}_{i2}^T\right]^T$$

where $\boldsymbol{\eta}_{i1}$ and $\boldsymbol{\eta}_{i2}$ are the 3-dimensional vectors containing the first and last three scaler elements of $\boldsymbol{\eta}_i$, respectively.

(a) For $i = n, \left[\boldsymbol{\eta}_n\right]_n = \hat{\varphi}_n\left[\boldsymbol{\psi}_n\right]_n$ {6M}
(b) For $i = n - 1, \ldots, 1\left[\boldsymbol{\eta}_i\right]_i$ {14M10A(n – 1)}

$$\left[\boldsymbol{\eta}_{i(b)}\right]_i = [\psi_{i1}]_i\hat{\varphi}_i + \mathbf{Q}_i\left[\boldsymbol{\eta}_{i+1,1}\right]_{i+1} : 11\text{M}\,7\text{A};$$
$$\left[\boldsymbol{\eta}_{i2}\right]_i = [\psi_{i2}]_i\hat{\varphi}_i + \left[\tilde{\mathbf{a}}_i\right]_i : 3\text{M}\,3\text{A}$$

C.12 Find $\tilde{\varphi}$: For $I = 1,\ldots, N$

$$\tilde{\varphi}_i = \frac{\hat{\varphi}_i}{\hat{m}_i} : 1\text{M} \qquad \{1\text{M}(\text{n})\}$$

C.13 Find $\ddot{\theta}$:

(a) For $i = 1, \ddot{\theta}_1 = \tilde{\varphi}_1$ {nil}
(b) For $i = 2, \ldots, n$, {4m 6A(n – 1)}

$$\left[\boldsymbol{\psi}_{i2}\right]_i^T\left[\bar{\boldsymbol{\mu}}_{i2}\right]_i : 2\text{M } 2\text{A}; \rightarrow \left[\boldsymbol{\psi}_{i1}\right]_i^T\left[\bar{\boldsymbol{\mu}}_{i1}\right]_i + () : 2\text{M } 3\text{A}; \rightarrow \ddot{\theta}_i = \tilde{\varphi}_i - () : 1\text{A}$$

C.14 Find the 6-dimensional vector $[\bar{\mu}_i]_i$ as

$$\bar{\mu}_i \equiv [\bar{\mu}_{i1}^T \; \bar{\mu}_{i2}^T]^T$$

where $\bar{\mu}_{i1}$ and $\bar{\mu}_{i2}$ are the 3-dimensional vectors containing first and last three elements of $\bar{\mu}_i$, respectively.

(a) For $i = 1$, $[\bar{\mu}_1]_1 = \mathbf{0}$ {nil}
(b) For $i = 1, \ldots, n-1$ {22M14A(n – 1)}

$$- [\mathbf{a}_{i-1}]_{i-1} \times [\bar{\mu}_{i-1,2}]_{i-1} : 6\mathrm{M}3\mathrm{A};$$

$$\rightarrow [\bar{\mu}_{i1}]_i = \mathbf{Q}_{i-1}^T \{ [\mu_{i-1,1}]_{i-1} + () \} : 8\mathrm{M}\,7\mathrm{A};$$

$$[\bar{\mu}_{i2}]_i = \mathbf{Q}_{i-1}^T [\mu_{i-1,2}]_{i-1} : 8\mathrm{M}\,4\mathrm{A}$$

C.15 The 6-dimensional vector $[\mu_i]_i$. The vector, μ_i, is defined as

$$\mu_i \equiv [\mu_{i1}^T \; \mu_{i2}^T]^T \qquad \{\mathrm{nil}\}$$

(a) $[\mu_1]_1$: For $i = 1$,

$$[\mu_{11}^T]_1 = \mathbf{0} : \mathrm{nil}; \; [\mu_{12}^T]_1 = \ddot{\theta}_1 [\mathbf{z}_1]_1 : \mathrm{nil}$$

(b) $[\mu_i]_i$: For $i = 1$,

$$[\mu_{i1}^T]_i = [\bar{\mu}_{i1}]_i : \mathrm{nil}; \; [\mu_{i2}^T]_i = \ddot{\theta}_i [\mathbf{z}_i]_i + [\bar{\mu}_{i2}]_i : \mathrm{nil}$$

The total complexity of the proposed algorithm for the forward dynamics algorithm is finally obtained by adding the values on the right inside the { and} as $(173n-128)$M, $(150n-133)$A, and $2n$T.

Appendix D
Derivation of Eq. (3.21a)

Using the expressions for the total kinetic energy, $T = \sum_{i=1}^{n} T_i$, where T_i is given by Eq. (3.17), the partial differentiations with respect to the set of *j*th independent generalized speeds and coordinates, i.e., $\dot{\mathbf{q}}_j$ and $\mathbf{q}_j$, respectively, as required in Eq. (3.20a) are obtained as

$$\frac{\partial T}{\partial \dot{\mathbf{q}}_j} = \sum_{i=1}^{n}\left[\int_0^{b_i} \rho_i \dot{\mathbf{r}}_i^T \frac{\partial \dot{\mathbf{r}}_i}{\partial \dot{\mathbf{q}}_j} \mathrm{d}\bar{b}_i + \int_0^{a_i} \rho_i \dot{\tilde{\mathbf{r}}}_i^T \frac{\partial \dot{\tilde{\mathbf{r}}}_i}{\partial \dot{\mathbf{q}}_j} \mathrm{d}\bar{a}_i + m_{pi} \dot{\mathbf{r}}_{pi}^T \frac{\partial \dot{\mathbf{r}}_{pi}}{\partial \dot{\mathbf{q}}_j} + \int_0^{a_i} \rho_i I_{pi} \dot{\beta}_i \frac{\partial \dot{\beta}_i}{\partial \dot{\mathbf{q}}_j} \mathrm{d}\bar{a}_i + (\mathbf{I}_{hi}\boldsymbol{\omega}_i)^T \frac{\partial \boldsymbol{\omega}_i}{\partial \dot{\mathbf{q}}_j}\right] \tag{D.1a}$$

$$\frac{\partial T}{\partial \mathbf{q}_j} = \sum_{i=1}^{n}\left[\int_0^{b_i} \rho_i \dot{\mathbf{r}}_i^T \frac{\partial \dot{\mathbf{r}}_i}{\partial \mathbf{q}_j} \mathrm{d}\bar{b}_i + \int_0^{a_i} \rho_i \dot{\tilde{\mathbf{r}}}_i^T \frac{\partial \dot{\tilde{\mathbf{r}}}_i}{\partial \mathbf{q}_j} \mathrm{d}\bar{a}_i + m_{pi} \dot{\mathbf{r}}_{pi}^T \frac{\partial \dot{\mathbf{r}}_{pi}}{\partial \mathbf{q}_j} + \int_0^{a_i} \rho_i I_{pi} \dot{\beta}_i \frac{\partial \dot{\beta}_i}{\partial \mathbf{q}_j} \mathrm{d}\bar{a}_i + (\mathbf{I}_{hi}\boldsymbol{\omega}_i)^T \frac{\partial \boldsymbol{\omega}_i}{\partial \mathbf{q}_j}\right] \tag{D.1b}$$

Substituting $\beta_i \equiv \bar{\mathbf{s}}_i^T \mathbf{c}_i$ and $\dot{\beta}_i \equiv \bar{\mathbf{s}}_i^T \dot{\mathbf{c}}_i$ from Eq. (3.5a), into Eqs. (D.1a–b), $\frac{\mathrm{d}}{\mathrm{d}t}\left(\frac{\partial T}{\partial \dot{\mathbf{q}}_j}\right)$ is obtained from Eq. (D.1a) as

$$\frac{d}{dt}\left(\frac{\partial T}{\partial \dot{\mathbf{q}}_j}\right) = \sum_{i=1}^{n}\left[\int_0^{b_i} \rho_i \left\{\ddot{\mathbf{r}}_i^T \frac{\partial \dot{\mathbf{r}}_i}{\partial \dot{\mathbf{q}}_j} + \dot{\mathbf{r}}_i^T \frac{d}{dt}\left(\frac{\partial \dot{\mathbf{r}}_i}{\partial \dot{\mathbf{q}}_j}\right)\right\} \mathrm{d}\bar{b}_i + \int_0^{a_i} \rho_i \left\{\ddot{\tilde{\mathbf{r}}}_i^T \frac{\partial \dot{\tilde{\mathbf{r}}}_i}{\partial \dot{\mathbf{q}}_j} + \dot{\tilde{\mathbf{r}}}_i^T \frac{d}{dt}\left(\frac{\partial \dot{\tilde{\mathbf{r}}}_i}{\partial \dot{\mathbf{q}}_j}\right)\right\} \mathrm{d}\bar{a}_i + m_{pi}\left\{\ddot{\mathbf{r}}_{pi}^T \frac{\partial \dot{\mathbf{r}}_{pi}}{\partial \dot{\mathbf{q}}_j} + \dot{\mathbf{r}}_{pi}^T \frac{d}{dt}\left(\frac{\partial \dot{\mathbf{r}}_{pi}}{\partial \dot{\mathbf{q}}_j}\right)\right\}\right.$$

P. V. Nandihal et al., *Dynamics of Rigid-Flexible Robots and Multibody Systems*, Intelligent Systems, Control and Automation: Science and Engineering 100,
https://doi.org/10.1007/978-981-16-2798-9

$$+\int_0^{a_i} \rho_i I_{pi}\bar{\mathbf{s}}_i^T \left\{\ddot{\mathbf{c}}_i\bar{\mathbf{s}}_i^T \frac{\partial \dot{\mathbf{c}}_i}{\partial \dot{\mathbf{q}}_j} + \dot{\mathbf{c}}_i\bar{\mathbf{s}}_i^T \frac{d}{dt}\left(\frac{\partial \dot{\mathbf{c}}_i}{\partial \dot{\mathbf{q}}_j}\right)\right\} \mathrm{d}\bar{a}_i$$

$$+\left\{(\mathbf{I}_{hi}\dot{\boldsymbol{\omega}}_i + \boldsymbol{\omega}_i \times \mathbf{I}_{hi}\boldsymbol{\omega}_i)^T \frac{\partial \boldsymbol{\omega}_i}{\partial \dot{\mathbf{q}}_j} + (\mathbf{I}_{hi}\boldsymbol{\omega}_i)^T \frac{d}{dt}\left(\frac{\partial \boldsymbol{\omega}_i}{\partial \dot{\mathbf{q}}_j}\right)\right\}\Bigg] \tag{D.2}$$

As shown in Stejskal and Valasek (1996) and others, it is evident that

$$\frac{\partial \dot{\mathbf{r}}_i}{\partial \dot{\mathbf{q}}_j} = \frac{\partial \mathbf{r}_i}{\partial \mathbf{q}_j};\ \frac{\partial \dot{\tilde{\mathbf{r}}}_i}{\partial \dot{\mathbf{q}}_j} = \frac{\partial \tilde{\mathbf{r}}_i}{\partial \mathbf{q}_j};\ \frac{\partial \dot{\mathbf{r}}_{pi}}{\partial \dot{\mathbf{q}}_j} = \frac{\partial \mathbf{r}_{pi}}{\partial \mathbf{q}_j};\ \text{and}\ \frac{\partial \dot{\mathbf{c}}_i}{\partial \dot{\mathbf{q}}_j} = \frac{\partial \mathbf{c}_i}{\partial \mathbf{q}_j} \tag{D.3}$$

Hence, the 2nd, 4th, 6th, and 8th terms on the right-hand side of Eq. (D.2) are given by

$$\begin{aligned}
&\frac{\mathrm{d}}{\mathrm{d}t}\left(\frac{\partial \dot{\mathbf{r}}_i}{\partial \dot{\mathbf{q}}_j}\right) = \frac{\mathrm{d}}{\mathrm{d}t}\left(\frac{\partial \mathbf{r}_i}{\partial \mathbf{q}_j}\right) = \frac{\partial \dot{\mathbf{r}}_i}{\partial \mathbf{q}_j};\ \frac{\mathrm{d}}{\mathrm{d}t}\left(\frac{\partial \dot{\tilde{\mathbf{r}}}_i}{\partial \dot{\mathbf{q}}_j}\right) = \frac{\mathrm{d}}{\mathrm{d}t}\left(\frac{\partial \tilde{\mathbf{r}}_i}{\partial \mathbf{q}_j}\right) = \frac{\partial \dot{\tilde{\mathbf{r}}}_i}{\partial \mathbf{q}_j}\\
&\frac{\mathrm{d}}{\mathrm{d}t}\left(\frac{\partial \dot{\mathbf{r}}_{pi}}{\partial \dot{\mathbf{q}}_j}\right) = \frac{\mathrm{d}}{\mathrm{d}t}\left(\frac{\partial \mathbf{r}_{pi}}{\partial \mathbf{q}_j}\right) = \frac{\partial \dot{\mathbf{r}}_{pi}}{\partial \mathbf{q}_j}\ \text{and}\ \frac{\mathrm{d}}{\mathrm{d}t}\left(\frac{\partial \dot{\mathbf{c}}_i}{\partial \dot{\mathbf{q}}_j}\right) = \frac{\mathrm{d}}{\mathrm{d}t}\left(\frac{\partial \mathbf{c}_i}{\partial \mathbf{q}_j}\right) = \frac{\partial \dot{\mathbf{c}}_i}{\partial \mathbf{q}_j}
\end{aligned} \tag{D.4}$$

Similar to Eq. (D.4) it can also be shown that

$$\frac{d}{dt}\left(\frac{\partial \boldsymbol{\omega}_i}{\partial \dot{\mathbf{q}}_j}\right) = \frac{\partial \boldsymbol{\omega}_i}{\partial \mathbf{q}_j} \tag{D.5}$$

Substituting Eqs. (D.4) and (D.5) into Eq. (D.2), and using the resulting expression, along with Eq. (D.1b), the left-hand side of Eq. (3.20a), is obtained as

$$\sum_{i=1}^{n}\Bigg[\int_0^{b_i} \rho_i \ddot{\mathbf{r}}_i^T \frac{\partial \dot{\mathbf{r}}_i}{\partial \dot{\mathbf{q}}_j}\mathrm{d}\bar{b}_i + \int_0^{a_i} \rho_i \ddot{\tilde{\mathbf{r}}}_i^T \frac{\partial \dot{\tilde{\mathbf{r}}}_i}{\partial \dot{\mathbf{q}}_j}\mathrm{d}\bar{a}_i + m_{pi}\ddot{\mathbf{r}}_{pi}^T \frac{\partial \dot{\mathbf{r}}_{pi}}{\partial \dot{\mathbf{q}}_j} + \int_0^{a_i} \rho_i I_{pi}\bar{\mathbf{s}}_i^T \ddot{\mathbf{c}}_i \bar{\mathbf{s}}_i^T \frac{\partial \dot{\mathbf{c}}_i}{\partial \dot{\mathbf{q}}_j}\mathrm{d}\bar{a}_i$$

$$+(\mathbf{I}_{hi}\dot{\boldsymbol{\omega}}_i + \boldsymbol{\omega}_i \times \mathbf{I}_{hi}\boldsymbol{\omega}_i)\frac{\partial \boldsymbol{\omega}_i}{\partial \dot{\mathbf{q}}_j}\Bigg] = \boldsymbol{\tau}_j \tag{D.6}$$

where cancelation of some terms happened. Next, substituting $\dot{\boldsymbol{r}}_i$, $\dot{\tilde{\boldsymbol{r}}}_i$, and $\dot{\boldsymbol{r}}_{pi}$ from Eq. (3.18b) into Eq. (D.6), one obtains

$$\sum_{i=1}^{n}\Bigg[\int_0^{b_i} \rho_i\left\{\ddot{\mathbf{r}}_i^T \frac{\partial \mathbf{v}_i}{\partial \dot{\mathbf{q}}_j} + \ddot{\mathbf{r}}_i^T \frac{\partial(\boldsymbol{\omega}_i \times \bar{b}_i\mathbf{z}_i)}{\partial \dot{\mathbf{q}}_j}\right\}\mathrm{d}\bar{b}_i$$

$$+\int_0^{a_i}\rho_i\left\{\ddot{\mathbf{r}}_i^T\frac{\partial \mathbf{v}_i}{\partial \dot{\mathbf{q}}_j}+\ddot{\mathbf{r}}_i^T\frac{\partial[\boldsymbol{\omega}_i\times\bar{\mathbf{r}}_i+\dot{\mathbf{u}}_i]}{\partial \dot{\mathbf{q}}_j}\right\}d\bar{a}_i$$

$$+m_{pi}\left\{\ddot{\mathbf{r}}_{pi}^T\frac{\partial \mathbf{v}_i}{\partial \dot{\mathbf{q}}_j}+\ddot{\mathbf{r}}_{pi}^T\frac{\partial[\boldsymbol{\omega}_i\times\bar{\mathbf{r}}_{pi}+\dot{\mathbf{u}}_{pi}]}{\partial \dot{\mathbf{q}}_j}\right\}+\int_0^{a_i}\rho_i I_{pi}\bar{\mathbf{s}}_i^T\ddot{\mathbf{c}}_i\bar{\mathbf{s}}_i^T\frac{\partial \dot{\mathbf{c}}_i}{\partial \dot{\mathbf{q}}_j}d\bar{a}$$

$$+\left.(\mathbf{I}_{hi}\dot{\boldsymbol{\omega}}_i+\boldsymbol{\omega}_i\times\mathbf{I}_{hi}\boldsymbol{\omega}_i)^T\frac{\partial\boldsymbol{\omega}_i}{\partial \dot{\mathbf{q}}_j}\right]=\boldsymbol{\tau}_j \tag{D.7}$$

Using the vector triple product rule (Strang 1980), $\mathbf{a}^T(\mathbf{b}\times\mathbf{c})=(\mathbf{c}\times\mathbf{a})^T\mathbf{b}$–**a**, **b** and **c** are any 3-dimensional Cartesian vectors, one can show that

$$\ddot{\mathbf{r}}_i^T\frac{\partial(\boldsymbol{\omega}_i\times\bar{b}_i\mathbf{z}_i)}{\partial \dot{\mathbf{q}}_j}=\bar{b}_i\ddot{\mathbf{r}}_i^T\left(\frac{\partial\boldsymbol{\omega}_i}{\partial \dot{\mathbf{q}}_j}\times\mathbf{z}_i+\boldsymbol{\omega}_i\times\frac{\partial\mathbf{z}_i}{\partial \dot{\mathbf{q}}_j}\right)=\bar{b}_i(\mathbf{z}_i\times\ddot{\mathbf{r}}_i)^T\frac{\partial\boldsymbol{\omega}_i}{\partial \dot{\mathbf{q}}_j} \tag{D.8a}$$

$$\begin{aligned}\ddot{\mathbf{r}}_i^T\frac{\partial(\boldsymbol{\omega}_i\times\bar{\mathbf{r}}_i+\dot{\mathbf{u}}_i)}{\partial \dot{\mathbf{q}}_j}&=\ddot{\mathbf{r}}_i^T\frac{\partial(\boldsymbol{\omega}_i\times\bar{\mathbf{r}}_i)}{\partial \dot{\mathbf{q}}_j}+\ddot{\mathbf{r}}_i^T\frac{\partial\dot{\mathbf{u}}_i}{\partial \dot{\mathbf{q}}_j}\\&=\ddot{\mathbf{r}}_i^T\left(\frac{\partial\boldsymbol{\omega}_i}{\partial \dot{\mathbf{q}}_j}\times\bar{\mathbf{r}}_i+\boldsymbol{\omega}_i\times\frac{\partial\bar{\mathbf{r}}_i}{\partial \dot{\mathbf{q}}_j}\right)+\ddot{\mathbf{r}}_i^T\frac{\partial\dot{\mathbf{u}}_i}{\partial \dot{\mathbf{q}}_j}\\&=(\bar{\mathbf{r}}_i\times\ddot{\mathbf{r}}_i)^T\frac{\partial\boldsymbol{\omega}_i}{\partial \dot{\mathbf{q}}_j}+\ddot{\mathbf{r}}_i^T\frac{\partial\dot{\mathbf{u}}_i}{\partial \dot{\mathbf{q}}_j}\end{aligned} \tag{D.8b}$$

and

$$\begin{aligned}\ddot{\mathbf{r}}_{pi}^T\frac{\partial\left(\boldsymbol{\omega}_i\times\bar{\mathbf{r}}_{pi}+\dot{u}_{pi}\right)}{\partial \dot{\mathbf{q}}_j}&=\ddot{\mathbf{r}}_{pi}^T\left(\frac{\partial\boldsymbol{\omega}_i}{\partial \dot{\mathbf{q}}_j}\times\bar{\mathbf{r}}_{pi}+\boldsymbol{\omega}_i\times\frac{\partial\bar{\mathbf{r}}_{pi}}{\partial \dot{\mathbf{q}}_j}\right)+\ddot{\mathbf{r}}_{pi}^T\frac{\partial\dot{\mathbf{u}}_{pi}}{\partial \dot{\mathbf{q}}_j}\\&=\left(\bar{\mathbf{r}}_{pi}\times\ddot{\mathbf{r}}_{pi}\right)^T\frac{\partial\boldsymbol{\omega}_i}{\partial \dot{\mathbf{q}}_j}+\ddot{\mathbf{r}}_{pi}^T\frac{\partial\dot{\mathbf{u}}_{pi}}{\partial \dot{\mathbf{q}}_j}\end{aligned} \tag{D.8c}$$

In Eqs. (D.8a–D.8c), $\partial\mathbf{z}_i/\partial\dot{\mathbf{q}}_j=\mathbf{O}$, $\partial\bar{\mathbf{r}}_i/\partial\dot{\mathbf{q}}_j=\mathbf{O}$, and $\partial\bar{\mathbf{r}}_{pi}/\partial\dot{\mathbf{q}}_j=\mathbf{O}$ are used as $\mathbf{z}_i$, $\bar{\mathbf{r}}_\mathbf{i}$, and $\bar{\mathbf{r}}_{pi}$ are functions of $\mathbf{q}_j$'s only, and not $\dot{\mathbf{q}}_j$'s. Moreover, using Eq. (3.4e) and substituting $\dot{\mathbf{u}}_i=\mathbf{S}_i\dot{\mathbf{d}}_i$, along with $\dot{\boldsymbol{u}}_{pi}=\mathbf{S}_i|_{a_i}\dot{\mathbf{d}}_i$ in Eqs. (D.8a–D.8c), $\ddot{\mathbf{r}}_i^T\frac{\partial\dot{\mathbf{u}}_i}{\partial\dot{\mathbf{q}}_j}$ and $\ddot{\mathbf{r}}_{pi}^T\frac{\partial\dot{\mathbf{u}}_{pi}}{\partial\dot{\mathbf{q}}_j}$ are rewritten as:

$$\ddot{\mathbf{r}}_i^T\frac{\partial\dot{\mathbf{u}}_i}{\partial \dot{\mathbf{q}}_j}=\ddot{\mathbf{r}}_i^T\mathbf{S}_i\frac{\partial\dot{\mathbf{d}}_i}{\partial \dot{\mathbf{q}}_j} \tag{D.9a}$$

$$\ddot{\mathbf{r}}_{pi}^T\frac{\partial\dot{\mathbf{u}}_{pi}}{\partial \dot{\mathbf{q}}_j}=\ddot{\mathbf{r}}_{pi}^T\mathbf{S}_i|_{a_i}\frac{\partial\dot{\mathbf{d}}_i}{\partial \dot{\mathbf{q}}_j} \tag{D.9b}$$

In Eq. (D.9b), $\mathbf{S}_i|_{a_i}$ is the shape function of the link evaluated at its tip, i.e., $\bar{a}_i=a_i$. Substituting Eqs. (D.9a–D.9b) into Eq. (D.7) yields

$$\sum_{i=1}^{n}\left[\int_0^{b_i}\rho_i\left\{\ddot{\mathbf{r}}_i^T\frac{\partial \mathbf{v}_i}{\partial \dot{\mathbf{q}}_j}+\bar{b}_i(\mathbf{z}_i\times\ddot{\mathbf{r}}_i)^T\frac{\partial \omega_i}{\partial \dot{\mathbf{q}}_j}\right\}\mathrm{d}\bar{b}_i\right.$$
$$+\int_0^{a_i}\rho_i\left\{\ddot{\tilde{\mathbf{r}}}_i^T\frac{\partial \mathbf{v}_i}{\partial \dot{\mathbf{q}}_j}+(\bar{\mathbf{r}}_i\times\ddot{\tilde{\mathbf{r}}}_i)^T\frac{\partial \omega_i}{\partial \dot{\mathbf{q}}_j}+\ddot{\tilde{\mathbf{r}}}_i^T\mathbf{S}_i\frac{\partial \dot{\mathbf{d}}_i}{\partial \dot{\mathbf{q}}_j}\right\}\mathrm{d}\bar{a}_i$$
$$+m_{pi}\left\{\ddot{\mathbf{r}}_{pi}^T\frac{\partial \mathbf{v}_i}{\partial \dot{\mathbf{q}}_j}+(\bar{\mathbf{r}}_{pi}\times\ddot{\mathbf{r}}_{pi})^T\frac{\partial \omega_i}{\partial \dot{\mathbf{q}}_j}+\ddot{\mathbf{r}}_{pi}^T\mathbf{S}_i|_{a_i}\frac{\partial \dot{\mathbf{d}}_i}{\partial \dot{\mathbf{q}}_j}\right\}$$
$$+\int_0^{a_i}\rho_i I_{pi}\bar{\mathbf{s}}_i^T\ddot{\mathbf{c}}_i\bar{\mathbf{s}}_i^T\frac{\partial \dot{\mathbf{c}}_i}{\partial \dot{\mathbf{q}}_j}\mathrm{d}\bar{a}_i$$
$$\left.+(\mathbf{I}_{hi}\dot{\omega}_i+\omega_i\times\mathbf{I}_{hi}\omega_i)^T\frac{\partial \omega_i}{\partial \dot{\mathbf{q}}_j}\right]=\mathbf{t}_j \quad \text{(D.9c)}$$

Now, introduce the following definitions:

$$\frac{\partial \mathbf{t}_i}{\partial \dot{\mathbf{q}}_j}\equiv\begin{bmatrix}\frac{\partial \mathbf{v}_i}{\partial \dot{\mathbf{q}}_j}\\ \frac{\partial \boldsymbol{\omega}_i}{\partial \dot{\mathbf{q}}_j}\\ \frac{\partial \dot{\mathbf{d}}_i}{\partial \dot{\mathbf{q}}_j}\\ \frac{\partial \dot{\mathbf{c}}_i}{\partial \dot{\mathbf{q}}_j}\end{bmatrix};\ \mathbf{w}_i^*\equiv\begin{bmatrix}\int_0^{b_i}\rho_i\ddot{\mathbf{r}}_i\mathrm{d}\bar{b}_i+\int_0^{a_i}\rho_i\ddot{\tilde{\mathbf{r}}}_i\mathrm{d}\bar{a}_i+m_{pi}\ddot{\mathbf{r}}_{pi}\\ \int_0^{b_i}\rho_i\bar{b}_i(\mathbf{z}_i\times\ddot{\mathbf{r}}_i)\mathrm{d}\bar{b}_i+\int_0^{a_i}\rho_i(\bar{\mathbf{r}}_i\times\ddot{\tilde{\mathbf{r}}}_i)\mathrm{d}\bar{a}_i+m_{pi}(\bar{\mathbf{r}}_{pi}\times\ddot{\mathbf{r}}_{pi})+\\ (\mathbf{I}_{hi}\dot{\boldsymbol{\omega}}_i+\boldsymbol{\omega}_i\times\mathbf{I}_{hi}\boldsymbol{\omega}_i)\\ \int_0^{a_i}\rho_i\mathbf{S}_i^T\ddot{\tilde{\mathbf{r}}}_i\mathrm{d}\bar{a}_i+m_{pi}\mathbf{S}_i|_{a_i}^T\ddot{\mathbf{r}}_{pi}\\ \int_0^{a_i}\rho_i I_{pi}\bar{\mathbf{s}}_i\ddot{\mathbf{c}}_i^T\bar{\mathbf{s}}_i\mathrm{d}\bar{a}_i\end{bmatrix} \quad \text{(D.10)}$$

Then, eq. (D.9c) is rewritten in compact form as

$$\left[\left(\frac{\partial \mathbf{t}_1}{\partial \dot{\mathbf{q}}_j}\right)^T\cdots\left(\frac{\partial \mathbf{t}_n}{\partial \dot{\mathbf{q}}_j}\right)^T\right]\begin{bmatrix}\mathbf{w}_1^*\\ \vdots\\ \mathbf{w}_n^*\end{bmatrix}=\boldsymbol{\tau}_j \quad \text{(D.11)}$$

which is the desired equation of motion (Eq. 3.21a) to obtain the generalized force corresponding to the vector of the jth generalized coordinates.

Appendix E
Computational Counts for Flexible Robots

In order to compute the joint accelerations, $\ddot{\mathbf{q}}$ of Eq. (3.31), using the recursive algorithm presented in Sect. 3.41, the following inputs are required: For $i = 1, \ldots, n$

(i) Constant Denavit–Hartenberg (DH) parameters of the system under study, i.e., α_i, a_i, and b_i (for revolute joints) or θ_i (for prismatic joints), as defined in Appendix AA.
(ii) Initial values for the variable DH parameter, i.e., θ_i, for the revolute joint, and b_i, for the prismatic joint, and other generalized coordinates associated with bending and torsion of the links, along with their first time derivatives.
(iii) Mass per unit length of each link, ρ_i.
(iv) Time history of input generalized forces, vector $\boldsymbol{\tau}$.
(v) Number of modes considered for the vibrations of the flexible links.

The computations associated with each step are given with the M, A, and T counts, i.e., multiplications/divisions, additions/subtractions, and trigonometric functions evaluations, respectively. They are mentioned inside {and}. For the purpose of evaluating the computational complexity, the payload mass and the centrifugal stiffness effect are considered to be absent. Moreover, all the links are assumed to vibrate in equal number of modes, namely m in bending and $\bar{m}$ in torsion, as mentioned in Sect. 3.4.1. The above assumptions are considered for the comparison of the final count with those available in the literature, e.g., Cyril (1988), Jain and Rodriguez (1992), where similar assumptions were taken. Due to the above assumptions, the degree of freedom of the flexible link system at hand reduces to, $\bar{n} \equiv n + \sum_{i=1}^{n_f} (2m_i + \bar{m}_i)$, where $n \equiv n_r + n_f$ is the total number of rigid and flexible links, n_r and n_f, respectively, or the joints preceding to the links. Moreover, the number 2 in the formula for $\bar{n}$ stands for Y and Z components of the spatial bending deflection of each flexible link. The number of modes considered in bending is m_i. The X-component of the deflection vanishes because axial stiffness has been neglected. However, the torsional deflection about X-axis is accounted for in which the number of modes considered is $\bar{m}_i$. To illustrate some of the computations of the intermediate steps in the sequence of the main steps, in-between computations are

P. V. Nandihal et al., *Dynamics of Rigid-Flexible Robots and Multibody Systems*,
Intelligent Systems, Control and Automation: Science and Engineering 100,
https://doi.org/10.1007/978-981-16-2798-9

also given after the colon ':', whereas a vector or a matrix represented in ith frame is denoted by $[\cdot]_i$. Furthermore, $\hat{m} \equiv 1 + m_f$, where $m_f \equiv 2m + \bar{m}$. Now, the calculation sequence is given below:

E.1 For $I = 1,\ldots, N$, find the 3-dimensional unit vectors, $[\hat{\boldsymbol{x}}_{i+1}]_i$ and $[z_i]_i$:

$$[\hat{\mathbf{x}}_{i+1}]_i = \tilde{\mathbf{Q}}_i[\hat{\mathbf{x}}_{i+1}]_{i+1}, \text{where } [\hat{\mathbf{x}}_{i+1}]_{i+1} \equiv \begin{bmatrix}1 & 0 & 0\end{bmatrix}^T, \text{and } [\mathbf{z}_i]_i \equiv \begin{bmatrix}0 & 0 & 1\end{bmatrix}^T \ \{\text{nil}\}.$$

The terms $\hat{\mathbf{x}}_{i+1}$ and $\mathbf{z}_i$ are, respectively, the unit vectors along $\hat{X}_{i+1}$- and Z_i-axes, and $\tilde{\mathbf{Q}}_i$ is the 3×3 rotation matrix that defines the orientation of the $(i + 1)^{st}$ frame with respect to the ith frame, as defined in Eq. (3.1) Matrix $\tilde{\mathbf{Q}}_i$ is evaluated as.

for $i = 1$,

$$\tilde{\mathbf{Q}}_i \equiv \mathbf{Q}_i \qquad \{4M\}$$

else for $i = 2, \ldots, n$

$$\tilde{\mathbf{Q}}_i \equiv \mathbf{Q}_i\hat{\mathbf{Q}}_{i-1} \qquad \{14M\,9A\,2T(n-1)\}$$

where $\mathbf{Q}_i$ is the rotation matrix associated with the rigid body motion of the ith link with respect to the $(i\text{-}1)^{st}$ one. Matrix $\hat{\mathbf{Q}}_{i-1}$ above represents the rotation matrix resulting from the angular motion of link $(i\text{-}1)^{st}$ due to its deformation. Matrix $\mathbf{Q}_i$ is given in step C.2 of Appendix C, and $\hat{\mathbf{Q}}_{i-1}$ is given by Eq. (3.2), which are reproduced here as

$$\mathbf{Q}_i \equiv \begin{bmatrix} c_i & -s_i\bar{c}_i & s_i\bar{s}_i \\ s_i & c_i\bar{c} & -c_i\bar{s}_i \\ 0 & \bar{s}_i & \bar{c}_i \end{bmatrix}; \hat{\mathbf{Q}}_{i-1} \equiv \begin{bmatrix} 1 & -\delta^z_{i-1} & \delta^y_{i-1} \\ \delta^z_{i-1} & 1 & -\delta^x_{i-1} \\ -\delta^y_{i-1} & \delta^x_{i-1} & 1 \end{bmatrix}$$

The items, δ^x_{i-1}, δ^y_{i-1}, and δ^z_{i-1} above are the angular slopes of the links due to the torsion about X-axis and bending about Y and Z axes, respectively. They are given by,

$$[\boldsymbol{\delta}_{i-1}] \equiv \begin{bmatrix} \delta^x_{i-1} \\ \delta^y_{i-1} \\ \delta^z_{i-1} \end{bmatrix} = \boldsymbol{\Delta}_{i-1}\mathbf{d}_{i-1} + \mathbf{C}_{i-1}\mathbf{c}_{i-1}$$

where the $3 \times 3\,m$ matrix, $\boldsymbol{\Delta}_{i-1}$, and the $3 \times \bar{m}_i$ matrix, $\mathbf{C}_{i-1}$, are, respectively, defined in Eqs. (3.10d) and (3.10e). The $\mathbf{d}_{i\text{-}1}$ and $\mathbf{c}_{i\text{-}1}$ are the vector of independent generalized coordinates, defined in Eqs. (3.4d) and (3.5b), respectively. Using Eqs. (3.4c), (3.5b), and (3.10d–e), the computations involved in evaluating $\boldsymbol{\delta}_{i-1}$ are as follows,

$$\delta^x_{i-1} = \frac{\partial}{\partial \bar{a}_{i-1}} \left[\bar{s}_{i-1,1} \cdots \bar{s}_{i-1,\bar{m}}\right] \Big|_{\bar{a}_{i-1}=a_{i-1}} \begin{bmatrix} c_{i-1,1} \\ \vdots \\ c_{i-1,\bar{m}} \end{bmatrix} \qquad \{\bar{m}\mathrm{M}(\bar{m}-1)\mathrm{A}(\mathrm{n}-1)\}$$

$$\delta^y_{i-1} = \frac{\partial}{\partial \bar{a}_{i-1}} \left[s^y_{i-1,1} \cdots {}^y_{i-1,m}\right] \Big|_{\bar{a}_{i-1}=a_{i-1}} \begin{bmatrix} d^y_{i-1,1} \\ \vdots \\ d^y_{i-1,m} \end{bmatrix} \qquad \{m\mathrm{M}(m-1)\mathrm{A}(\mathrm{n}-1)\}$$

$$\delta^z_{i-1} = \frac{\partial}{\partial \bar{a}_{i-1}} \left[s^z_{i-1,1} \cdots {}^z_{i-1,m}\right] \Big|_{\bar{a}_{i-1}=a_{i-1}} \begin{bmatrix} d^z_{i-1,1} \\ \vdots \\ d^z_{i-1,m} \end{bmatrix} \qquad \{m\mathrm{M}(m-1)\mathrm{A}(\mathrm{n}-1)\}$$

in which $\bar{s}_{i,1} \cdots \bar{s}_{i,\bar{m}}$ are the shape functions of the beam in torsion and $s^y_{i,1} \cdots s^y_{i,m}$ and $s^z_{i,1} \cdots s^z_{i,m}$ are, respectively, the shape functions of the beam with respect to the $\hat{Y}$ and $\hat{Z}$ axes. Finally, the multiplication of $\tilde{\mathbf{Q}}_\mathbf{i}$ with $\left[\hat{\mathbf{x}}_{i+1}\right]_{i+1}$ does not require any computation, as $\tilde{\mathbf{Q}}_i\left[\hat{\mathbf{x}}_{i+1}\right]_{i+1}$ is nothing but the first column of $\tilde{\mathbf{Q}}_i$. Hence, {nil} is put in the beginning of step E.1.

E.2 Find the integrals of the shape functions:

The shape functions of the beam along $\hat{Y}$ and $\hat{Z}$ axes, namely, $s^y_{i,j}$ and $s^z_{i,j}$, are, respectively, as appeared in Eq. (3.4a). The evaluation of the integrals of shape functions of the beam along $\hat{Y}$-axis, $s^y_{i,j}$, is shown below, which are expected to be done offline and, hence, their counts are put as {nil}.

$$\left[\mathbf{K}^y_{1,i}\right]_{i+1} \equiv \int_0^{a_i} \mathbf{S}^y_i \, d\bar{a}_i = \int_0^{a_i} \begin{bmatrix} 0 & \ldots 0 \\ s^y_{i,1} & \ldots s^y_{i,m} \\ 0 & \ldots 0 \end{bmatrix} d\bar{a}_i \qquad \{\text{nil}\}$$

$$\left[\mathbf{K}^y_{2,i}\right]_{i+1} \equiv \int_0^{a_i} \left[\mathbf{S}^y_i\right]^T \mathbf{S}^y_i \, d\bar{a}_i = \int_0^{a_i} \begin{bmatrix} 0 & \ldots 0 \\ \left(s^y_{i,1}\right)^2 & \ldots \left(s^y_{i,m}\right)^2 \\ 0 & \ldots 0 \end{bmatrix} d\bar{a}_i \qquad \{\text{nil}\}$$

$$\left[\mathbf{K}^y_{3,i}\right]_{i+1} \equiv \int_0^{a_i} \bar{a}_i \mathbf{S}^y_i \, d\bar{a}_i = \int_0^{a_i} \begin{bmatrix} 0 & \ldots 0 \\ \bar{a}_i s^y_{i,1} & \ldots \bar{a}_i s^y_{i,m} \\ 0 & \ldots 0 \end{bmatrix} d\bar{a}_i \qquad \{\text{nil}\}$$

$$\left[\mathbf{K}^y_{4,i}\right]_{i+1} \equiv \mathbf{S}^y_i \Big|_{\bar{a}_i=a_i} = \begin{bmatrix} 0 & \ldots 0 \\ s^y_{i,1} & \ldots s^y_{i,m} \\ 0 & \ldots 0 \end{bmatrix}_{\bar{a}_i=a_i} \qquad \{\text{nil}\}$$

Similarly, the integrals of the shape functions of the beam along Z_i-axis, $K^z_{1,i}$, $K^z_{2,i}$, $K^z_{3,i}$, and $K^z_{4,i}$ are also evaluated without any computation. For the torsional-shaped functions, the following computations are performed:

$$\left[\mathbf{K}_{5,i}\right]_{i+1} \equiv \int_0^{a_i} \bar{\mathbf{s}}_i\bar{\mathbf{s}}_i^T d\bar{a}_i = \int_0^{a_i} \begin{bmatrix} \bar{s}_{i1}^2 & \bar{s}_{i1}\bar{s}_{i2} & \cdots & \bar{s}_{i1}\bar{s}_{i\bar{m}} \\ & \ddots & & \vdots \\ & & \ddots & \vdots \\ \text{syms} & & & \bar{s}_{i\bar{m}}^2 \end{bmatrix} d\bar{a}_i \qquad \{\text{nil}\}$$

E.3 Considering the frame rotations, the integrals of shape functions evaluated above are now represented in the *i*th frame, for $I = 1, \ldots, N$, as

$$\left[\mathbf{K}_{1,i}^y\right]_i \equiv \tilde{\mathbf{Q}}_f\left[\mathbf{K}_{1,i}^y\right]_{i+1} \qquad \{3m\mathrm{M}\,(n)\}$$

Similarly, for the other integrals of the shape function matrices, i.e.,

$$\left[\mathbf{K}_{2,\mathrm{i}}^y\right]_i, \left[\mathbf{K}_{3,\mathrm{i}}^y\right]_i, \left[\mathbf{K}_{4,\mathrm{i}}^y\right]_i, \left[\mathbf{K}_{1,\mathrm{i}}^z\right]_i, \left[\mathbf{K}_{2,\mathrm{i}}^z\right]_i, \left[\mathbf{K}_{3,\mathrm{i}}^z\right]_i, \left[\mathbf{K}_{4,\mathrm{i}}^z\right]_i, \left[\mathbf{K}_{5,\mathrm{i}}\right]_i \quad \{(21m + 3\bar{m})\mathrm{M(n)}\}$$

E.4 Find the Mass Matrix, $[\mathbf{M}_i]_i$, Which is Denoted as

$$\mathbf{M}_i \equiv \begin{bmatrix} \mathbf{M}_{ivv} & \mathbf{M}_{ivw} & \mathbf{M}_{ivd} & \mathbf{O} \\ & \mathbf{M}_{iww} & \mathbf{M}_{iwd} & \mathbf{O} \\ & & \mathbf{M}_{idd} & \mathbf{O} \\ \text{syms} & & & \mathbf{M}_{icc} \end{bmatrix}$$

where $\mathbf{M}_{i\mathrm{vv}}$, $\mathbf{M}_{i\mathrm{vw}}$, and $\mathbf{M}_{iww}$ are the 3×3 matrices, whereas $\mathbf{M}_{ivd}$ and $\mathbf{M}_{iwd}$ are the $3 \times 2m$ matrices. Finally, $\mathbf{M}_{idd}$ is the $2m \times 2m$ matrix, and $\mathbf{M}_{icc}$ is the $\bar{m} \times \bar{m}$ diagonal matrix. The expressions for the above matrices are available from Eq. (3.24a). The computations involved in their evaluations are given below as, for $i = 1, \ldots, n$,

(a) The 3×3 matrix, $[\mathbf{M}_{ivv}]_i$, is obtained as:

$$[\mathbf{M}_{ivv}]_i = \rho_i(a_i + b_i)\mathbf{1} \qquad \{\text{nil}\}$$

where$\rho_i(a_i + b_i)$ is constant for a link and expected to be computed offline. Hence, it is not counted.

(b) The 3×3 matrix, $[\mathbf{M}_{ivw}]_i$, is obtained as:

$$[\mathbf{M}_{ivw}]_i = -\int_0^{b_i} \rho_i\left(\left[\bar{\mathbf{b}}_i\right]_i \times \mathbf{1}\right)d\bar{b}_i$$

$$-\int_0^{a_i} \rho_i([\bar{\mathbf{r}}_i]_i \times \mathbf{1})d\bar{a}_i \quad \{(12 + 2m)\mathrm{M}(5 + 2m)\mathrm{A}(n)\}$$

Substituting, $\bar{\mathbf{b}}_i = \bar{b}_i\mathbf{z}_i$ and $\bar{\mathbf{r}}_i = b_i\mathbf{z}_i + \bar{a}_i\mathbf{x}_{i+1} + \mathbf{u}_i$, the step-by-step details of the computations involved are shown below:

$$\int_0^{b_i} \rho_i\left([\bar{b}_i\mathbf{z}_i]_i \times \mathbf{1}\right)\mathrm{d}\bar{b}_i + \int_0^{a_i} \rho_i([b_i\mathbf{z}_i]_i \times \mathbf{1})\mathrm{d}\bar{a}_i = \tilde{C}_{1i}\begin{bmatrix} 0 & 1 & 0 \\ -1 & 0 & 0 \\ 0 & 0 & 0 \end{bmatrix} \qquad : 0\mathrm{M}$$

where $\tilde{C}_{1i} \equiv \rho_i b_i\left(\frac{b_i}{2} + a_i\right)$ is computed offline, which is similar to C_{1i} for the rigid body systems in Step C.3.

$$\int_0^{a_i} \rho_i\left([\bar{a}_i\mathbf{x}_{i+1}]_i \times \mathbf{1}\right)\mathrm{d}\bar{a}_i = \tilde{C}_{2i}\begin{bmatrix} 0 & q_{31i} & -q_{21i} \\ -q_{31i} & 0 & q_{11i} \\ q_{21i} & -q_{11i} & 0 \end{bmatrix} \qquad : 3\mathrm{M}$$

where $\tilde{C}_{2i} \equiv \rho_i \frac{a_i^2}{2}$ is also computed offline. Moreover, the term, $q_{sti,}$ is nothing but the (s, t) element of matrix $\tilde{Q}_i$. Next, the integral, $\int_0^{a_i} \rho_i([\mathbf{u}_i]_i \times \mathbf{1})\mathrm{d}\bar{a}_i$, is evaluated as,

$$[\mathbf{u}_i]_i = \tilde{\mathbf{Q}}_i[\mathbf{u}_i]_{i+1} \qquad : 6\mathrm{M}\ 3\mathrm{A}$$

in which,

$$[\mathbf{u}_i]_{i+1} \equiv [\mathbf{S}_i]_{i+1}[\mathbf{d}_i]_i = \begin{bmatrix} 0 & \sum_0^m s_{i,j}^y d_{i,j}^y & \sum_0^m s_{i,j}^z d_{i,j}^z \end{bmatrix}^T \qquad : 2m\mathrm{M}(2m-2)\mathrm{A}$$

(c) The 3×3 matrix, $[\mathbf{M}_{iww}]_i$, is obtained as

$$[\mathbf{M}_{iww}]_i = -\int_0^{b_i} \rho_i\bar{b}_i^2\{[\mathbf{z}_i]_i \times ([\mathbf{z}_i]_i \times \mathbf{1})\}\mathrm{d}\bar{b}_i - \int_0^{a_i} \rho_i\{[\bar{\mathbf{r}}_i]_i \times ([\bar{\mathbf{r}}_i]_i \times \mathbf{1})\}\mathrm{d}\bar{a}_i$$

$$\{(54 + 6m)\mathrm{M}(23 + 6m)\mathrm{A}(n)\}$$

The computational count is obtained similar to that of $[\mathbf{M}_{ivw}]_i$.The details of computations involved in finding $[\mathbf{M}_{iww}]_i$ is omitted here to avoid perplexity.

(d) The $3 \times 2m$ matrix, $[\mathbf{M}_{ivd}]_i$, is obtained as

$$[\mathbf{M}_{ivd}]_i = \int_0^{a_i} \rho_i[\mathbf{S}_i]_i\mathrm{d}\bar{a}_i = \rho_i\left[\mathbf{K}_{1,i}^y\ \mathbf{K}_{1,i}^z\right]_i \qquad \{6m\mathrm{M}(n)\}$$

(e) The $3 \times 2m$ matrix, $[\mathbf{M}_{iwd}]_i$, is obtained as

$$[\mathbf{M}_{iwd}]_i = \int_0^{a_i} \rho_i [\bar{\mathbf{r}}_i]_i \times [\mathbf{S}_i]_i d\bar{a}_i \qquad \{12m\text{M}\, 6m\text{A}(n)\}$$

Note here that vector $\rho_i[\bar{\mathbf{r}}_i]_i$ has already been calculated, above, while evaluating $[\mathbf{M}_{iww}]_i$.

(g) The $2m \times 2m$ matrix, $[\mathbf{M}_{idd}]_i$, is obtained as

$$[\mathbf{M}_{idd}]_i = \rho_i \begin{bmatrix} \mathbf{K}^y_{i,2} & \mathbf{O} \\ \mathbf{O} & \mathbf{K}^z_{i,2} \end{bmatrix}_i \qquad \{6m\text{M}(n)\}$$

where the evaluation of the block matrices are shown in Step E.2.

(h) The $\bar{m} \times \bar{m}$-dimensional diagonal matrix, $\mathbf{M}_{icc}$, is obtained as

$$\mathbf{M}_{icc} = I_{pi}\left[\mathbf{K}_{5,i}\right]_i \qquad \{\text{nil}\}$$

Similar to items (d) and (g) above, $\mathbf{M}_{icc}$, does not require any online computations.

Thus, the computation of mass matrix, $[\mathbf{M}_i]_i$, in E.4, along with other associated matrices in E.1–E.3 involves $\left[(80 + 58m + 4\bar{m})n - m_f - 10\right]$ M and $\left[(34 + 16m + \bar{m})n - m_f - 6\right]$ A and 2nT.

E.5 Find the Matrix, $\left[\hat{\mathbf{M}}_i\right]_i$. For the purpose of showing the computational details of $\hat{\boldsymbol{M}}_i$, it is expressed similar to $\boldsymbol{M}_i$ in step E.4 as

$$\hat{\mathbf{M}}_i \equiv \begin{bmatrix} \hat{\mathbf{M}}_{ivv} & \hat{\mathbf{M}}_{ivw} & \hat{\mathbf{M}}_{ivd} & \mathbf{O} \\ & \hat{\mathbf{M}}_{iww} & \hat{\mathbf{M}}_{iwd} & \mathbf{O} \\ & & \hat{\mathbf{M}}_{idd} & \mathbf{O} \\ \text{syms} & & & \hat{\mathbf{M}}_{icc} \end{bmatrix}$$

For the sake of clarity in representation, the $3 \times 2m$ matrix $\hat{\mathbf{M}}_{ivd}$ is denoted as:

$$\hat{\mathbf{M}}_{ivd} \equiv \left[\hat{\mathbf{M}}^y_{ivd}\ \hat{\mathbf{M}}^z_{ivd}\right]$$

where $\hat{\mathbf{M}}^y_{ivd}$ and $\hat{\mathbf{M}}^z_{ivd}$ are the $3 \times m$ submatrices of $\hat{\mathbf{M}}_{ivd}\hat{\mathbf{M}}$ corresponding to the vibrations in $\hat{Y}$ and $\hat{Z}$ directions, respectively. Similarly, the $3 \times 2m$ matrix, $\hat{\mathbf{M}}_{iwd}$, is denoted as $\hat{\mathbf{M}}_{iwd} \equiv \left[\hat{\mathbf{M}}^y_{iwd}\ \hat{\mathbf{M}}^z_{iwd}\right]$. Now,

For $i = n$,

$$\left[\hat{\mathbf{M}}_i\right]_i = [\mathbf{M}_i]_i \qquad \{\text{nil}\}$$

For $i = n$-1, …, 1,

(a) $\left[\hat{\mathbf{M}}_{ivv}\right]_i = [\mathbf{M}_{ivv}]_i + \tilde{\mathbf{Q}}_i\left[\bar{\mathbf{M}}_{i+1,vv}\right]_{i+1}\tilde{\mathbf{Q}}_i^T$ {45M 33A(n − 1)}

(b) $\left[\hat{\mathbf{M}}_{ivw}\right]_i = [\mathbf{M}_{ivw}]_i + \tilde{\mathbf{Q}}_i\left[\mathbf{\Gamma}_{1,i+1}\right]_{i+1}\tilde{\mathbf{Q}}_i^T$ {54M 42A (n − 1)}

where

$$\left[\mathbf{\Gamma}_{1,i+1}\right]_{i+1} \equiv \left[\bar{\mathbf{M}}_{ivw}\right]_{i+1} - \left[\bar{\mathbf{M}}_{i+1,vv}\right]_{i+1}\left([\bar{\mathbf{r}}_{ai}]_{i+1} \times 1\right) \quad \{15\text{M } 18\text{A } (\text{n} - 1)\}$$

in which the computations involved in evaluating $[\bar{\mathbf{r}}_{ai}]_{i+1} \times 1$ are nil as it is a mere representation scheme. Note that the 3 × 3 matrix, $\mathbf{\Gamma}_{1,i+1}$, is introduced here for clarity in representation of the subsequent equations.

(c) The 3 × 3 matrix, $\hat{\mathbf{M}}_{iww}$, is computed as

$$\left[\hat{\mathbf{M}}_{iww}\right]_i = [\mathbf{M}_{iww}]_i + \tilde{\mathbf{Q}}_i\Big\{\left([\bar{\mathbf{r}}_{ai}]_{i+1} \times 1\right)^T\Gamma_{1,i+1} - \left[\bar{\mathbf{M}}_{i+1,vw}\right]_{i+1}\left([\bar{\mathbf{r}}_{ai}]_{i+1} \times 1\right) + \left[\bar{\mathbf{M}}_{i+1,ww}\right]_{i+1}\Big\}\tilde{\mathbf{Q}}_i^T$$

{57M 48A(n − 1)}

(d) The 3 × 2m matrix, $\hat{\mathbf{M}}_{ivd}$, is computed as,

$$\left[\hat{\mathbf{M}}_{ivd}^y\right]_i = \left[\mathbf{M}_{ivd}^y\right]_i + \tilde{\mathbf{Q}}_i\left[\mathbf{\Gamma}_{2,i+1}^y\right]_{i+1} \quad \{9m\text{M } 3m\text{A}(\text{n} - 1)\}$$

$$\left[\hat{\mathbf{M}}_{ivd}^z\right]_i = \left[\mathbf{M}_{ivd}^z\right]_i + \tilde{\mathbf{Q}}_i\left[\mathbf{\Gamma}_{2,i+1}^z\right]_{i+1} \quad \{9m\text{M } 3m\text{A}(\text{n} - 1)\}$$

where similar to $\hat{\mathbf{M}}_{ivd}^y$ and $\hat{\mathbf{M}}_{ivd}^z$, the 3 × m matrices, $\mathbf{M}_{ivd}^y$ and $\mathbf{M}_{ivd}^z$, are the submatrices of $\mathbf{M}_{ivd}$ corresponding to the vibrations in Y-direction and Z-direction, respectively. Moreover, the 3 × m matrices, $\mathbf{\Gamma}_{2,i+1}^y$ and $\mathbf{\Gamma}_{2,i+1}^z$, are introduced here for improving clarity, i.e.,

$$\left[\Gamma_{2,i+1}^y\right]_{i+1} \equiv \left[\bar{\mathbf{M}}_{i+1,vv}\right]_{i+1}\left[\mathbf{K}_{4,i}^y\right]_{i+1} + \left[\bar{\mathbf{M}}_{i+1,vw}\right]_{i+1}\left[\bar{\delta}_i^y\right]_{i+1} \quad \{12m\text{M } 9m\text{A}(\text{n} - 1)\}$$

$$\left[\Gamma_{2,i+1}^z\right]_{i+1} \equiv \left[\bar{\mathbf{M}}_{i+1,vv}\right]_{i+1}\left[\mathbf{K}_{4,i}^z\right]_{i+1} + \left[\bar{\mathbf{M}}_{i+1,vw}\right]_{i+1}\left[\bar{\delta}_i^z\right]_{i+1} \quad \{12m\text{M } 9m\text{A}(\text{n} - 1)\}$$

in which,

$$\left[\bar{\boldsymbol{\delta}}_i^y\right]_{i+1} = \tilde{\mathbf{Q}}_i^T\left[0 \; \boldsymbol{\delta}_i^y \; 0\right]^T \quad \{3m\text{M}(\text{n} - 1)\}$$

$$\left[\bar{\boldsymbol{\delta}}_i^z\right]_{i+1} = \tilde{\mathbf{Q}}_i^T\left[0 \; 0 \; \boldsymbol{\delta}_i^z\right]^T \quad \{3m\text{M}(\text{n} - 1)\}$$

(e) The 3 × 2m matrix, $\hat{\mathbf{M}}_{iwd}$, is computed as

$$\hat{\mathbf{M}}_{iwd} \equiv \left[\hat{\mathbf{M}}_{iwd}^y \; \hat{\mathbf{M}}_{iwd}^z\right]$$

where

$$\left[\hat{\mathbf{M}}^y_{iwd}\right]_i = \left[\mathbf{M}^y_{iwd}\right]_i + \tilde{\mathbf{Q}}_i\left\{\left([\bar{\mathbf{r}}_{ai}]_{i+1} \times 1\right)^T\left[\Gamma^y_{2,i+1}\right]_{i+1} + \left[\Gamma^y_{3,i+1}\right]_{i+1}\right\}$$

{15*m*M 15*m*A(n − 1)}

$$\left[\hat{\mathbf{M}}^z_{iwd}\right]_i = \left[\mathbf{M}^z_{iwd}\right]_i + \tilde{\mathbf{Q}}_i\left\{\left([\bar{\mathbf{r}}_{ai}]_{i+1} \times 1\right)^T\left[\Gamma^z_{2,i+1}\right]_{i+1} + \left[\Gamma^z_{3,i+1}\right]_{i+1}\right\}$$

{15*m*M 15*m*A(n − 1)}

where similar to $\hat{\mathbf{M}}^y_{iwd}$ and $\hat{\mathbf{M}}^z_{iwd}$, the $3 \times m$ matrices $\mathbf{M}^y_{iwd}$ and $\mathbf{M}^z_{iwd}$, are the submatrices of $\mathbf{M}_{iwd}$ corresponding to the vibrations in *Y*-direction and *Z*-direction, respectively. The $3 \times m$ matrix, $\boldsymbol{\Gamma}^y_{3,i+1}$ and $\boldsymbol{\Gamma}^z_{3,i+1}$, is defined as

$$\left[\Gamma^y_{3,i+1}\right]_{i+1} \equiv \left[\bar{\mathbf{M}}_{i+1,vw}\right]^T_{i+1}\left[\mathbf{K}^y_{4,i}\right]_{i+1} + \left[\bar{\mathbf{M}}_{i+1,ww}\right]_{i+1}\left[\bar{\delta}^y_i\right]_{i+1} \quad \{12m\mathrm{M}\, 9m\mathrm{A}(\mathrm{n}-1)\}$$

$$\left[\Gamma^z_{3,i+1}\right]_{i+1} \equiv \left[\bar{\mathbf{M}}_{i+1,vw}\right]_{i+1}\left[\mathbf{K}^z_{4,i}\right]_{i+1} + \left[\bar{\mathbf{M}}_{i+1,ww}\right]_{i+1}\left[\bar{\delta}^y_i\right]_{i+1} \quad \{12m\mathrm{M}\, 9m\mathrm{A}(\mathrm{n}-1)\}$$

(f) The $2m \times 2m$ matrix $\hat{\mathbf{M}}_{idd}$, is computed as

$$\hat{\mathbf{M}}_{idd} \equiv \begin{bmatrix} \hat{\mathbf{M}}^y_{idd} & \mathbf{O} \\ \mathbf{O} & \hat{\mathbf{M}}^z_{idd} \end{bmatrix}$$

where the $m \times m$-dimensional diagonal submatrices, $\hat{\mathbf{M}}^y_{idd}$ and $\hat{\mathbf{M}}^z_{idd}$, are the terms of $\hat{\mathbf{M}}_{idd}$ corresponding to the vibrations in *Y*-direction and *Z*-direction, respectively. Now,

$$\left[\hat{\mathbf{M}}^y_{idd}\right]_i = \left[\mathbf{M}^y_{idd}\right]_i + \left[\mathbf{K}^y_{4,i}\right]^T_{i+1}\left[\Gamma^y_{2,i+1}\right]_{i+1} + \left[\bar{\delta}^y_i\right]^T_{i+1}\left[\Gamma^y_{3,i+1}\right]_{i+1} \quad \{4m\mathrm{M}\, 4m\mathrm{A}(\mathrm{n}-1)\}$$

$$\left[\hat{\mathbf{M}}^z_{idd}\right]_i = \left[\mathbf{M}^z_{idd}\right]_i + \left[\mathbf{K}^z_{4,i}\right]^T_{i+1}\left[\Gamma^z_{2,i+1}\right]_{i+1} + \left[\bar{\delta}^z_i\right]^T_{i+1}\left[\Gamma^z_{3,i+1}\right]_{i+1} \quad \{4m\mathrm{M}\, 4m\mathrm{A}(\mathrm{n}-1)\}$$

(g) Finally,

$$\left[\hat{\mathbf{M}}_{icc}\right]_i = \left[\mathbf{M}_{icc}\right]_i \qquad \{\text{nil}\}$$

E.6 Find the $(6 + m_f) \times (1 + m_f)$ matrix, $\left[\hat{\boldsymbol{\Psi}}_i\right]_i$, which is as follows:

$$\hat{\boldsymbol{\Psi}}_i \equiv \hat{\mathbf{M}}_i\mathbf{P}_i \qquad \{\text{nil}\}$$

Note that the computations required for the determination of $\left[\hat{\boldsymbol{\Psi}}_i\right]_i$ are associated with $\mathbf{P}_i$ only. Matrix $\mathbf{P}_i$, as given by Eq. (3.10b), involves algebraic operations with

1, **O**, or $\mathbf{z}_i$, only which do not require any computations. Hence, the computations required for the determination of $\left[\hat{\boldsymbol{\psi}}_i\right]_i$ are nil.

E.7 Find the $(1 + m_f) \times (1 + m_f)$ Matrix, $\left[\hat{\mathbf{I}}_i\right]_i$, for $I = n,\ldots,1$, as

$$\hat{\mathbf{I}}_i \equiv \mathbf{P}_i^T \hat{\boldsymbol{\Psi}}_i \qquad \{\text{nil}\}$$

Similar to $\left[\hat{\boldsymbol{\psi}}_i\right]_i$, the evaluation of $\left[\hat{\mathbf{I}}_i\right]_i$ requires operations associated with $\mathbf{P}_i$ only and, hence, the computations required for $\left[\hat{\mathbf{I}}_i\right]_i$ are nil.

E.8 Find the $(1 + m_f) \times (1 + m_f)$ matrix of, $\left[\hat{\mathbf{I}}_i\right]_i^{-1}$, for $I = n,\ldots,1$, as

$$\left[\hat{\mathbf{I}}_i\right]_i^{-1} \text{(inverse of } \hat{\mathbf{I}}_i\text{)} \qquad \left\{\left(\frac{\hat{m}^3}{6} + \frac{\hat{m}^2}{2} + \frac{\hat{m}}{3}\right)\mathrm{M}\left(\frac{\hat{m}^3}{6} + \frac{\hat{m}^2}{2} + \frac{\hat{m}}{3}\right)\mathrm{A}(n)\right\}$$

E.9 Find the $(6 + m_f) \times (1 + m_f)$ matrix, $\left[\boldsymbol{\psi}_i\right]_i$, for $I = n,\ldots,1$ as

$$[\boldsymbol{\Psi}_i]_i \equiv \left[\hat{\boldsymbol{\Psi}}_i\right]_i \left[\hat{\mathbf{I}}_i\right]_i^{-1} \qquad \left\{(6 + m_f)\hat{m}^2\mathrm{M}\,(6 + m_f)m_f\hat{m}\;\mathrm{A}\,(\mathrm{n})\right\}$$

E.10 Find the $(6 + m_f) \times (6 + m_f)$ Matrix, $\left[\bar{\mathbf{M}}_i\right]_i$, for $I = n,\ldots,2$ as

$$\left[\bar{\mathbf{M}}_i\right]_i = \left[\hat{\mathbf{M}}_i\right]_i - \left[\hat{\boldsymbol{\psi}}_i\right]_i [\Psi_i]_i^T$$
$$\left\{\frac{\hat{m}(6 + m_f)(7 + m_f)}{2}\mathrm{M}\frac{\hat{m}(6 + m_f)(7 + m_f)}{2}\mathrm{A}(\mathrm{n} - 1)\right\}$$

which is required in E.5.

E.11 Find the block element of the $\bar{n}$-dimensional vector, $\boldsymbol{\varphi}_i$, i.e., $\boldsymbol{\varphi}_i$: for $I = 1, \ldots, N$

$$\boldsymbol{\varphi}_i = \boldsymbol{\tau}_i - \boldsymbol{\tau}_i^q \qquad \{m_f\mathrm{A}(\mathrm{n})\}$$

where vector $\boldsymbol{\tau}_i^q$ is the $(1 + 2m + \bar{m})$-dimensional vector of the terms corresponding to the Coriolis and centrifugal, and gravity terms which can be evaluated using any recursive inverse dynamics algorithm while $\ddot{\mathbf{q}}_i = \mathbf{0}$. Now, to find $\boldsymbol{\tau}_i = \boldsymbol{\tau}_i^E + \boldsymbol{\tau}_i^s$, as defined after Eq. (3.20a), $\boldsymbol{\tau}_i^E$ is the input, whereas $\boldsymbol{\tau}_i^s$ is evaluated using Eq. (3.20d–g) as,

$$\boldsymbol{\tau}_i^s \equiv \begin{bmatrix} 0 \\ \boldsymbol{\tau}_i^{sy} \\ \boldsymbol{\tau}_i^{sz} \\ \boldsymbol{\tau}_i^{sx} \end{bmatrix}$$

Comparing with Eq. (3.20d), only scalar zero will appear in the above expression as deformation along X-axis is neglected for the purpose of calculating computational complexity. The m-dimensional vector $\boldsymbol{\tau}_i^{sy}$ is obtained as,

$$\boldsymbol{\tau}_i^{sy} \equiv \begin{bmatrix} \tau_{i1}^{sy} & \cdots & \tau_{im_i}^{sy} \end{bmatrix}^T \qquad \{m^2\mathrm{M}(m-1)m\mathrm{A(n)}\}$$

where the elements, τ_{ij}^{sy}, are defined in Eq. (3.20f), which are evaluated offline. However, multiplications with $\mathbf{d}_i^y$ are counted. Similarly m-dimensional vector $\boldsymbol{\tau}_i^{sz}$ and $\bar{m}$-dimensional vector $\boldsymbol{\tau}_i^{sx}$ are evaluated below,

$$\boldsymbol{\tau}_i^{sz} = \begin{bmatrix} \tau_{i1}^{sz} & \cdots & \tau_{im_i}^{sz} \end{bmatrix}^T \qquad \{m^2\mathrm{M}(m-1)m\mathrm{A(n)}\}$$

$$\boldsymbol{\tau}_i^{sx} = \begin{bmatrix} \tau_{i1}^{sx} & \cdots & \tau_{i\bar{m}_i}^{sx} \end{bmatrix}^T \qquad \{\bar{m}^2\mathrm{M}(\bar{m}-1)\bar{m}\mathrm{A(n)}\}$$

Then,

$$\boldsymbol{\tau}_i = \boldsymbol{\tau}_i^E + \boldsymbol{\tau}_i^s \qquad \{m_f\mathrm{A(n)}\}$$

E.12 Find $\hat{\boldsymbol{\varphi}}$:

For the purpose of computational counts, the vector, $\boldsymbol{\varphi}_i$, is denoted as

$$\boldsymbol{\varphi}_i \equiv \begin{bmatrix} \varphi_{i,\theta} & \boldsymbol{\varphi}_{i,dy}^T & \boldsymbol{\varphi}_{i,dz}^T & \boldsymbol{\varphi}_{i,c}^T \end{bmatrix}^T$$

(a) For $i = n$, $[\hat{\boldsymbol{\varphi}}_i]_i = [\boldsymbol{\varphi}_\mathrm{i}]_i$ {nil}

For clarity in representation, vector $\hat{\boldsymbol{\varphi}}_i$ is also represented like $\boldsymbol{\varphi}_i$ above. Computations are now given as.

(b) For $i = n-1,\ldots,1$:

$$\hat{\varphi}_{i,\theta} = \varphi_{i,\theta} + [\mathbf{z}_i]_i^T \left\{ \tilde{\mathbf{Q}}_i \left((\bar{\mathbf{r}}_{ai} \times 1)\left[\bar{\boldsymbol{\eta}}_{i+1,\theta 1}\right]_{i+1} + \left[\bar{\boldsymbol{\eta}}_{i+1,2}\right]_{i+1,\theta 2} \right) \right\} \quad \{15\mathrm{M}\ 13\mathrm{A}\ (\mathrm{n}-1$$

$$\left[\hat{\boldsymbol{\varphi}}_{i,dy}\right]_i = \left[\boldsymbol{\varphi}_{i,dy}\right]_i + \left\{\left[\mathbf{K}^y_{4,i}\right]^T_{i+1}\left[\bar{\boldsymbol{\eta}}_{i+1,\theta 1}\right]_{i+1} + \left[\bar{\delta}^y_i\right]^T_{i+1}\left[\bar{\boldsymbol{\eta}}_{i+1,\theta 2}\right]_{i+1}\right\}$$
$\{4m\mathrm{M}\,4m\mathrm{A}(\mathrm{n}-1)\}$

$$\left[\hat{\boldsymbol{\varphi}}_{i,dz}\right]_i = \left[\boldsymbol{\varphi}_{i,dz}\right]_i + \left\{\left[\mathbf{K}^z_{4,i}\right]^T_{i+1}\left[\bar{\boldsymbol{\eta}}_{i+1,\theta 1}\right]_{i+1} + \left[\bar{\delta}^z_i\right]^T_{i+1}\left[\bar{\boldsymbol{\eta}}_{i+1,\theta 2}\right]_{i+1}\right\}$$
$\{4m\mathrm{M}\,4m\mathrm{A}(\mathrm{n}-1)\}$

$$\left[\hat{\boldsymbol{\varphi}}_{i,c}\right]_i = \left[\boldsymbol{\varphi}_{i,c}\right]_i \ \{\text{nil}\}$$

in which the components of $(6+m_f)$-dimensional vector $\bar{\boldsymbol{\eta}}_{i+1}$ for the ith link are obtained from E.13 below.

E.13 Find $(6+m_f)$-dimensional vector $\left[\bar{\boldsymbol{\eta}}_i\right]_i$. For the purpose of counting the number of computations involved, vector $\bar{\boldsymbol{\eta}}_i$ is denoted, in a manner similar to the vector $\boldsymbol{\varphi}_i$, as

$$\bar{\boldsymbol{\eta}}_i \equiv \left[\bar{\boldsymbol{\eta}}_{i,\theta}\ \bar{\boldsymbol{\eta}}^T_{i,dy}\ \bar{\boldsymbol{\eta}}^T_{i,dz}\ \bar{\boldsymbol{\eta}}^T_{i,c}\right]^T$$

where the 6-dimensional vector, $\bar{\boldsymbol{\eta}}_{i,\theta} \equiv \left[\bar{\boldsymbol{\eta}}^T_{i,\theta 1}\ \bar{\boldsymbol{\eta}}^T_{i,\theta 2}\right]^T$. Vectors $\bar{\boldsymbol{\eta}}_{i,dy}$, $\bar{\boldsymbol{\eta}}_{i,dz}$ and $\bar{\boldsymbol{\eta}}_{i,c}$ are m- and $\bar{m}$-dimensional, respectively.

(a) For $i = n$,

$$\left[\boldsymbol{\eta}_n\right]_n = \left[\boldsymbol{\psi}_n\right]_n\left[\hat{\boldsymbol{\varphi}}_n\right]_n \qquad \{(6+m_f)\hat{m}\mathrm{M}(6+m_f)m_f\mathrm{A}\}$$

(b) For $i = n-1,\ldots,1$, $\left[\bar{\boldsymbol{\eta}}_i\right]_i$ is obtained by first obtaining an intermediate $(6+m_f)$-dimensional vector $\boldsymbol{\eta}_i$, denoted similar to $\bar{\boldsymbol{\eta}}_i$, as

$$\left[\boldsymbol{\eta}_i\right]_i \equiv \left[\boldsymbol{\eta}_{i,\theta}\ \boldsymbol{\eta}^T_{i,dy}\ \boldsymbol{\eta}^T_{i,dz}\ \boldsymbol{\eta}^T_{i,c}\right]^T = [\psi_i]_i\left[\hat{\varphi}_i\right]_i$$
$$\left\{(6+m_f)\hat{m}\mathrm{M}(6+m_f)m_f\mathrm{A}(n-1)\right\}$$

(c) The elements of $\left[\bar{\boldsymbol{\eta}}_i\right]_i$ are then obtained as

$$\left[\bar{\boldsymbol{\eta}}_{i,\theta 1}\right]_i = \left[\boldsymbol{\eta}_{i,\theta 1}\right]_i + \tilde{\mathbf{Q}}_i\left[\bar{\boldsymbol{\eta}}_{i+1,\theta}\right]_{i+1} \quad \{9\mathrm{M}\,9\mathrm{A}(n-1)\}$$
$$\left[\bar{\boldsymbol{\eta}}_{i,\theta 2}\right]_i = \left[\boldsymbol{\eta}_{i,\theta 2}\right]_i + \underline{\left\{\tilde{\mathbf{Q}}_i\left((\tilde{\mathbf{r}}_{ai}\times 1)\left[\bar{\boldsymbol{\eta}}_{i+1,\theta 1}\right]_{i+1} + \left[\bar{\boldsymbol{\eta}}_{i+1,\theta 2}\right]_{i+1}\right)\right\}}\{1\mathrm{A}(n-1)\}$$

Since $\bar{\boldsymbol{\eta}}_{i,dy}$, $\bar{\boldsymbol{\eta}}_{i,dz}$, and $\bar{\boldsymbol{\eta}}_{i,c}$ are not required in calculations above, they are not evaluated.

Note that the underlined terms involved in the computation of $\bar{\boldsymbol{\eta}}_{i,\theta 2}$ have already been computed while evaluating the vector $\hat{\boldsymbol{\varphi}}_{i,d}$ above, and the computations involved in them are not counted here.

E.14 Find $\tilde{\boldsymbol{\varphi}}$: For $I = 1, \ldots n$

$$[\tilde{\boldsymbol{\varphi}}_i]_i = \left[\hat{\mathbf{I}}_i\right]_i^{-1} [\hat{\boldsymbol{\varphi}}_i]_i \qquad \{\hat{m}^2 \mathrm{M}\, m_f \hat{m} \mathrm{A(n)}\}$$

Again, for the clarity of representation, the $(1 + m_f)$-dimensional vector $\tilde{\boldsymbol{\varphi}}$ is represented as:

$$\tilde{\boldsymbol{\varphi}}_i \equiv \left[\tilde{\varphi}_{i,\theta}\ \tilde{\boldsymbol{\varphi}}_{i,d}^T\ \tilde{\boldsymbol{\varphi}}_{i,c}^T\right]^T$$

where, $\tilde{\varphi}_{i,\theta}$ is a scalar, and $\tilde{\boldsymbol{\varphi}}_{i,d}$ and $\tilde{\boldsymbol{\varphi}}_{i,c}$ are the 2 m- and $\bar{m}$-dimensional vectors, respectively.

The backward recursion loop ends here and the forward recursion, i.e., the determination of $\ddot{\mathbf{q}}_i$ starts next.

E.15 Find $\ddot{\mathbf{q}}_i$: For $I = 1, \ldots n$

(a) For $i = 1$,

$$\ddot{\mathbf{q}}_i = \tilde{\boldsymbol{\varphi}}_i \qquad \{\mathrm{nil}\}$$

(b) For $i = 2, \ldots, n$, $\ddot{\mathbf{q}}_i$ is obtained by first evaluating an intermediate $(6 + m_f)$-dimensional vector $\bar{\boldsymbol{\mu}}_i$ denoted as

$$\bar{\boldsymbol{\mu}}_i \equiv \left[\bar{\boldsymbol{\mu}}_{i,\theta}\ \bar{\boldsymbol{\mu}}_{i,d}^T\ \bar{\boldsymbol{\mu}}_{i,c}^T\right]^T$$

where the 6-dimensional vector, $\bar{\boldsymbol{\mu}}_{i,\theta} \equiv \left[\bar{\boldsymbol{\mu}}_{i,\theta 1}^T\ \bar{\boldsymbol{\mu}}_{i,\theta 2}^T\right]^T$, and vectors $\bar{\boldsymbol{\mu}}_{i,d}$ and $\bar{\boldsymbol{\mu}}_{i,c}$ are 2 m- and $\bar{m}$-dimensional vectors. Moreover, $\bar{\boldsymbol{\mu}}_{i,d}$ is denoted as $\bar{\boldsymbol{\mu}}_{i,d} \equiv \left[\bar{\boldsymbol{\mu}}_{i,dy}^T\ \bar{\boldsymbol{\mu}}_{i,dz}^T\right]^T$. Now,

$$[\bar{\mu}_{i,\theta 1}]_i = \tilde{\mathbf{Q}}_{i-1}^T\Big([\mu_{i-1,\theta 1}]_{i-1} - (\bar{\mathbf{r}}_{ai} \times 1)[\mu_{i-1,\theta 2}]_{i-1}$$
$$+ [\mathbf{K}_{4,i-1}^y]_i [\mu_{i-1,dy}]_{i-1} + [\bar{\delta}_{i-1}^y]_i [\mu_{i-1,dz}]_{i-1}\Big)$$
$$\{(15 + 2m)\mathrm{M}(12 + 2m)\mathrm{A}(\mathrm{n} - 1)\}$$

$$[\bar{\mu}_{i,\theta 2}]_i = \tilde{\mathbf{Q}}_{i-1}^T\Big([\mu_{i-1,\theta 2}]_{i-1} + [\mathbf{K}_{4,i-1}^z]_i [\mu_{i-1,dy}]_{i-1} + [\bar{\delta}_{i-1}^z]_i [\mu_{i-1,dz}]_{i-1}\Big)$$
$$\{(9 + 2m)\mathrm{M}(9 + 2m)\mathrm{A}(\mathrm{n} - 1)\}$$

Since, $\bar{\boldsymbol{\mu}}_{i,d}$, and $\bar{\boldsymbol{\mu}}_{i,c}$ are not required in calculations they are not evaluated. The $(1+m_f)$-dimensional vector $\ddot{\mathbf{q}}_i$ is then obtained from

$$[\ddot{\mathbf{q}}_i]_i = [\tilde{\boldsymbol{\varphi}}_i]_i - [\boldsymbol{\psi}_i]_i^T [\bar{\boldsymbol{\mu}}_i]_i \qquad \{6\hat{m}\text{M}6\hat{m}\text{A}(\text{n}-1)\}$$

Again, in the subsequent steps, the $(1+m_f)$-dimensional vector, $\ddot{\mathbf{q}}_i$, is represented as

$$\ddot{\mathbf{q}}_i \equiv \left[\ddot{\theta}_i \; \ddot{\mathbf{d}}_{iy}^T \; \ddot{\mathbf{d}}_{iz}^T \; \ddot{\mathbf{c}}_i^T\right]^T$$

θ_i is a scalar and $\mathbf{d}_{iy}$, $\mathbf{d}_{iz}$, and $\mathbf{c}_i$ are the m- and $\bar{m}$-dimensional vectors.

E.16 Find the $(6+m_f)$-dimensional vector, $[\boldsymbol{\mu}_i]_i$. It is denoted, as

$$\boldsymbol{\mu}_i \equiv \left[\boldsymbol{\mu}_{i\theta 1}^T \; \boldsymbol{\mu}_{i\theta 2}^T \; \boldsymbol{\mu}_{idy}^T \; \boldsymbol{\mu}_{idz}^T \; \boldsymbol{\mu}_{ic}^T\right]^T$$

where vectors $\boldsymbol{\mu}_{i\theta 1}$ and $\boldsymbol{\mu}_{i\theta 2}$ are the 3-dimensional vectors, whereas $\boldsymbol{\mu}_{idy}$ and $\boldsymbol{\mu}_{idz}$ are the m-dimensional vectors. The last term, $\boldsymbol{\mu}_{ic}$, is the $\bar{m}$-dimensional vector.

(a) For $i = 1$,

$$\boldsymbol{\mu}_i \equiv \mathbf{P}_i \ddot{\mathbf{q}}_i \qquad \{\text{nil}\}$$

Since the evaluation of $\boldsymbol{\mu}_i$ requires operations associated with $\mathbf{P}_i$ only, the computations required for it are nil.

(b) For $i = 2, \ldots, n$,

$$\begin{aligned}
[\boldsymbol{\mu}_{i\theta 1}]_i &= [\bar{\boldsymbol{\mu}}_{i\theta 1}]_i \\
[\boldsymbol{\mu}_{i\theta 2}]_i &= [\mathbf{z}_i]_i \ddot{\theta}_i + [\bar{\boldsymbol{\mu}}_{i\theta 2}]_i \qquad \{1\text{A}(n-1)\} \\
[\boldsymbol{\mu}_{idy}]_i &= [\ddot{\mathbf{d}}_{iy}]_i;\; [\boldsymbol{\mu}_{idz}]_i = [\ddot{\mathbf{d}}_{iz}]_i;\; [\boldsymbol{\mu}_{ic}]_i = [\ddot{\mathbf{c}}_i]_i
\end{aligned}$$

The total complexity of the proposed forward dynamics algorithm for a general all flexible link robot is finally obtained by adding the above number of computations shown within { and}, which is

$$\left[(2m^2 + \bar{m}^2 + 180m + 4\bar{m} + 299 + M_{m1})n - (124m + \bar{m} + 229 + M_{m2})\right]\text{M};$$
$$\left[(6m^2 + 2\bar{m}^2 + 4m\bar{m} + 122m + 8\bar{m} + 220 + A_{m1})n - (94m + \bar{m} + 192 + A_{m2})\right]\text{A};$$

and $2n$ T.

where M_{m1}, M_{m2}, A_{m1}, and A_{m2} are the functions of the number of modes considered for vibration, i.e.,

$$M_{m1} \equiv \frac{\hat{m}^3}{6} + \left(2m + \bar{m} + \frac{15}{2}\right)\hat{m}^2 + \left(\frac{4m^2 + \bar{m}^2 + 4m\bar{m} + 30m + 15\bar{m}}{2} + \frac{100}{3}\right)\hat{m}$$

$$M_{m2} \equiv \left(27 + \frac{4m^2 + \bar{m}^2 + 4m\bar{m} + 26m + 13\bar{m}}{2}\right)\hat{m};$$

$$A_{m1} \equiv \frac{\hat{m}^3}{6} + \frac{\hat{m}^2}{2} + \left(6m^2 + \frac{3}{2}\bar{m}^2 + 6m\bar{m} + 27m + \frac{27}{2}\bar{m} + \frac{82}{3}\right)\hat{m};$$

$$A_{m2} \equiv \left(2m^2 + \frac{\bar{m}^2}{2} + 2m\bar{m} + 13m + \frac{13}{2}\bar{m} + 27\right)\hat{m}$$

where $\hat{m} = 1 + m_f$ and $m_f = 2m + \bar{m}$.

Appendix F
Modeling of Four-Bar Mechanism in Recurdyn Software

The comparison of the results obtained using an in-house algorithm and a Multibody Dynamics Software (MBDS) is not trivial, mainly in the case of flexible multibody systems. This appendix presents a systematic procedure adopted to compare the forward dynamics results of different closed-loop systems of Chaps. 5 and 6. The steps were explained by modeling *RRF* 4-bar mechanism in RecurDyn. RecurDyn (Recursive dynamics) from Function bay Inc. is a popular dynamic analysis software for modeling and simulation of both rigid and flexible multibody systems.

F.1 Modeling of a 4-Bar Mechanism

In this section, the steps carried out for modeling an *RRF* 4-bar mechanism in RecurDyn, shown in Fig. F.1, are discussed. The *RRF* 4-bar mechanism has two rigid links and a flexible link. Since the fourth link is grounded, it need not be modeled. These links are connected by revolute joints. The modeling of the rigid link, flexible link, and modeling of joints are discussed in the following subsections.

F.1.1 Modeling of a Rigid Link

The mechanism has two rigid links. These links are modeled as a cylinder. This option is in the *body* modeling section of the *professional* tab of RecurDyn, as shown in Fig. F.2. This asks the two endpoints and the radius. The mass and inertia properties are automatically calculated by the density of the preassigned material and the geometric properties. The material preassigned here is *steel.*

P. V. Nandihal et al., *Dynamics of Rigid-Flexible Robots and Multibody Systems*, Intelligent Systems, Control and Automation: Science and Engineering 100, https://doi.org/10.1007/978-981-16-2798-9

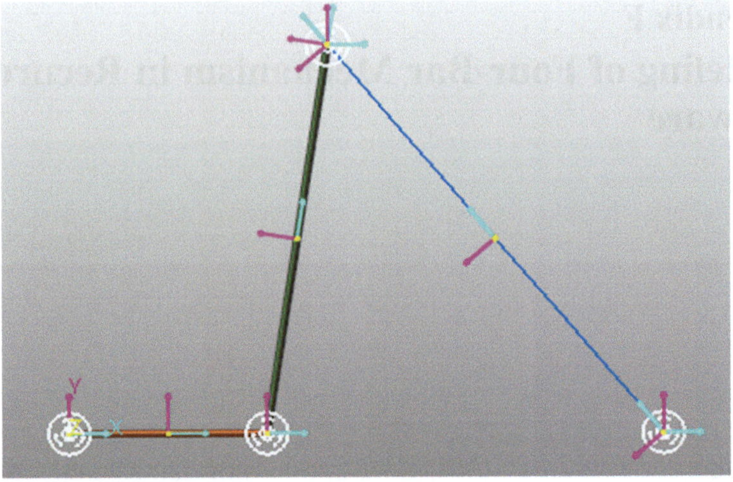

Fig. F.1 The *RRF* 4-bar mechanism

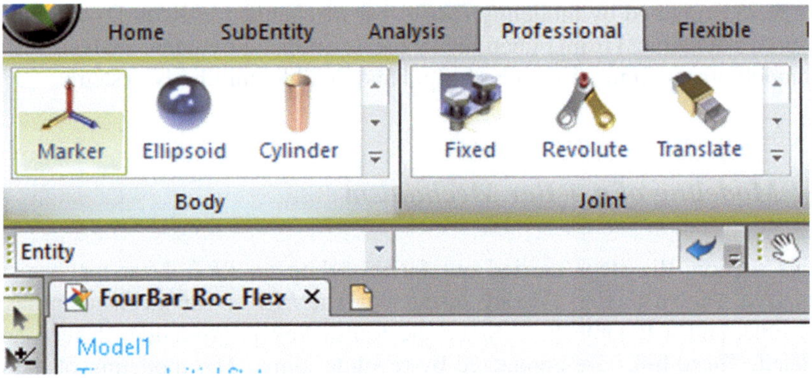

Fig. F.2. RecurDyn menu

F.1.2 Modeling of a Joint

The different joints are provided in the *professional* tab of the RecurDyn for multi-body analysis. To model the revolute joint for the *RRF* 4-bar mechanism, the *Revolute* option (shown in Fig F.2) from the *Joint* section is selected. It asks 1st body, 2nd body, and point at which joint is to be created. Providing these informations, the joint is modeled.

F.1.3 Modeling of a Flexible Link

The flexible link can be modeled by creating a line element and meshing it with a 1-dimensional beam element by providing the geometric parameters. To do this in RecurDyn, the link is modeled using *general* option in *body* section *professional* tab. In *general* option, a line is created using *Outline* option. Then, the created line is taken to meshing environment by right clicking the line and selecting the *Mesh* option. Once that is done, the mesher tab will automatically open as shown in Fig. F.3. The beam properties are defined as shown in Fig F.4 using the properties option in

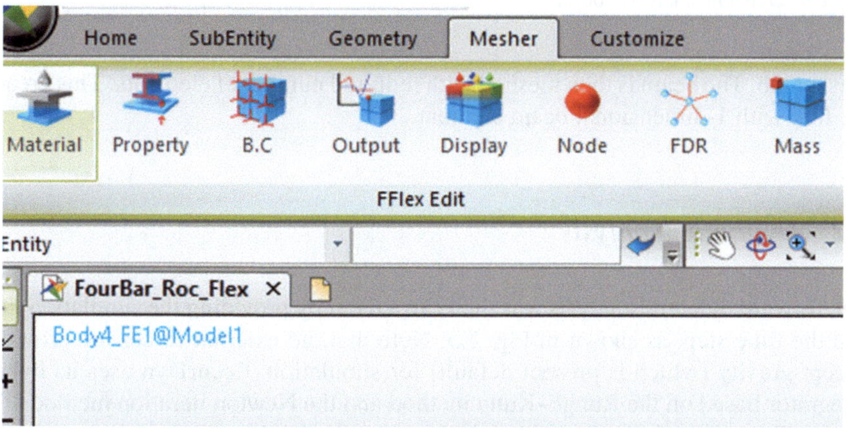

Fig. F.3 RecurDyn mesher

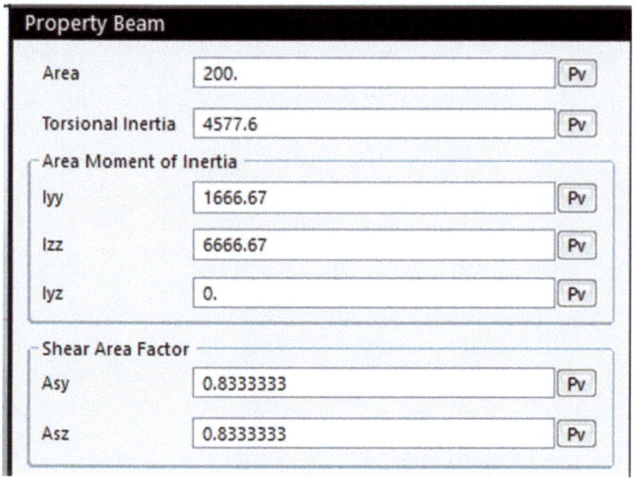

Fig. F.4 Beam properties

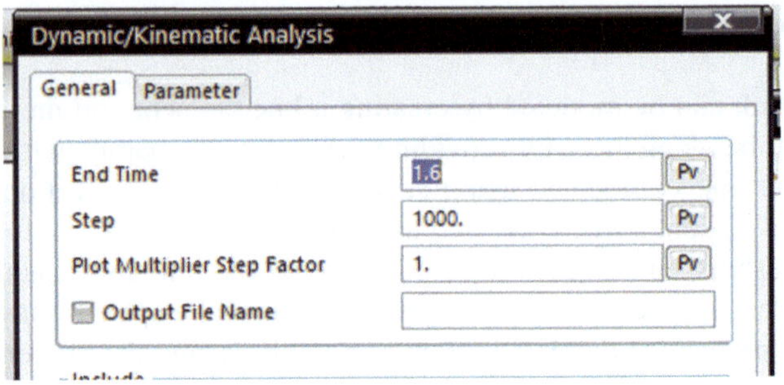

Fig. F.5 Dynamic analysis options

mesher tab. The beam is then meshed with required number of elements. This creates the link with 1-dimensional beam element.

F.2 Dynamic Analysis

The forward dynamic analysis was then carried out by providing the simulation time and the time step, as shown in Fig. F.5. Note that, no external forces are provided except gravity (which is present default) for simulation. RecurDyn uses its hybrid integrator based on the Runge–Kutta method and the Newton iteration method. The simulation was performed, and the results were obtained.

Bibliography

Angeles J (2003) Fundamentals of robotic mechanical system. Springer Verlag, New York

Bhangale P (2004) Dynamics based modeling, computational complexity and architecture selection of robot manipulators, Ph.D. Thesis, Indian Institute of Technology Delhi, India

Cyril X (1988) In: Dynamics of flexible-link manipulators. McGill University, Canada

Denavit J, Hartenberg RS (1955) A kinematic notation for lower-pair mechanisms based on matrices. ASME J Applied Mechanics 77:445–450

Jain A, Rodriguez G (1992) Recursive flexible multibody system dynamics using spatial operators. J Guidance Control Dynam 15 (6):1453–1466

Saha SK (1999a) Analytical expression for the inverted inertia matrix of serial robots. Int J Robot Res 18(1):116–124.

Strang G (1980) Linear algebra and its applications. Harcourt, Brace, Jovanovich Pub., Florida

Stejskal V, Valášek M (1996) Kinematics and dynamics of machinery. M. Dekker

Walker MW, Orin DE (1982) Efficient dynamic computer simulation of robotic mechanisms. J Dyn Syst Meas Control Trans Asme V 104(N3):205–211.

Yoshikawa T (1998) Foundations of robotic analysis and control. Prentice hall of India Pvt. Ltd, New Delhi

P. V. Nandihal et al., *Dynamics of Rigid-Flexible Robots and Multibody Systems*,
Intelligent Systems, Control and Automation: Science and Engineering 100,
https://doi.org/10.1007/978-981-16-2798-9

MIX
Papier aus verantwortungsvollen Quellen
Paper from responsible sources
FSC® C105338

If you have any concerns about our products,
you can contact us on
ProductSafety@springernature.com

In case Publisher is established outside the EU,
the EU authorized representative is:
Springer Nature Customer Service Center GmbH
Europaplatz 3, 69115 Heidelberg, Germany

Printed by Libri Plureos GmbH
in Hamburg, Germany